I have endeavoured to indicate tonight in broad outline ~~though~~ the early history of correlation which has now a most extensive literature. It is a long step from Francis Galton's 'reversion' in sweet peas to the full theory of multiple correlation, which we now know to be identical with the spherical trigonometry of high-dimensioned space, the total correlation coefficients being the cosines of the edges of the polyhedra and the partial correlation coefficients the cosines of the polyhedral angles. But to find the correlation of the health of a child with the number of people per room while you render neutral its age, the health of its parents, the wages of its father, and the habits of its mother, is no less a vital a problem than Galton's correlation of character in parent and offspring. It requires indeed more mathematics, but the mathematics are not there for the joy of the analyst but because they are essential to the solution. It is the transition from the mill as pestle & mortar to the mill with steam driven steel grain crushing rollers. But the inventor of milling was the person who bruised grain between two stones, and Galton was the man who discovered the highway across this new country with what he aptly terms "its easy descents to different goals."

Karl Pearson.

Facsimile of the last pages of the MS. of Karl Pearson's paper on the History of Correlation read before the Society of Biometricians and Mathematical Statisticians, June 14, 1920 (see *Biometrika*, XIII, p. 45)

# TABLES

OF

## THE ORDINATES AND
## PROBABILITY INTEGRAL OF THE DISTRIBUTION
## OF THE CORRELATION COEFFICIENT IN
## SMALL SAMPLES

BY

## F. N. DAVID

ISSUED BY THE *BIOMETRIKA* OFFICE
UNIVERSITY COLLEGE, LONDON

CAMBRIDGE
AT THE UNIVERSITY PRESS
1954

CAMBRIDGE UNIVERSITY PRESS
Cambridge, New York, Melbourne, Madrid, Cape Town, Singapore,
São Paulo, Delhi, Dubai, Tokyo

Cambridge University Press
The Edinburgh Building, Cambridge CB2 8RU, UK

Published in the United States of America by Cambridge University Press, New York

www.cambridge.org
Information on this title: www.cambridge.org/9780521124126

First published 1938
Reprinted 1954
This digitally printed version 2009

*A catalogue record for this publication is available from the British Library*

ISBN 978-0-521-07169-7 Hardback
ISBN 978-0-521-12412-6 Paperback

# CONTENTS

# EDITOR'S PREFACE

The coefficient of correlation has played an important part during the last fifty years in the development of the theory of mathematical statistics and its applications. In a field where many have worked three names may be linked specially with the development of the theory of $r$: those of Francis Galton, Karl Pearson and R. A. Fisher. Galton's pioneer work may seem to us easy now, but it is just such simple first steps which are often the most difficult to make and yet the most far-reaching in their consequences. It was Pearson who was largely responsible for the development of the theory of correlation from these first foundations, and for demonstrating beyond question that this new concept "brought psychology, anthropology, medicine and sociology into the field of mathematical treatment". Finally, at almost his first venture into the field of statistical theory, Fisher deduced the probability distribution of $r$ in samples from a normal population, and at the same time drew attention to that valuable conception by which a sample may be represented as a point in multiple space.

The importance of Fisher's 1914 result was at once appreciated by Pearson, who, with his characteristic eagerness to put theory into numbers, was already early in 1915 planning the "Cooperative Study", which presented in tables of ordinates, in frequency constants and in photographs of models the varied forms assumed by the distribution of $r$. To improve on the tabled ordinates had probably long been Pearson's plan, but it was not until 1931 that he suggested to Miss David the computation of the tables of the probability integral now published in this volume. In the meantime Fisher had suggested the very useful logarithmic transformation of $r$, which enables the probability integral to be obtained from tables of the normal function with an accuracy sufficient for most common purposes. Nevertheless, in Pearson's view, however useful an approximation might be, it was desirable that for a statistical measure as important as the coefficient of correlation there should be available fundamental tables which would enable the probability integral to be determined with a considerable degree of mathematical accuracy for any size of sample, $n$, and any population correlation, $\rho$.

While accepting this view the present editor is well aware that the ideal of a mathematically accurate, all-embracing table, has not been attained. Karl Pearson's hope that it would be possible to interpolate accurately up to $n = 1600$ from a framework of tables at $n = 25, 50, 100, 200, 400, 800$ and $1600$ was not fulfilled, and it has been necessary to rest content with a more restricted objective.

Miss David has loyally completed a difficult task whose solution in the region of high $n$ and $\rho$ proved increasingly elusive. She has made two special contributions of her own: (1) a scheme of charts based on the tables which, for the range they cover, provide a more comprehensive and useful picture of the relation between $r$, $\rho$ and $n$ than is elsewhere available; (2) an introduction in which she has tried, with, I believe, no little measure of success, to link together a number of illustrative examples with a simple statement of the principles by which the theory of probability may be used as a guide in drawing inferences from observation.

Recent developments of theory, coming in response to a need for the solution of new types of problem, have laid emphasis on a technique which deals with regression rather than correlation coefficients and considers the apportionment of the covariance or product sum, $\Sigma (x-\bar{x})(y-\bar{y})$, into parts that may be associated with different factors determining variation. But however important these new conceptions may be, the product-moment coefficient of correlation is likely always to have an essential part to play in the application of statistical method. When two variables are approximately normally correlated—and there are good reasons for supposing that there is considerable latitude in the stringency of this approximation—the coefficient by itself provides a completely adequate measure of the intensity of association. Lying between $-1$ and $+1$ and being independent of any units of measurement, in fact a pure measure of correlation, $r$ has a direct and simple appeal. In providing tables and charts which deal with the relationship between $r$, $\rho$ and $n$ we are not, I feel certain, merely toying with an interesting historical relic of the past.

E. S. PEARSON

DEPARTMENT OF STATISTICS
UNIVERSITY COLLEGE
LONDON
*February* 1938

*Editorial Note to* 1954 *re-issue.* We are indebted to Dr A. K. Gayen for pointing out certain errors in R. A. Fisher's original formulae (7), (8) and (9) (p. viii) and for computing the consequent corrections to the columns headed Approximation I and II in Table VIII (p. xxxii).

# INTRODUCTION

## SECTION I. INTRODUCTORY

Throughout the following pages it will be assumed that we are dealing with two correlated random variables $x$ and $y$ for which the joint probability law is

$$p(x,y) = \frac{1}{2\pi\sigma_1\sigma_2\sqrt{1-\rho^2}} e^{-\frac{1}{2(1-\rho^2)}\left\{\frac{(x-\xi_1)^2}{\sigma_1^2} - \frac{2\rho(x-\xi_1)(y-\xi_2)}{\sigma_1\sigma_2} + \frac{(y-\xi_2)^2}{\sigma_2^2}\right\}}. \qquad \ldots\ldots(1)$$

The probability laws of $x$ and $y$ thus each follow the normal distribution, i.e.

$$p(x) = \frac{1}{\sqrt{2\pi}\sigma_1} e^{-\frac{1}{2\sigma_1^2}(x-\xi_1)^2}; \quad p(y) = \frac{1}{\sqrt{2\pi}\sigma_2} e^{-\frac{1}{2\sigma_2^2}(y-\xi_2)^2}. \qquad \ldots\ldots(2)$$

We shall be concerned with the relation between the correlation coefficient

$$r = \frac{\sum\limits_{i=1}^{n}(x_i-\bar{x})(y_i-\bar{y})}{\sqrt{\sum\limits_{i=1}^{n}(x_i-\bar{x})^2 \sum\limits_{i=1}^{n}(y_i-\bar{y})^2}},^{*} \qquad \ldots\ldots(3)$$

calculated from a sample of size $n$, randomly drawn from an infinite population represented by the normal bivariate distribution law (1), and the correlation coefficient, $\rho$, of that population. The tables which follow give the ordinates and areas of the curves representing the sampling distribution of $r$ for differing values of $n$ and $\rho$.

The form of the distribution of $r$ for $\rho = 0$ and any $n$ was first given by "Student"[1], while for the general distribution for any $n$ and $\rho$ we are indebted to R. A. Fisher[2]. "Student's" results were obtained by empirical means, but R. A. Fisher's proof, dependent upon geometrical argument and analogy, is mathematically rigorous.[†]

*Distribution of $r$ for any $n$, $\rho = 0$.*

$$p(r \mid n, \rho = 0) = \frac{\Gamma\left(\dfrac{n-1}{2}\right)}{\sqrt{\pi}\,\Gamma\left(\dfrac{n-2}{2}\right)}(1-r^2)^{\frac{n-4}{2}}. \qquad \ldots\ldots(4)$$

*Distribution of $r$ for any $n$ and any $\rho$.*

$$p(r \mid n, \rho) = \frac{(1-\rho^2)^{\frac{n-1}{2}}}{\pi(n-3)!}(1-r^2)^{\frac{n-4}{2}} \frac{d^{n-2}}{d(r\rho)^{n-2}}\left(\frac{\arccos(-\rho r)}{\sqrt{1-\rho^2 r^2}}\right). \qquad \ldots\ldots(5)$$

Among other papers on the form of the distribution we may notice one by H. E. Soper[3] and one published from the Department of Applied Statistics entitled "A Cooperative Study"[4]. In the latter paper tables were included of the ordinates of the distribution of $r$ for given values of $n$ and $\rho$. From these tables it was possible to calculate by quadrature the area under the curves and therefore the probability integral of $r$.

---

* $\bar{x}$ and $\bar{y}$ denote the sample means, i.e. $\quad \bar{x} = \dfrac{1}{n}\sum\limits_{i=1}^{n} x_i, \quad \bar{y} = \dfrac{1}{n}\sum\limits_{i=1}^{n} y_i.$

† An alternative proof of the general distribution is given in the Appendix. While it follows closely the lines of Professor Fisher's proof the distribution is reached by means of algebraic transformations. It is hoped that the proof in this form will be of use to those who are unable to visualize a solid figure. For an elegant proof using characteristic functions see S. Kullback, *Annals of Mathematical Statistics*, v, 4 (1934), pp. 263–305.

This method of quadrature was the only one by which a precise measure of the significance of $r$ could be obtained until the publication of an important paper by R. A. Fisher[5], which contains the now well-known transformation of $r$ referred to below.

Fisher showed that by a suitable transformation of $r$ and $\rho$ the distribution curves of $r$ could be transformed approximately into normal curves.

Write
$$z' = \tfrac{1}{2} \log\left(\frac{1+r}{1-r}\right); \quad \zeta = \tfrac{1}{2} \log\left(\frac{1+\rho}{1-\rho}\right). \qquad \ldots\ldots(6)$$

From $p\,(z' \mid n, \zeta)$ we obtain the moments in series:

$$\text{Mean } (z') = \zeta + \frac{\rho}{2\,(n-1)}\left\{1 + \frac{5+\rho^2}{4\,(n-1)} + \ldots\right\}, \qquad \ldots\ldots(7)$$

$$\sigma_{z'}^2 = \frac{1}{n-1}\left\{1 + \frac{4-\rho^2}{2\,(n-1)} + \frac{22-6\rho^2-3\rho^4}{6\,(n-1)^2} + \ldots\right\}, \qquad \ldots\ldots(8)$$

$$\left.\begin{array}{l} \beta_1 = \dfrac{\rho^6}{(n-1)^3} + \ldots, \\[3mm] \beta_2 = 3 + \dfrac{2}{n-1} + \dfrac{4+2\rho^2-3\rho^4}{(n-1)^2} + \ldots \end{array}\right\}. \qquad \ldots\ldots(9)$$

It is seen that, provided $n$ be of reasonable size, $z'$ is approximately normally distributed. Therefore instead of calculating the areas of the distribution curves of $r$, the problem can be put into terms of the area of the normal curve. This transformation is simple and, as will be seen later, it gives accurate results over the whole range of values of $\rho$ and $r$.

The idea of constructing tables of the probability integral of $r$ was suggested to the present writer by Karl Pearson in 1931. It was his desire to complete the series of extensive tables, calculated in his laboratories, associated with the fundamental tests of statistical sampling theory.* It is true that R. A. Fisher's $z'$-transformation is sufficient for most practical purposes, and is very simple to apply when $n$ is large enough for us to take the expectation of $z'$ as $\zeta$ and the standard error as $1/\sqrt{n-3}$. Nevertheless it is the function of a basic table, such as that given here, to form a standard against which the adequacy of approximations may be judged. This table should also be of some permanent value in providing a point of departure for the construction of useful working tools such as the charts which have been included in this volume.

## SECTION II. CONSTRUCTION OF THE TABLES

Separate tables are given for sizes of sample from $n = 3$ to $n = 25$. For each size of sample ten distributions of $r$ have been tabulated for values of $\rho = \cdot0, \cdot1, \cdot2, \ldots \cdot9$, each probability integral and ordinate being accurate to five decimal places. Tables are also given for $n = 50, 100, 200$ and $400$ and a method of logarithmic interpolation is employed to obtain any intervening $n$. It was originally intended to include also tables for size of sample $n = 800$ and $n = 1600$. A small portion of each of these tables was calculated, but the addition of these extra values did not improve the results of logarithmic interpolation. In view of the labour involved (no ordinates of the curves had been calculated previously), it was decided to omit them.

When these tables were begun some time was spent in searching for a suitable method or methods by which the probability integral might be obtained. It would be superfluous to state these in detail, but it may be mentioned that the fitting of Pearson curves to the ordinates already tabled was carried out. This method had to be rejected, since the results obtained did not reach the required standard of accuracy. In 1932 F. Garwood[6] gave exact formulae for the probability integrals, but his results were expressed in terms of the ordinates, which themselves were only tabulated to five decimal places, and therefore the

---

\* Most notable among these are the Tables of the Incomplete Gamma- and the Incomplete Beta-Functions.

areas calculated from his formulae had also to be rejected. Quadrature was the only method which gave accuracy, and it proved the simplest to use. A suitable quadrature formula was found and this was used throughout.

*Case $\rho = 0$.*

The probability law of $r$ for any $n$, when $\rho$ is zero, may be written

$$p(r \mid n, \rho = 0) = \frac{\Gamma\left(\dfrac{n-1}{2}\right)}{\sqrt{\pi}\,\Gamma\left(\dfrac{n-2}{2}\right)} (1-r^2)^{\frac{n-4}{2}}. \qquad \ldots\ldots(10)$$

Write $(1+r) = 2\omega$; then

$$\frac{\displaystyle\int_{-1}^{r} p(r \mid n, \rho = 0)\, dr}{\displaystyle\int_{-1}^{+1} p(r \mid n, \rho = 0)\, dr} = I_{\frac{1}{2}(1+r)}\left(\frac{n-2}{2}, \frac{n-2}{2}\right). \qquad \ldots\ldots(11)$$

Or, alternatively, write $(1-r^2) = z$; then

$$\frac{\displaystyle\int_{-1}^{r} p(r \mid n, \rho = 0)\, dr}{\displaystyle\int_{-1}^{+1} p(r \mid n, \rho = 0)\, dr} = \begin{cases} \frac{1}{2} I_{1-r^2}\left(\dfrac{n-2}{2}, \dfrac{1}{2}\right) & r - \text{ve}, \\[2ex] 1 - \frac{1}{2} I_{1-r^2}\left(\dfrac{n-2}{2}, \dfrac{1}{2}\right) & r + \text{ve}. \end{cases} \qquad \ldots\ldots(12)$$

It is seen that the probability integral of (10) can be transformed into either of two incomplete $B$-function ratios. Quadrature was used to obtain the probability integral and the results were checked from the $B$-function tables.

*Case $0 < \rho < +1$.*

The method of quadrature used throughout to construct the tables is that due to Gregory (see for e.g. (7)).

If $z_0, z_1, \ldots z_p$ represent the ordinates of the curve, $f(x)$, at equal distances $h$ apart, and $\Delta$ has its usual meaning as a difference symbol, then

$$\frac{1}{h}\int_{z_0}^{z_p} f(x)\, dx = [\tfrac{1}{2}z_0 + z_1 + z_2 + \ldots + z_{p-2} + z_{p-1} + \tfrac{1}{2}z_p] - \tfrac{1}{12}[\Delta z_{p-1} - \Delta z_0] - \tfrac{1}{24}[\Delta^2 z_{p-2} + \Delta^2 z_0]$$
$$- \tfrac{19}{720}[\Delta^3 z_{p-3} - \Delta^3 z_0] - \tfrac{3}{160}[\Delta^4 z_{p-4} + \Delta^4 z_0] - \tfrac{863}{60480}[\Delta^5 z_{p-5} - \Delta^5 z_0] - \tfrac{275}{24192}[\Delta^6 z_{p-6} + \Delta^6 z_0] - \ldots \text{etc.}$$
$$\ldots\ldots(13)$$

Formula (13) as quoted appears to entail much laborious calculation. If, however, we replace the differences by ordinates, i.e. if we write

$$\Delta z_{p-1} = z_p - z_{p-1}, \quad \Delta^2 z_{p-2} = z_p + z_{p-2} - 2z_{p-1},$$

and so on, then the formula is at once simplified and can be written in such a way that a minimum of calculation is necessary.

*Gregory's formula (i)* (up to and including sixth differences).

$$\frac{1}{h}\int_{z_0}^{z_p} f(x)\, dx = [\tfrac{1}{2}z_0 + z_1 + \ldots + z_{p-1} + \tfrac{1}{2}z_p] \qquad + 0\cdot471{,}429\ [z_{p-3} + z_3]$$
$$- 0\cdot195{,}776\ [z_p + z_0] \qquad - 0\cdot260{,}607\ [z_{p-4} + z_4]$$
$$+ 0\cdot460{,}384\ [z_{p-1} + z_1] \qquad + 0\cdot082{,}474\ [z_{p-5} + z_5] \qquad \ldots\ldots(14\,a)$$
$$- 0\cdot546{,}536\ [z_{p-2} + z_2] \qquad - 0\cdot011{,}367\ [z_{p-6} + z_6]$$

*Gregory's formula (ii)* (up to and including eighth differences).

$$\frac{1}{h}\int_{z_0}^{z_p} f(x)\,dx = [\tfrac{1}{2}z_0 + z_1 + \dots + z_{p-1} + \tfrac{1}{2}z_p] \qquad\qquad -1\cdot140{,}564\ \ [z_{p-4}+z_4]$$

$$-0\cdot213{,}025\ ^-[z_p+z_0] \qquad\qquad +0\cdot720{,}944\ \ [z_{p-5}+z_5]$$

$$+0\cdot589{,}020\ \ [z_{p-1}+z_1] \qquad\qquad -0\cdot297{,}855\ ^-[z_{p-6}+z_6] \qquad \dots\dots(14\,b)$$

$$-0\cdot964{,}015\ ^-[z_{p-2}+z_2] \qquad\qquad +0\cdot072{,}497\ \ [z_{p-7}+z_7]$$

$$+1\cdot240{,}890\ \ [z_{p-3}+z_3] \qquad\qquad -0\cdot007{,}893\ \ [z_{p-8}+z_8]$$

The simplicity of either of these formulae can be seen at a glance. Given the ordinates of the distribution of $r$ for a fixed $\rho$ and $n$, we first ascertain the chordal areas from $z_0$ to each succeeding ordinate. The corrective term involving $z_0, z_1, \dots z_6$, or $z_0, z_1, \dots z_8$, is next calculated, this being constant for the given distribution. There is now left only one calculation for each probability integral required. Either formula $(14\,a)$ or $(14\,b)$ was used throughout. The aim has been to provide tables of the probability integral which would be correct to five decimal places and it is hoped that this has been achieved. It is possible that in some cases the figure in the fifth decimal place may be one unit wrong. If this occurs it will be due to the fact that the ordinates from which the probability integral is obtained are themselves only correct to five decimal places.

The tables of ordinates published in "A Cooperative Study" were the foundation of the present tables. The values of the probability integral corresponding to these tables of ordinates were completed in 1934, but when examples came to be worked out from them, it was found that the differences were too large to allow of accurate interpolation. Accordingly more ordinates, and therefore more values of the probability integral, were calculated as it was found they were needed, and this accounts for the somewhat "lopsided" character of the table. Not all the ordinates which were calculated have been included. It is felt that the tables of ordinates as printed will give a sufficiently reasonable idea of the shape of the curves, and that interpolation between the ordinates is adequate for all practical purposes.

### Calculation of Ordinates

Two methods were adopted for the calculation of the ordinates which were interpolated between those already published in "A Cooperative Study". It is known that

$$y_3 = \frac{1-\rho^2}{\pi\sqrt{1-r^2}}\left(\frac{1}{1-\rho^2 r^2} + \frac{\rho r \arccos(-\rho r)}{(1-\rho^2 r^2)^{\frac{3}{2}}}\right),$$

$$y_4 = \frac{(1-\rho^2)^{\frac{3}{2}}}{\pi}\left(\frac{3\rho r}{(1-\rho^2 r^2)^2} + \frac{(1+2\rho^2 r^2)\arccos(-\rho r)}{(1-\rho^2 r^2)^{\frac{5}{2}}}\right),$$

where $y_{n_1}$ is the ordinate of the curve $p(r\,|\,n, \rho)$ for $n = n_1$.

Hence by repeated application of the recurrence formula

$$y_{n+2} = \frac{2n-1}{n-1}\kappa_1 y_{n+1} + \frac{n-1}{n-2}\kappa_2 y_n,$$

where

$$\kappa_1 = \frac{\rho r \sqrt{1-\rho^2}\sqrt{1-r^2}}{1-\rho^2 r^2}, \qquad \kappa_2 = \frac{(1-\rho^2)(1-r^2)}{1-\rho^2 r^2},$$

it was possible to obtain the ordinates for succeeding values of $n$, ranging from 5 to 25, for a given $r$ and $\rho$. Where isolated ordinates were required the following formula was used:

$$y_n = \frac{n-2}{\sqrt{n-1}}(1-\rho^2)^{\frac{3}{2}}\chi(\rho, r)\left[1 + \frac{\phi_1}{n-1} + \frac{\phi_2}{(n-1)^2} + \frac{\phi_3}{(n-1)^3} + \frac{\phi_4}{(n-1)^4} + \dots\right],$$

where
$$\log \chi (\rho, r) = -(n-1)\log \chi_1 - \log \chi_2,$$

$$\chi_1 = \frac{1-\rho r}{\{(1-\rho^2)(1-r^2)\}^{\frac{1}{2}}}, \qquad \chi_2 = \frac{\sqrt{2\pi}\{(1-\rho^2)(1-r^2)\}^{\frac{3}{2}}}{(1-\rho r)^{\frac{5}{2}}},$$

and
$$\phi_1 = \frac{r\rho+2}{8}, \qquad \phi_2 = \frac{(3r\rho+2)^2}{128}, \qquad \phi_3 = \frac{5\{15\,(r\rho)^3 + 18\,(r\rho)^2 - 4\,(r\rho) - 8\}}{1024},$$

$$\phi_4 = \frac{3675\,(r\rho)^4 + 4200\,(r\rho)^3 - 2520\,(r\rho)^2 - 3360\,(r\rho) - 336}{32768}.$$

This last formula is only approximate, and was originally intended to be used for $n > 25$, but by checking against the true results obtained from the recurrence formula, it was found to be reasonably accurate, even for $n$ as low as 10.

### Checks

The differences of each table of the probability integral were found on completion, and this process was sufficient to detect any gross errors which were made. A closer check presented and still presents more difficulty. In the main, panel-area formulae were used to check differences between successive tabled areas.

*Panel Area between $z_{-1}$ and $z_0$ (5 ordinates).*

$$\int_{z_{-1}}^{z_0} f(x)\,dx = \frac{h}{720}[-19z_{-2} + 346z_{-1} + 456z_0 - 74z_1 + 11z_2]. \qquad \ldots\ldots(15a)$$

*Panel Area between $z_0$ and $z_1$ (6 ordinates).*

$$\int_{z_0}^{z_1} f(x)\,dx = \frac{h}{1440}[475z_0 + 1427z_1 - 798z_2 + 482z_3 - 173z_4 + 27z_5]. \qquad \ldots\ldots(15b)$$

*Panel Area between $z_0$ and $z_1$ (8 ordinates).*

$$\int_{z_0}^{z_1} f(x)\,dx = \frac{h}{120960}[36799z_0 + 139849z_1 - 121797z_2 + 123133z_3 - 88547z_4 + 41499z_5 - 11351z_6 + 1375z_7].$$
$$\ldots\ldots(15c)$$

In cases where there was disagreement between the results obtained from formulae (14) and formulae (15), it was usually found to be because formulae (15) did not include enough differences, and the result obtained from (14) was allowed to stand, after being carefully checked by both formulae (14). It is, however, too much to hope that tables of this size will be completely free from error, and the writer will be grateful for notification when any errors are found.

## SECTION III. INTERPOLATION. METHODS AND ILLUSTRATION

For very many purposes modern statistical method calls for little more knowledge of a probability integral than a tabulation of the various "significance levels", and for those who wish only to use these levels, charts are published at the end of this volume. Sometimes, however, it is necessary to obtain a probability by interpolation, so the interpolation formulae which have proved most accurate will be stated briefly. These formulae are general adaptations of those already given by Karl Pearson[8]. The calculation is straightforward but in most cases somewhat laborious.

Suppose that for a given sample of size $n$, it is desired to find the probability integral $z_{\theta\chi}$, for given $\rho_\chi$ and $r_\theta$. It is simplest to set out the ordinates used in a diagram (see Fig. 1).

*Using first differences.*

$$z_{\theta\chi} = \phi\psi z_{0,0} + \phi\chi z_{0,1} + \theta\psi z_{1,0} + \theta\chi z_{1,1}. \qquad \ldots\ldots(16)$$

*Up to and including third differences.*

$$z_{\theta\chi} = \{1 + \tfrac{1}{2}(\theta\phi + \chi\psi)\}\{\phi(\psi z_{0,0} + \chi z_{0,1}) + \theta(\psi z_{1,0} + \chi z_{1,1})\}$$
$$- \tfrac{1}{6}\theta\phi\{(1+\phi)(\psi z_{-1,0} + \chi z_{-1,1}) + (1+\theta)(\psi z_{2,0} + \chi z_{2,1})\}$$
$$- \tfrac{1}{6}\chi\psi\{(1+\psi)(\phi z_{0,-1} + \theta z_{1,-1}) + (1+\chi)(\phi z_{0,2} + \theta z_{1,2})\}. \qquad \dots\dots(17)$$

*Illustration.*

Given $n = 20$, $\rho = \cdot277$, $r = \cdot185$, what is the value, by interpolation, of

$$\int_{-1}^{+\cdot185} p(r \mid n = 20,\ \rho = \cdot277)\,dr\ ? \qquad \dots\dots(18)$$

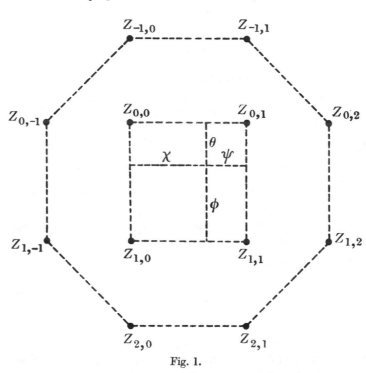

Fig. 1.

In order to test the accuracy of the interpolation formulae (16) and (17), ordinates of the distribution of $r$ for $n = 20$ and $\rho = \cdot277$ were calculated, and (18) found by quadrature. This was found to be $\cdot33007$. The values of $z$ required for interpolation are:

|  | $\rho = \cdot1$ | $\rho = \cdot2$ | $\rho = \cdot3$ | $\rho = \cdot4$ |
|---|---|---|---|---|
| $r = \cdot10$ |  | $\cdot32570$ | $\cdot18188$ |  |
| $r = \cdot15$ | $\cdot57957$ | $\cdot40536$ | $\cdot24280$ | $\cdot11854$ |
| $r = \cdot20$ | $\cdot66144$ | $\cdot49060$ | $\cdot31482$ | $\cdot16617$ |
| $r = \cdot25$ |  | $\cdot57806$ | $\cdot39668$ |  |

Here $\theta = \cdot7$ and $\chi = \cdot77$. Applying (16) we see that

$$\int_{-1}^{+\cdot185} p(r \mid n = 20,\ \rho = \cdot277)\,dr = \cdot336785,$$

and using (17),

$$\int_{-1}^{+\cdot185} p(r \mid n = 20,\ \rho = \cdot277)\,dr = \cdot33021.$$

Linear interpolation as in (16) is seen to be a rough approximation, the interpolated value differing from the true value in the third decimal place; but the result obtained from (17) is reasonable, there being a difference of only ·00014 between the true and interpolated values.

Formula (17) is for use in the middle of the table. We may quote one further formula, which will be of use when interpolating on the border of the table. For this, the scheme will be:

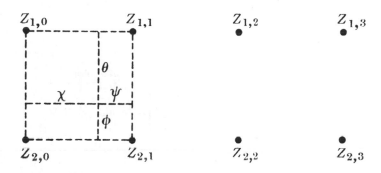

Fig. 2.

and $z_{\theta\chi}$ is obtained by substitution in the following:

$$
\begin{aligned}
z_{\theta\chi} = {} & \phi\{1+\tfrac{1}{3}\theta(1+\phi)-\tfrac{1}{6}\theta(1+\theta)\}\{\psi z_{0,0}+\chi z_{1,1}\}+\theta\{1+\tfrac{1}{3}\phi(1+\theta)-\tfrac{1}{6}\phi(1+\phi)\}\{\psi z_{2,0}+\chi z_{2,1}\} \\
& -\tfrac{1}{6}\chi\psi\{(4+\psi)[\phi z_{1,0}+\theta z_{2,0}]-3(3+\psi)[\phi z_{1,1}+\theta z_{2,1}]\}-\tfrac{1}{6}\chi\psi\{3(2+\psi)[\phi z_{1,2}+\theta z_{2,2}]-(1+\psi)[\phi z_{1,3}+\theta z_{2,3}]\} \\
& -\tfrac{1}{6}\theta\phi\{(1+\phi)[\psi z_{0,0}+\chi z_{0,1}]+(1+\theta)[\psi z_{3,0}+\chi z_{3,1}]\}.
\end{aligned}
$$
......(19)

Using (19) on the same figures as before, the tabulated values required are:

|  |  |  |  |
|---|---|---|---|
| ·32570 | ·18188 |  |  |
| ·40536 | ·24280 | ·11854 | ·04383 |
| ·49060 | ·31482 | ·16617 | ·06696 |
| ·57806 | ·39668 |  |  |

and we get

$$\int_{-1}^{+\cdot185} p\left(r \mid n=20,\, \rho=\cdot277\right)dr = \cdot 32979,$$

differing from the true value by ·00028.

These three formulae should be enough for interpolation into all parts of the table where the specific size of sample is given.

### Possibility of Graphical Interpolation

Formulae (17) and (19), while giving accurate results, often need laborious calculation in application. Graphical interpolation is possible if the probability integral is required to be correct to two units only. Let us consider the possibility of finding (18) graphically. Plot $r$ along the abscissa, as ordinates take the corresponding values of the probability integral, and plot the two curves $\rho = \cdot 2$ and $\rho = \cdot 3$, for size of sample $n = 20$. Provided these be drawn on a reasonable scale it is possible to read off the probability integral

Diagram showing Graphical Interpolation for size of sample $n = 20$

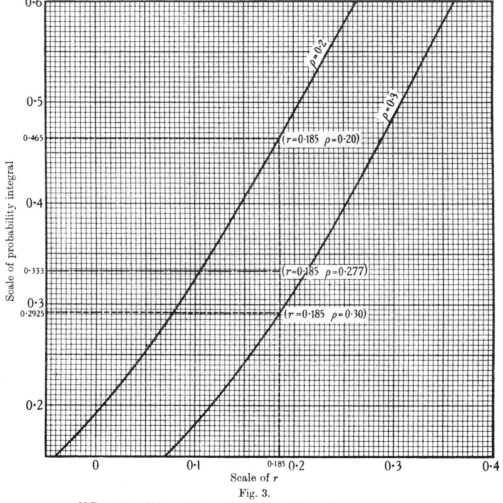

Fig. 3.

N.B.   $\cdot 465 - \cdot 2925 = \cdot 1725$,   $\cdot 1725 \times \cdot 23 = \cdot 0397$,   $\cdot 0397 + \cdot 2925 = \cdot 3332$.

for $\rho = \cdot 277$ and $r = \cdot 185$. The probability integral obtained by this method was $\cdot 333$, which is slightly more accurate than the result obtained by linear interpolation by formula (16). This result is sufficiently good if a rough guide is all that is required, and hence it is suggested that this method be used when a quick approximation to the probability integral is desired.

### Logarithmic Interpolation for $n > 25$

The interpolation formulae discussed previously in this section are chiefly applicable to the first part of the table, where the size of sample is from $n = 3$ to $n = 25$. The second part of the table gives the ordinates and areas for samples of sizes 50, 100, 200 and 400. We shall now consider interpolation for the probability integral in this second part.

Consider Lagrange's mid-point formula for the graduation of four ordinates at equal distance, $h$, apart (9). Suppose the four ordinates to be $z_0$, $z_1$, $z_2$, $z_3$. Then if $z_x$, the required ordinate, be situated, as shown in Fig. 4, at a distance $hx$ from the foot of the first ordinate, $z_0$,

$$z_x = z_0 + \tfrac{1}{6}\{x^3[(z_3 - z_2) - 2(z_2 - z_1) + (z_1 - z_0)]$$
$$- 3x^2[(z_3 - z_2) - 3(z_2 - z_1) + 2(z_1 - z_0)]$$
$$+ x[2(z_3 - z_2) - 7(z_2 - z_1) + 11(z_1 - z_0)]\}. \qquad \ldots\ldots(20)$$

We may use this formula in order to construct tables for large values of $n$. The method is as follows. Using a logarithmic scale for $n$ it is seen that the probability integral is tabled at five equidistant values of this new variable, corresponding to an argument interval of log 2, i.e. at

$$\log 25, \quad \log 25 + \log 2, \quad \log 25 + 2\log 2, \quad \log 25 + 3\log 2, \quad \log 25 + 4\log 2.$$

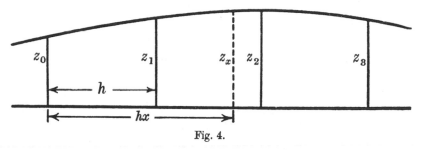

Fig. 4.

If then, for example, it is wished to interpolate for $n = 160$, the values of $z_0$, $z_1$, $z_2$, $z_3$ to be inserted in (20) will be the values of the probability integral at $n = 50, 100, 200$ and $400$, and

$$x = (\log 160 - \log 50)/\log 2 = 1\!\cdot\!678,072.$$

Both for illustration and as a test of the accuracy of (20), the probability integral for $\rho = \cdot8$ and $n = 160$ is given below for several values of $r$. The exact result obtained by quadrature from ordinates calculated from the formulae of p. x is also given.

Table I.  *Probability integral for $\rho = \cdot8$*

| $n$ \ $r$ | 50 | 100 | 200 | 400 | 160 (by four-point) | 160 (exact) |
|---|---|---|---|---|---|---|
| ·60 | ·00237 | ·00005⁻ | — | — | ·00000 | ·00000 |
| ·65 | ·01146 | ·00067 | — | — | ·00000 | ·00003 |
| ·70 | ·04945 | ·01023 | ·00054 | — | ·00168 | ·00172 |
| ·75 | ·17732 | ·10000 | ·03643 | ·00582 | ·05422 | ·05388 |
| ·80 | ·47693 | ·48387 | ·48864 | ·49200 | ·48729 | ·48729 |
| ·85 | ·84835⁺ | ·93506 | ·98547 | ·99909 | ·97323 | ·97402 |
| ·90 | ·99374 | ·99986 | 1·00000 | 1·00000 | 1·00000 | 1·00000 |

It will be seen that the agreement between the exact values and the results of applying (20) is fairly good. These results are again used in Section V, where they are compared with the results of applying the $z'$-transformation.

If the probability integral is required for a value of $\rho$ lying between the tabled values, it will be necessary to construct two such tables as that given above for the tabled values of $\rho$ lying immediately above and immediately below the desired value. These obtained, the method of graphical interpolation suggested on p. xiv will give the required value correct to two decimal places. The writer is of the opinion that the increase in accuracy would be very little if four tables such as the above were calculated and interpolation by second differences employed.

It may be that the probability integral for a given $\rho$, $r$ and large $n$ will be wanted quickly and approximately. To this end diagrams, numbered I–X, have been constructed and will be found at the end of this introduction. A separate diagram is given for each value of $\rho$ which is tabled. $n$ is plotted along the abscissa and curves are drawn of the probability integral for given values of $r$. Hence for a given $n$ and $\rho$, the values of the probability integral for $r$ at an argument of $\cdot 05$ can be read off immediately. For a value of $\rho$ lying between the tabled values the probability integral for the $\rho$ lying immediately above and immediately below may be found, and the method of p. xiv employed.

## SECTION IV. USE OF THE TABLES AND ILLUSTRATIONS

In whatever field he is working the applied mathematician is concerned with bridging a gap between a conceptual mathematical model and the data of his experience. Thus the mathematical statistician has need to consider how a precise but abstract theory of probability may be employed most usefully to draw inferences from observation. There may be differences of opinion as to the best methods of answering some of the questions discussed below, but there should be general agreement on the importance of defining with precision the terms of the questions asked and the principles adopted in answering them. For this reason it has seemed well to introduce the illustrations of the use of tables and charts given below with a somewhat formal statement of the guiding principles which the writer has followed in their solution. At the same time no claim is made that the types of problem illustrated are exhaustive, nor that the approach to their solution is unique; the problems discussed might certainly have been formulated in a different way, the same tables or charts being used in their solution.

The problems of practical statistics which call for the introduction of the theory of probability will almost always be found at the root to be concerned with the relation between what may be termed the collective character* or characters of a sample, and the collective character or characters of the population from which the sample has been randomly drawn. It is rarely possible to determine a collective character directly from a knowledge of the population, for in most cases the populations studied are either infinite or very large, and even if it were possible the question would arise as to whether we are justified in spending much time and labour in so doing. In practice we take our randomly drawn sample, and use it to obtain information about the population we are studying. Provided the sample is of reasonable size we may do this with a fair degree of accuracy.

Thus if the collective character under consideration in the population is the coefficient of correlation, $\rho$, between two variable characteristics $x$ and $y$, we may wish to obtain answers to such questions as the following:

1. Are the observed data in a sample consistent with the hypothesis that in the population

   (i) $\rho \geqslant \rho_0$, (ii) $\rho \leqslant \rho_0$, (iii) $\rho = \rho_0$,

where $\rho_0$ is some specified value?

2. How may the observed data in a sample be used to the best advantage in order to calculate limits $\rho_b$ and $\rho_a$, such that the statements

   (i) $\rho_b \leqslant \rho \leqslant +1$, (ii) $-1 \leqslant \rho \leqslant \rho_a$, (iii) $\rho_b \leqslant \rho \leqslant \rho_a$,

regarding the unknown value of $\rho$ in the population sampled may be made with given degrees of confidence?

3. Suppose we have $k$ $(k \geqslant 2)$ independent randomly drawn samples. Are the observed data consistent with the hypothesis that in the $k$ populations sampled the coefficients of correlation, $\rho_1, \rho_2, \ldots \rho_k$, are all equal to

   (i) a specified value, $\rho_0$,

   (ii) a common but unspecified value, $\rho$?

---

\* As far as the writer is aware, the term "collective character" was first used by J. Neyman in his paper "On two different aspects of the Representative Method", *J. Roy. Statist. Soc.* xcvii (1934), 561. He obtained it by translating a Russian phrase. Applied respectively to the sample and the population it is used instead of the terms "statistic" and "parameter".

Questions 1 and 3 are concerned with the testing of a statistical hypothesis. We ask whether the data are consistent with a specified hypothesis or not; if we decide to say "No", we risk one form of error, that of rejecting the hypothesis when it is true. In any precise test it should be possible to fix what is commonly called the significance level so as to control this risk of error, say $\epsilon$, at some prescribed figure, e.g. $\epsilon = \cdot 01$ or $\cdot 05$. If, on the other hand, we say "Yes, the data appear consistent with the hypothesis", it may nevertheless happen that the hypothesis tested is false, and that through the inadequacy of the data we have failed to detect the fact that some alternative hypothesis is true. These two types of error cannot altogether be avoided in testing a statistical hypothesis; if the test is arranged to reduce the risk of the first it will increase that of the second, and *vice versa*. Some illustration of this is given below. Question 2 leads to the problem of interval estimation. Here again two considerations will be taken into account; for example, in question 2 (iii) we must consider both (a) the risk that the interval $\rho_b$ to $\rho_a$ fails to cover the unknown value of $\rho$ in the population sampled, and (b) the breadth of the interval $\rho_b$ to $\rho_a$, which for the given risk we should like to be as narrow as possible. In this connexion the definition of narrowness will have to be discussed.

It is necessary to emphasize again that in the work which follows we assume that the samples have been randomly drawn from some population in which the variates $x$ and $y$ under consideration follow the normal bivariate distribution given in equation (1) above. That the results will be approximately true even when the distribution is far removed from normal is suggested by the empirical work of E. S. Pearson* and others[10, 11], but so far it has not been established mathematically.

### *Question* 1

A sample consisting of $n$ pairs of observations is available.

(1) Is $\rho \geqslant \rho_0$? Here the admissible hypotheses alternative to that tested are that $\rho < \rho_0$.

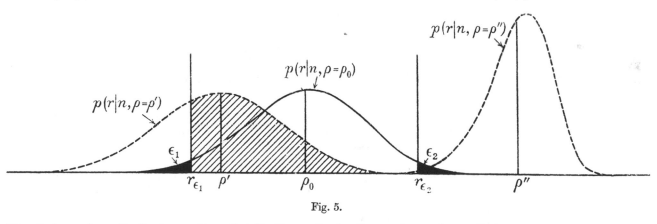

Fig. 5.

Fig. 5 shows hypothetical sampling distributions of $r$ for three values of $\rho$, and size of sample equal to $n$. We want to test if $\rho \geqslant \rho_0$ and, bearing in mind the alternative hypotheses, we may suggest the following rule:

Reject the hypothesis tested if $r < r_{\epsilon_1}$, where

$$\epsilon_1 = \int_{-1}^{r_{\epsilon_1}} p\,(r \mid n, \rho_0)\,dr. \qquad \qquad \ldots\ldots(21)$$

Accordingly, if $\rho > \rho_0$, e.g. equals $\rho''$, then the chance of rejecting this hypothesis when it is true will be $< \epsilon_1$. If $\rho = \rho_0$ the chance of rejection when the hypothesis is true is exactly equal to $\epsilon_1$. On the other hand,

---

* E. S. Pearson, *Biometrika,* xxi (1929), 356–60: "The results suggest that the normal bivariate surface can be mutilated and distorted to a remarkable degree without affecting the frequency distribution of $r$."

if in fact $\rho = \rho'$ and the hypothesis tested were false, the risk of failing to detect this is measured by the shaded area in Fig. 5 under $p\,(r\,|\,n,\,\rho = \rho')$ to the right of the ordinate at $r_{\epsilon_1}$. Clearly in the case illustrated, if $\rho = \rho'$, we should be almost as likely as not to fail to detect the fact that the hypothesis $\rho \geqslant \rho_0$ was false.

(ii) Is $\rho \leqslant \rho_0$? The admissible alternative hypotheses will be that $\rho > \rho_0$, and in a way similar to the above we may set up the following rule:

Reject $H_0$, the hypothesis tested, if $r > r_{\epsilon_2}$, where

$$\epsilon_2 = \int_{r_{c_2}}^{+1} p\,(r\,|\,n,\,\rho_0)\,dr. \qquad\qquad \ldots\ldots(22)$$

The four charts provided at the end of this volume can be used to obtain the limits $r_{\epsilon_1}$ and $r_{\epsilon_2}$, for values of $\epsilon_1$ and $\epsilon_2$ equal to ·005, ·01, ·025 and ·05, and for varying values of $n$ and $\rho$. If it is desired to take a different probability level from those given, it will be possible to find the necessary figures from the main tables.

(iii) Is $\rho = \rho_0$? Here the admissible alternative hypotheses to that tested will be that $-1 < \rho < \rho_0$ (e.g. $\rho = \rho'$), and $\rho_0 < \rho < +1$ (e.g. $\rho = \rho''$). The test we may set up will be a combination of those used in questions 1 (i) and 1 (ii). We may postulate as our rule:

Reject the hypothesis tested if $r > r_{\epsilon_2}$ or $r < r_{\epsilon_1}$, where $\epsilon_1$ and $\epsilon_2$ are defined as above and

$$\epsilon_1 = \epsilon_2 = \tfrac{1}{2}\epsilon, \qquad\qquad \ldots\ldots(23)$$

where $\epsilon$ will be the chance of rejecting the hypothesis tested when it is true. It is easily seen that this risk of rejection will be the same if $\epsilon_1$ is not equal to $\epsilon_2$ so long as

$$\epsilon_1 + \epsilon_2 = \epsilon. \qquad\qquad \ldots\ldots(24)$$

Neyman and Pearson [12] have shown that in certain cases if we take

$$\epsilon_1 = \epsilon_2,$$

we shall be less likely to reject the hypothesis tested when it is false, than when it is true. A test leading to such consequences they have termed biased.* Such a situation would arise if in Fig. 5, for any curves of the system $p\,(r\,|\,n,\,\rho)$ having $\rho$ either below or above $\rho_0$, the proportional area included between ordinates at $r_{\epsilon_1}$ and $r_{\epsilon_2}$ was greater than $1-\epsilon$. It was decided to test whether the rejection limits obtained from the distribution of $r$ by taking equal tail areas were biased. Following the procedure of Neyman and Pearson an unbiased test may be obtained by solving for $r_{\epsilon_1}$ and $r_{\epsilon_2}$ from the two following equations:

$$\int_{r_{\epsilon_1}}^{r_{\epsilon_2}} p\,(r\,|\,n,\,\rho)\,dr = 1-\epsilon, \qquad\qquad \ldots\ldots(25)$$

$$\frac{d}{d\rho}\int_{r_{\epsilon_1}}^{r_{\epsilon_2}} p\,(r\,|\,n,\,\rho)\,dr = 0. \qquad\qquad \ldots\ldots(26)$$

This solution has been investigated and a note on unbiased limits for $r$ has already been published [13]. We may state here the conclusion which was reached. It was found that the rejection limits obtained by taking unbiased limits for $r$ differed very little from the limits which were obtained by taking equal tail areas from the $r$-distribution, and that for all practical purposes these two sets of limits could be regarded as coincident. This result is not surprising when it is remembered that by R. A. Fisher's $z'$-transformation the distribution curves of $r$ are transformed approximately into a series of normal curves. There is a slight error introduced by this transformation, and this is the reason why the unbiased limits and the equal tail-area limits are not quite coincident, but this error is usually so small as to be negligible.

---

* The word "biased" throughout the next few pages will be used in Neyman and Pearson's sense.

*Illustration. Question 1. Case (i).*

In the production of a certain aluminium die-casting previous experience over a long period of time had shown a manufacturer that two measurements of quality, namely tensile strength ($x$), measured in pounds per square inch, and hardness ($y$), measured in terms of Rockwell's $E$, were both approximately normally distributed. Since the determination of $x$ involves the destruction of the casting, it is desired to use the character $y$ as a measure of strength in its place, and for this purpose it is considered essential that the correlation, $\rho$, between the value of $x$ and $y$ in the same specimen should be at least as high as $+\cdot80$. Tests of $x$ and $y$ on 25 specimens are available, giving a sample correlation, $r$, of $+\cdot641$. Should it be concluded that the correlation between the characteristics in this type of casting is insufficient to justify the use of $y$ in predicting strength?

In statistical terminology we see that it is necessary to test the hypothesis that $\rho \geqslant +\cdot80$, the admissible alternative hypotheses being that $\rho < +\cdot80$.

Using the tables we find that $P\{r \leqslant \cdot641 \mid n = 25, \rho = \cdot80\} = \cdot04584$.* From Charts I, II, III, IV, respectively, we see that using the

| | | | | | |
|---|---|---|---|---|---|
| $\cdot05$ limit we reject the hypothesis tested if $r$ is less than $\cdot65$, | | | | | |
| $\cdot025$ | ,, | ,, | $r$ | ,, | $\cdot605$, |
| $\cdot01$ | ,, | ,, | $r$ | ,, | $\cdot55$, |
| $\cdot005$ | ,, | ,, | $r$ | ,, | $\cdot51$. |

It is therefore seen that, if in the sampled material the correlation between the two characters was $\cdot80$ or more, so small a value of $r$ as that observed would be expected to occur through chance sampling fluctuations less than once in 20 times, when testing 25 specimens. On the assumption that the specimens tested are a random selection from the material, the manufacturer would feel very doubtful whether the correlation between $x$ and $y$ was high enough for his purpose, although he might be well advised to examine a further sample before rejecting the hypothesis, $\rho \geqslant +\cdot80$.

*Illustration. Question 1. Case (iii).*

Let us suppose that in dealing with the same material as in Case (i) the manufacturer had found from past experience that $\rho = +\cdot63$, and as a routine test for control of quality he proposed that in the future random samples of 20 specimens should be drawn from each batch of several hundred castings, $x$ and $y$ measured, and the correlation between them calculated. What control limits $r_1$ and $r_2$ should he specify in order that he may be reasonably certain of detecting whether $\rho$ for the batch had altered appreciably from $+\cdot63$?

We shall assume (i) that the batch is so large that we may regard the sample as being drawn from an infinite population, (ii) that within the batch the quality of the material is homogeneous. By making $r_1$ and $r_2$ close to $\rho_0 = +\cdot63$, the manufacturer would reduce the risk of passing material for which $\rho$ was much greater or much less than $+\cdot63$. This he might do if he were trying to establish a rigorous control. But if $r_1$ and $r_2$ are too close to $\rho_0 = +\cdot63$, he would run a second risk in that he would often reject material which was really satisfactory. This would be possible owing to the wide variation in $r$ for samples of 20. Therefore, before setting up his two limits $r_1$ and $r_2$ he must decide upon the risk he is willing to undertake. We shall suppose that he is content with the control if the limits chosen entail a risk of rejecting satisfactory material 5 times in 100, i.e. 5 times in 100 he will run the risk of rejecting the hypothesis $\rho = \rho_0 = +\cdot63$, when it is actually true.

We obtain the limits $r_1$ and $r_2$ from Chart II, which shows that

$$P\{(\cdot275 \leqslant r \leqslant \cdot845) \mid n = 20, \rho = +\cdot63\} = \cdot05.$$

The two limits for $r$ that he should therefore set up will be $r_1 = +\cdot275$, $r_2 = +\cdot845$.

* This useful shorthand is often found in statistical papers. Put into words it means "the probability that $r$ is less than or equal to $\cdot641$, given that $n = 25$ and $\rho = \cdot80$, is equal to $\cdot04584$".

*Question* 2. *To determine a confidence interval $\rho_b$ to $\rho_a$ for $\rho$.*

Certain aspects of the problem of estimation have recently been advanced considerably by the work of R. A. Fisher[14] on fiducial probability and J. Neyman[15] on interval estimation. The procedure and terminology developed by the latter will be followed, but from the practical point of view there is no substantial difference between the results reached by either approach. For the sake of clarity the problem has been divided into three classes.

*Case* (i). *To determine $\rho_b$ so that we may make the statement $\rho_b \leqslant \rho \leqslant +1$ with a given degree of confidence.* If we turn to Chart I, with which is associated a confidence coefficient of ·90, and as an illustration consider the lower of the two curves marked $n = 10$, we know that, whatever be the value of $\rho$ in the sampled population, we shall expect in repeated sampling to find that 5 out of 100 samples of 10 give a value of $r$ falling below this curve. Thus if we consider the set of points $(r, \rho)$, which we may meet when drawing samples of 10 from normal bivariate populations, we should expect 5 per cent. of these points to lie below the lower curve, and 95 per cent. above it. It follows that if in general, in a given sample of size $n = n_0$,

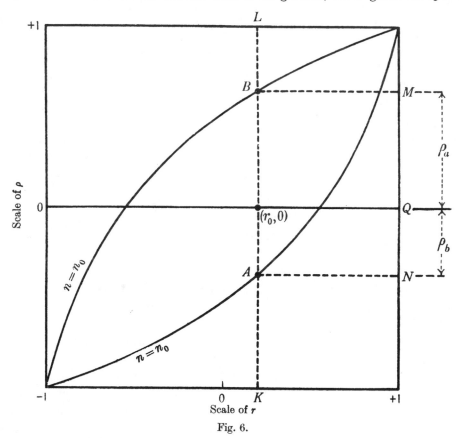

Fig. 6.

we find the sample correlation coefficient $r = r_0$, we may adopt the following procedure to determine the limit $\rho_b$:

Plot the point $(r_0, \rho = 0)$ and draw a line parallel to the axis of $\rho$ through this point. Suppose this line cuts the lower curve for $n = n_0$ at the point $A$. Draw the line $AN$ perpendicular to the axis of $\rho$, cutting the axis in the point $N$. Then $QN = \rho_b$, and we may make the statement

$$\rho_b \leqslant \rho \leqslant +1 \qquad\qquad \ldots\ldots(27)$$

with a degree of confidence measured by the confidence coefficient of ·95. We may express this by saying that 95 times in 100 we shall expect the interval $\rho_b$ to $+1\cdot00$ to cover the true population value. The confidence coefficient represents the probability that the statement (27) will be correct if we follow this procedure. If Charts II, III and IV are used the confidence coefficients will be $\cdot975$, $\cdot99$ and $\cdot995$, respectively.

*Case (ii). To determine $\rho_a$ so that we may make the statement $-1 \leqslant \rho \leqslant \rho_a$ with a given degree of confidence.* In case (ii), given $r_0$ as in case (i), continue the line through $(r_0, \rho=0)$, which is parallel to the axis of $\rho$, until it cuts the upper of the two curves for $n = n_0$. Suppose the point of intersection to be $B$, and draw a line $BM$, perpendicular to the axis of $\rho$, and cutting this axis in the point $M$. Then $MQ = \rho_a$ and we may associate with the statement

$$-1 \leqslant \rho \leqslant \rho_a \qquad \qquad \ldots\ldots(28)$$

a confidence coefficient of $\cdot95$ if Chart I is used.

*Case (iii). To determine $\rho_a$ and $\rho_b$ so that we may make the statement $\rho_b \leqslant \rho \leqslant \rho_a$ with a given degree of confidence.* Using Chart I, and the pair of curves for $n = n_0$, where $n_0$ is the specified size of sample, we know that of the set of points $(r, \rho)$ that may be met in our statistical experience when randomly drawing samples of size $n_0$, 90 per cent. will fall within the lozenge-shaped belt between the two curves, 5 per cent. above the upper curve, and 5 per cent. below the lower curve. Hence if for an observed value of $r = r_0$ in a sample we determine, as in Fig. 6, the points $A$ and $B$, $M$ and $N$, and therefore $\rho_a$ and $\rho_b$, where $\rho_b = NQ$ and $\rho_a = MQ$, we may associate with the statement

$$\rho_b \leqslant \rho \leqslant \rho_a \qquad \qquad \ldots\ldots(29)$$

a confidence coefficient of $\cdot90$. Intervals with confidence coefficients of $\cdot95$, $\cdot98$ and $\cdot99$ can similarly be obtained from Charts II, III and IV, respectively.

In the present type of problem it is clear that we could obtain an infinite variety of belts provided that (24) holds. We should have the same risk of error in making statement (29) if the two curves of Fig. 6 were based on those values of $r_{\epsilon_1}$ and $r_{\epsilon_2}$ obtained by a consideration of (21), (22) and (24) instead of those obtained from (21), (22) and (23). For instance, a belt with confidence coefficient $\cdot90$ could be obtained by taking $\epsilon_1 = \cdot02$, $\epsilon_2 = \cdot08$, as well as with $\epsilon_1 = \epsilon_2 = \cdot05$.

One such type of belt has been discussed in the note on unbiased limits for $r$, to which we have already referred [13]. It has been shown by J. Neyman [15] that, in the case of certain skew distributions, such "unbiased" belts have definite theoretical advantages over those obtained by taking equal tail areas. In the case of the $r$-distribution, however, evidence given in the paper [13] suggests that the equal tail-area belt and the unbiased belt may be considered as coincident for all practical purposes.

*Illustration.*

The width of span ($x$) and length of forearm ($y$) of 20 males have been measured, and the correlation between these two variates is found to be $+\cdot550$. Assuming that width of span and length of forearm are both approximately normally distributed, what interval will cover the correlation coefficient between $x$ and $y$ in the population?

The sample correlation coefficient is $+\cdot550$. Using Chart I and the pair of curves for $n = 20$, we see that we may make the statement
$$\cdot21 \leqslant \rho \leqslant \cdot76,$$
with a degree of confidence measured by a confidence coefficient of $\cdot90$. This is equivalent to saying that in repeated sampling nine times out of ten we expect the interval $\cdot21$ to $\cdot76$ to cover the true population value.

We may decide that the risk of the interval failing to cover the true value once in ten times is too great. Accordingly we would turn to Chart II and make the statement
$$P\{\cdot160 \leqslant \rho \leqslant \cdot785\} = \cdot95.$$
Here the risk of failing to cover the true population value by the interval $\cdot160$ to $\cdot785$ is reduced to $\cdot05$, but it should be noticed that in reducing the risk of error we have increased the breadth of the interval.

We should naturally like the interval for $\rho$ to be as narrow as possible, but we can make this interval narrow only by increasing the risk of being wrong. It is therefore necessary to balance the consequences entailed by a wrong decision against the advantages of the narrow interval.

The sample we are discussing was in fact randomly drawn from a population of 10,000 males. The correlation between width of span and length of forearm for this population was found to be ·758.

### Question 3

In attempting to use any single comprehensive criterion to test hypotheses regarding the values of unknown parameters in more than one population, we find that little progress has so far been made in the development of such tests except in the case where the sample estimates of the unknown population parameters are normally distributed. If, for example, we have samples from two populations with correlations $\rho_1$ and $\rho_2$, the classical method of testing the hypothesis that $\rho_1$ equals $\rho_2$ would be to calculate the ratio of the difference between the sample correlations to an estimate of the standard error of this difference, i.e.

$$\frac{r_1 - r_2}{\sqrt{\dfrac{(1 - r_1^2)^2}{n_1 - 1} + \dfrac{(1 - r_2^2)^2}{n_2 - 1}}}, \qquad\qquad ......(30)$$

and refer this ratio to the normal probability scale. This procedure is adequate when dealing with large samples, provided the hypothetical common value of $|\rho|$ is not too near unity. It may however be extremely inaccurate in other cases. The sampling distribution of (30) is unknown, and its form would be extraordinarily difficult to determine, since $r_1$ and $r_2$ follow different non-normal distributions, depending on the value of $n_1$ and $n_2$ and that of the unknown common $\rho$. Further, even if the sampling distribution were obtainable, the test might be less efficient in detecting true differences between $\rho_1$ and $\rho_2$ than other tests which might be devised.

R. A. Fisher's $z'$-transformation has the great practical advantage that it provides, instead of $r$, a quantity $z'$ which is approximately normally distributed, and whose standard error is practically independent of $\rho$. Thus when we have samples from several populations, we may test hypotheses regarding the population $\rho$'s by applying to the sample $z''$s the appropriate tests from "normal theory". Since, however, the distribution is not precisely normal or independent of $\rho$, some approximation is entailed by such a procedure and it is therefore important to note that for one type of problem, such as Question 3 (i), the present tables provide an exact test.

### Question 3 (i)

$k$ samples, containing respectively $n_1, n_2, \ldots n_k$ observations of two correlated variables $x$ and $y$, have been drawn from $k$ normal bivariate populations, with unknown correlation coefficients $\rho_1, \rho_2, \ldots \rho_k$. Are the data consistent with the hypothesis that

$$\rho_1 = \rho_2 = \ldots = \rho_k = \rho_0, \qquad\qquad ......(31)$$

where $\rho_0$ is some specified value?

In order to determine the best form of test to use, it is necessary to decide upon the kind of alternatives to (31) that appear most likely, having regard to general considerations of the type of problem dealt with. If the $\rho$'s are not all equal to $\rho_0$ we might have grounds for thinking (a) that they would have some other common value differing from $\rho_0$, (b) that they might be unequal but all $>\rho_0$ (or $<\rho_0$), (c) that they might assume any different and unequal values whatsoever. In practice it is evident that we cannot be *certain* of the class of admissible hypotheses alternate to those tested; nevertheless we shall generally have a fairly clear idea of the form of departure from (31) which it is most important for us not to overlook. Consequently we shall prefer to use that test which is most likely to detect such a departure if it exists.

Consider first the alternative (c) mentioned above. Since $\rho_0$ is specified in (31), for each sample we may obtain the probability integral of $r$, given $n$, for $\rho = \rho_0$. For the $t$th sample ($t = 1, 2, \ldots k$), define $\pi_t$ as follows:

$$\left. \begin{array}{l} \pi_t = 2 \displaystyle\int_{-1}^{r_t} p\,(r \mid n = n_t, \rho = \rho_0)\,dr, \text{ if } r_t \leqslant \text{median value of } r^* \\[3mm] \pi_t = 2 \displaystyle\int_{r_t}^{+1} p\,(r \mid n = n_t, \rho = \rho_0)\,dr, \text{ if } r_t > \text{median value of } r \end{array} \right\} \qquad \ldots\ldots(32)$$

Hence $0 \leqslant \pi_t \leqslant 1$. Further, if the hypothesis tested is true, the $k$ values of $\pi_t$ will be independent and each equally likely to assume any value between 0 and 1. We may now follow the suggestion of R. A. Fisher [16] and Karl Pearson [17] and take

$$P = \pi_1 \pi_2 \ldots \pi_k \qquad\qquad \ldots\ldots(33)$$

as a criterion to test the hypothesis (31). It will be seen that $P$ has a maximum value of unity and tends to zero as the values of $r_t$ diverge from $\rho_0$.

As Fisher has stated, and as follows also from Karl Pearson's work, if the hypothesis tested be true

$$\chi^2 = -2 \log_e (P) \qquad\qquad \ldots\ldots(34)$$

is distributed as in the standard $\chi^2$ distribution

$$p\,(\chi^2) = \frac{1}{2^{f/2}\,\Gamma\!\left(\dfrac{f}{2}\right)} (\chi^2)^{f/2-1} e^{-\frac{1}{2}\chi^2} \qquad\qquad \ldots\ldots(35)$$

with degrees of freedom, $f$, equal to $2k$. Thus a simple test of the hypothesis (31) is available, provided that the values of the $\pi_t$'s can be calculated. The present tables and charts make such calculations possible.

In considering alternative (b), we see that the hypothesis tested is that $\rho_t \leqslant \rho_0$, and the admissible alternative hypotheses will assume that there is at least one population, though which one cannot be specified, for which $\rho_t > \rho_0$. In this case we define $\pi_t$ as

$$\pi_t = \int_{r_t}^{+1} p\,(r \mid n = n_t, \rho = \rho_0)\,dr, \qquad\qquad \ldots\ldots(36)$$

so that $0 \leqslant \pi_t \leqslant 1$ ($t = 1, 2, \ldots k$), and if the hypothesis tested is true the $\pi_t$'s will have the same properties as before. Thus the criterion $P$ will be as in (33) and will be related by the transformation (34) to the $\chi^2$ distribution with $2k$ degrees of freedom. It will be noted that $P$ will now approach zero as the differences $r_t - \rho_0$ increase, i.e. as it becomes less likely that the hypothesis $\rho_t \leqslant \rho_0$ is true.

Karl Pearson suggested a slightly different procedure to test the significance of (33), the hypothesis of alternative (b). He wrote

$$P\{\lambda \geqslant \lambda_n\} = P_{\lambda_n} = I\left(n-1, \; -\frac{\log_{10} \lambda_n}{\sqrt{n}\,\log_{10} e}\right), \qquad\qquad \ldots\ldots(37)$$

where $I\,(p, u)$ is the function given in *Tables of the Incomplete Gamma-Function*, $\lambda_n$ is the criterion (33) and $n$ is the number of samples which are tested. Tables of $P_{\lambda_n}$ were calculated for different values of $-\log_{10} \lambda_n$ and $n$, and may therefore be used to test hypothesis (31). This process is, however, essentially the same as that proposed by R. A. Fisher. Since

$$1 - P\{\chi^2 > \chi_0^2\} = 1 - P_{\chi^2} \text{ (say)} = I\left(\tfrac{1}{2}(N-3), \; \frac{\tfrac{1}{2}\chi_0^2}{\sqrt{\tfrac{1}{2}(N-1)}}\right), \qquad\qquad \ldots\ldots(38)$$

if we calculate $P_{\chi^2}$ for $N = 2n+1$ and $\chi^2 = -\dfrac{2 \log_{10} \lambda_n}{\log_{10} e}$, then from (37) and (38)

$$P\{\lambda \geqslant \lambda_n\} = 1 - P_{\chi^2}. \qquad\qquad \ldots\ldots(39)$$

---

* For very small samples median $r$ will of course differ somewhat from $\rho_0$.

It should be understood that these two tests are precise, in the sense that the sampling distribution of the criterion used is known if $H_0$, the hypothesis tested, be true, so that the risk of rejecting $H_0$ when true is exactly controlled. How far such tests are biased, in the sense of Neyman and Pearson's terminology, or whether more powerful tests could be found which would be more likely to detect departures of the $\rho$'s from $\rho_0$, are theoretical problems requiring further investigation.

*Example I.* The following example is quoted by Tippett[18] from a paper by Tschepourkowsky (1905). The table contains the correlation coefficient between cephalic index and upper face form for samples of skulls belonging to thirteen races. It is desired to test the hypothesis that there is no association between cephalic index and upper face form. Accordingly, in our terminology we shall test the hypothesis that

$$\rho_1 = \rho_2 = \ldots = \rho_{13} = 0, \qquad \ldots \ldots (40)$$

where alternatively the $\rho$'s have any other values between $-1$ and $+1$.

Table II

| Race | Number of skulls measured | Correlation coefficient | $\int_{-1}^{r_t} p(r \mid n, \rho = 0)$ | $\pi_t$ |
|---|---|---|---|---|
| Australians | 66 | $+\cdot089$ | $\cdot761$ | $\cdot478$ |
| Negroes | 77 | $+\cdot182$ | $\cdot946$ | $\cdot108$ |
| Duke of York Islanders | 53 | $-\cdot093$ | $\cdot255$ | $\cdot510$ |
| Malays | 60 | $-\cdot185$ | $\cdot079$ | $\cdot158$ |
| Fijians | 32 | $+\cdot217$ | $\cdot883$ | $\cdot334$ |
| Papuans | 39 | $-\cdot255$ | $\cdot060$ | $\cdot120$ |
| Polynesians | 44 | $+\cdot002$ | $\cdot505$ | $\cdot990$ |
| Alfourous | 19 | $-\cdot302$ | $\cdot104^5$ | $\cdot209$ |
| Micronesians | 32 | $-\cdot251$ | $\cdot083$ | $\cdot166$ |
| Copts | 34 | $-\cdot147$ | $\cdot203$ | $\cdot406$ |
| Etruscans | 47 | $-\cdot021$ | $\cdot445$ | $\cdot890$ |
| Europeans | 80 | $-\cdot198$ | $\cdot039$ | $\cdot078$ |
| Ancient Thebans | 152 | $-\cdot067$ | $\cdot207^5$ | $\cdot415$ |

The values of $\pi_t$ were calculated directly from the *Tables of the Incomplete Beta-Function*, by means of the relations expressed in (11) and (12). Similar results would have been obtained by application of the Lagrangian interpolation formula to the tables of $r$.

Here we see that

$$\log P = \log \prod_{t=1}^{13} \pi_t = -7\cdot174,045$$

$$\chi^2 = -2\log_e P = 33\cdot0376, \qquad f = 2k = 26,$$

$$f = 26, \qquad \chi^2 = 35\cdot563, \quad P_{\chi^2} = \cdot10,$$

$$f = 26, \qquad \chi^2 = 31\cdot795, \quad P_{\chi^2} = \cdot20,$$

and hence the probability of getting a larger $\chi^2$ than the one we have obtained will lie between $\cdot1$ and $\cdot2$.

Tippett approaches the example in another way. Using equations (6) and (8) of Fisher's $z'$-transformation he finds the quantity

$$\chi^2 = \sum_{t=1}^{13} (n_t - 3)(z_t')^2 = 17\cdot26,$$

and refers this to the $\chi^2$ tables with degrees of freedom $f = k = 13$. The tables give

$$f = 13, \qquad \chi^2 = 19\cdot812, \quad P_{\chi^2} = \cdot10,$$

$$f = 13, \qquad \chi^2 = 16\cdot985, \quad P_{\chi^2} = \cdot20.$$

Since here he is testing the hypothesis $\rho = 0$, the error introduced by the transformation is very slight. It is seen that the result is comparable with the previous one obtained by using the probability integral of $r$.

The results of either of these tests would therefore give us no clear grounds for rejecting the hypothesis that there is no association between cephalic index and upper face form.

*Example II.* Matuszewski and Supinska [19] carried out a series of experiments with streptococcus. They measured on samples of different types of bacteria the rate of increase per hour of the number of bacteria, and the amount of acid in $10^{-10}$ mg. produced by one cell in one hour. The results of their experiments were given in tabular form, and rough plotting seemed to suggest that the assumption of normality would be justifiable. The present writer worked out a series of correlation coefficients, which are given below in Table III.

We shall test the hypothesis* that

$$\rho_1 = \rho_2 = \rho_3 = \ldots = \rho_{11} = 0, \qquad \ldots \ldots (41)$$

i.e. that there is no association between rate of increase of the bacteria and the amount of acid produced by one cell, the alternative hypotheses being that the $\rho$'s may have any other values between $+1$ and $-1$.

Using the tables of $r$ we obtain $\int_{-1}^{r} p(r \mid n, \rho)\, dr$ for each sample and hence $\pi_t$, using (32).

### Table III

| Type of bacteria | | Number of experiments with the same culture | Correlations between rate of increase and amount of acid | $\int_{-1}^{r} p(r)\, dr$ | $\pi_t$ | $\log_{10} \pi_t$ |
|---|---|---|---|---|---|---|
| Streptococcus Lactis | 3 | 0 | $-\cdot3945$ | $\cdot2195$ | $\cdot4389$ | $\bar{1}\cdot64237$ |
| ,,    ,, | 4 | 6 | $+\cdot6268$ | $\cdot9085^+$ | $\cdot1830$ | $\bar{1}\cdot26245$ |
| ,,    ,, | 5 | 7 | $+\cdot8276$ | $\cdot9892$ | $\cdot0215^+$ | $\bar{2}\cdot33244$ |
| Streptococcus Cremoris | 6 | 6 | $-\cdot1973$ | $\cdot3539$ | $\cdot7078$ | $\bar{1}\cdot84991$ |
| Streptococcus Lactis | 7 | 6 | $+\cdot5015^+$ | $\cdot8446$ | $\cdot3108$ | $\bar{1}\cdot49248$ |
| ,,    ,, | 8 | 6 | $-\cdot4498$ | $\cdot1854$ | $\cdot3708$ | $\bar{1}\cdot56914$ |
| ,,    ,, | 9 | 5 | $-\cdot0878$ | $\cdot5557$ | $\cdot8886$ | $\bar{1}\cdot94871$ |
| ,,    ,, | 10 | 5 | $-\cdot6396$ | $\cdot1226$ | $\cdot2452$ | $\bar{1}\cdot38952$ |
| ,,    ,, | 11 | 6 | $-\cdot0167$ | $\cdot4875^-$ | $\cdot9749$ | $\bar{1}\cdot98896$ |
| ,,    ,, | 12 | 5 | $-\cdot3717$ | $\cdot2690$ | $\cdot5379$ | $\bar{1}\cdot73070$ |
| ,,    ,, | 13 | 5 | $-\cdot4953$ | $\cdot1981$ | $\cdot3961$ | $\bar{1}\cdot59780$ |

$$\chi^2 = -2\log_e P = 23\cdot926, \qquad f = 2k = 22,$$
$$P_{\chi^2} = \cdot35.$$

An alternative method of testing hypothesis (41) is that used by Tippett in the previous example. We shall note further on in the text that a rough approximation to equation (8) is to assume that $z'$ is approximately normally distributed with standard deviation equal to $1/\sqrt{n-3}$. The quantity $z'\sqrt{n-3}$ will therefore be normally distributed with unit standard deviation in populations where $\rho$ is zero. Hence if we have $k$ samples and consider the expression

$$\chi^2 = \sum_{t=1}^{k} (n_t - 3)\, z_t'^2,$$

we see that this will be distributed as $\chi^2$ with $k$ degrees of freedom. The alternative method by which we proceed is therefore clear. Converting each $r_t$ of Table III to $z_t'$ by means of the relation (6), we finally obtain

$$\chi^2 = \sum_{t=1}^{11} (n_t - 3)\, z_t'^2 = 11\cdot5416, \qquad f = k = 11,$$
$$P_{\chi^2} = \cdot40.$$

* It would, of course, be wrong to consider that the truth of any hypothesis tested is proved or disproved when it is based on such scanty data. The only inference we may draw would be that the hypothesis may be true, or alternatively, may be false, but that further experimentation would be necessary to confirm it.

This result is comparable with that of the other method. In either method we therefore find no reason to reject the hypothesis that there is no association between the rate of increase of the bacteria and the amount of acid produced by one cell in one hour.

### Question 3 (ii)

This type of problem differs from the previous one in that now $\rho_0$ is not specified and the hypothesis to be tested is that

$$\rho_1 = \rho_2 = \ldots = \rho_k. \quad\quad \ldots\ldots(42)$$

In such cases no exact test is known, but two lines of procedure are possible:

(a) If $k$ is large enough, we may obtain from the sample correlation coefficients, $r_t$, some form of weighted estimate, say $r_0$, of the unknown hypothetical common $\rho_0$, and using this for $\rho_0$ apply the same methods as were described in dealing with Question 3 (i).

In a recent paper Karl Pearson[17] suggested using as $r_0$ an approximation to the maximum likelihood estimate of the unknown $\rho_0$. The method of approximation was as follows: He calculated the weighted mean of the first four powers of the sample correlation coefficients, i.e.

$$\mu_v = \frac{\sum\limits_{t=1}^{n} n_t r_t^v}{N}, \quad \text{where} \quad N = \sum\limits_{t=1}^{n} n_t, \quad\quad \ldots\ldots(43)$$

and substituted them in the following equations:

$$\left.\begin{aligned}
\rho_1 &= \mu_1 \\
\rho_2 &= \mu_1 + \rho_1(\mu_2 - \rho_1^2) \\
\rho_3 &= \mu_1 + \rho_2(\mu_2 - \rho_2^2) + \rho_1^2(\mu_3 - \rho_1^3) \\
\rho_4 &= \mu_1 + \rho_3(\mu_2 - \rho_3^2) + \rho_2^2(\mu_3 - \rho_2^3) + \rho_1^3(\mu_4 - \rho_1^4)
\end{aligned}\right\} \quad\quad \ldots\ldots(44)$$

$\rho_4$ was his final approximation to the common $\rho$. The process could be extended to $\rho_5$ and $\rho_6$ and so on, but moments higher than $\mu_4$ would then have to be calculated. This procedure has been tried on several examples and found to give quite a reasonable value for $\rho$.*

Since $r_0$ is a function of $r_1, r_2, \ldots r_k$, the expressions $\pi_t$ of (32) or (36) will not now be independent, and consequently the test based on the transformation

$$\chi^2 = -2\log_e(P)$$

will no longer be accurate, in the sense that in repeated sampling this $\chi^2$ would not follow the distribution (35).

(b) As an alternative R. A. Fisher's $z'$-transformation may be used. If it may be assumed that when (42) is true, $z_t'$ is distributed normally about a common but unknown mean for all values of $t$, with standard deviation $1/\sqrt{n_t - 3}$,† then

$$\chi^2 = \sum_{t=1}^{k}(n_t - 3)(z_t' - \bar{z}')^2, \quad\quad \ldots\ldots(45)$$

where

$$\bar{z}' = \frac{\sum\limits_{t=1}^{k}(n_t - 3)z_t'}{\sum\limits_{t=1}^{k}(n_t - 3)}, \quad\quad \ldots\ldots(46)$$

* It is probable that this procedure was in Karl Pearson's mind when he wrote that the present tables would "largely assist the investigator to determine whether a series of correlation coefficients of samples may be assumed to have a common origin". See also *Biometrika*, xxv, 395.

† This approximation for the standard deviation is derived from equation (8). If we neglect the terms containing $\rho$ and higher powers of $\rho$ we get

$$\sigma_{z'}^2 = \frac{3n^2 + 8}{3(n-1)^2},$$

which is approximately equal to $1/(n-3)$.

will be distributed as $\chi^2$ with degrees of freedom $f = k-1$. Unless all the $n_t$ are equal, it will be seen from (7) that the expected values of the $z'_t$ are not precisely the same, so that an approximation is involved in this test also.

*Example I.* The following example is taken from a recent paper by E. S. Pearson and S. S. Wilks [20].

Table IV. *Racial Correlation Coefficients for equal small samples of 20 taken from 30 races*

| $r_t$ | $\int_{-1}^{r_t} p(r\mid n,\rho)\,dr$ | $\pi_t$ | $\log \pi_t$ | $r_t$ | $\int_{-1}^{r_t} p(r\mid n,\rho)\,dr$ | $\pi_t$ | $\log \pi_t$ | $r_t$ | $\int_{-1}^{r_t} p(r\mid n,\rho)\,dr$ | $\pi_t$ | $\log \pi_t$ |
|---|---|---|---|---|---|---|---|---|---|---|---|
| +·097 | ·2085 | ·4170 | $\bar{1}$·62014 | +·219 | ·3853 | ·7706 | $\bar{1}$·58580 | +·178 | ·3100 | ·6200 | $\bar{1}$·79239 |
| +·198 | ·3507 | ·7014 | $\bar{1}$·84597 | −·152 | ·0331 | ·0662 | $\bar{2}$·82086 | +·763 | ·9979 | ·0042 | $\bar{3}$·62325 |
| +·576 | ·9418 | ·1164 | $\bar{1}$·06595 | +·319 | ·5637 | ·8726 | $\bar{1}$·94082 | +·101 | ·2127 | ·4254 | $\bar{1}$·62880 |
| −·015 | ·1008 | ·2016 | $\bar{1}$·30449 | +·310 | ·5473 | ·9054 | $\bar{1}$·95684 | +·449 | ·7877 | ·4246 | $\bar{1}$·62798 |
| +·173 | ·3115 | ·6230 | $\bar{1}$·79449 | +·019 | ·1277 | ·2554 | $\bar{1}$·40722 | +·245 | ·4300 | ·8600 | $\bar{1}$·93450 |
| +·764 | ·9980 | ·0040 | $\bar{3}$·60206 | +·445 | ·7816 | ·4368 | $\bar{1}$·64028 | +·360 | ·6385 | ·7230 | $\bar{1}$·85914 |
| −·037 | ·0858 | ·1716 | $\bar{1}$·23452 | +·410 | ·7256 | ·5488 | $\bar{1}$·73941 | +·592 | ·9460 | ·1080 | $\bar{1}$·03342 |
| +·667 | ·9823 | ·0354 | $\bar{2}$·54900 | [+·946 | 1·0000 | ·0000 | < $\bar{6}$　]* | −·515 | ·0003 | ·0006 | $\bar{4}$·77815 |
| +·014 | ·1234 | ·2468 | $\bar{1}$·39235 | +·018 | ·1268 | ·2536 | $\bar{1}$·40415 | +·023 | ·1311 | ·2622 | $\bar{1}$·41863 |
| −·112 | ·0472 | ·0944 | $\bar{2}$·97497 | +·160 | ·2921 | ·5842 | $\bar{1}$·76656 | +·259 | ·4458 | ·8916 | $\bar{1}$·95017 |

Samples of 20 skulls are randomly drawn from each of 30 different races, and the correlation between head length and head breadth is calculated for each sample. The question which we may ask is: Are the data consistent with the hypothesis

$$\rho_1 = \rho_2 = \ldots = \rho_{30} = \rho, \qquad\qquad \ldots\ldots(47)$$

where $\rho$ is not specified?

We assume that if the hypothesis (47) is not true the $\rho$'s may assume any different and unequal values whatever. Since $\rho$ is not specified it is necessary to obtain an estimate of the common correlation coefficient, $r_0$, from the data. Various methods may be devised. Here Karl Pearson's maximum likelihood estimate is used, and to this end equations (44) are employed. Successive approximations give

$$\rho_1 = ·2490, \quad \rho_2 = ·2726, \quad \rho_3 = ·2761, \quad \rho_4 = ·2774.$$

These results seem to suggest that we may well take $r_0 = ·277$. Using the table for $n = 20$ we obtain by interpolation the $\pi_t$'s of equation (32),

$$\log_{10} P = -20·7077,*$$

and hence from (34)

$$\chi^2 = -2\log_e(P) = 95·362, \qquad f = 58,\dagger$$

$$P_{\chi^2} < ·0001.$$

We therefore reject the hypothesis (47) and decide that it is most unlikely that each of the 30 races have the same correlation between head length and head breadth.

E. S. Pearson and S. S. Wilks also decided to reject (47) but they employed a different procedure. Using (46), they found

$$\bar{z}' = ·24913.$$

Hence using (45)

$$\chi^2 = \sum_{t=1}^{30} (n_t - 3)(z'_t - \bar{z}')^2 = 96·01, \qquad f = k - 1 = 29,$$

$$P_{\chi^2} < ·000,030.$$

We see that using either procedure we should reject hypothesis (47).

* The value $+·946$ was used to obtain the maximum likelihood estimate of $\rho$, but was omitted from the calculations which follow, since the purpose of the example is purely illustrative. It is clear that had this coefficient been included, the effect would have been to increase the value of $\chi^2$, and hence make the hypothesis (47) even more improbable.

† It is not obvious what the exact degrees of freedom will be in this case. We know that if we are testing the hypothesis $\rho = \rho_0$, where $\rho_0$ is some fixed value, the degrees of freedom will be $2k$. Without further theoretical work it is not possible to say what the effect will be on the number of degrees of freedom of calculating the weighted means of the first four powers of the $r$'s. The writer is aware that $2k$ is not correct in this case but offers it as an approximation until the problem is solved correctly.

Karl Pearson, using his $P_{\lambda n}$ test, found that in only 3·7 per cent. of cases would a more improbable result than the one he reached be obtained, and he therefore decided also to reject hypothesis (47). It is of interest to note why his result differs so greatly from those obtained by the two previous methods.

Pearson defines his $\pi_t$ as
$$\pi_t = \int_{-1}^{r_t} p\,(r \mid n, \rho)\,dr,$$

and his criterion is
$$P = \prod_{t=1}^{k} \pi_t.$$

If we consider the class of admissible alternative hypotheses we see that here Pearson is really testing the hypothesis $\rho \geqslant r_0 = ·277$ with the alternative that at least one $\rho_t < ·277$. This is equivalent to the alternative (b) discussed under Question 3 (i), while the two previous methods are equivalent to the alternative (c), where we assume that the $\rho_t$'s may take any values between minus unity and plus unity. The discrepancy between the results of Karl Pearson and of E. S. Pearson and S. S. Wilks is therefore not important, because we see that they are really using methods designed to test different hypotheses.

In the case of two samples of size $n_1$ and $n_2$, respectively, it would be quite unjustifiable to attempt to estimate a common $\rho$ for the two populations, and to carry out the first procedure suggested on p. xxiii above, i.e. by calculating a $\pi_1$ and $\pi_2$ based on the estimate of $\rho$. The second procedure involving the use of the $z'$-transformation may however still be used. Since the degrees of freedom for $\chi^2$ are now $k-1 = 1$, the second method of procedure reduces to the following:

Calculate
$$z'_t = \tfrac{1}{2} \log \frac{1 + r_t}{1 - r_t}$$

for $t = 1, 2$, and test the hypothesis
$$\rho_1 = \rho_2 \qquad \qquad \qquad ……(48)$$

by finding the ratio
$$\frac{z'_1 - z'_2}{\sqrt{\dfrac{1}{n_1 - 3} + \dfrac{1}{n_2 - 3}}} \qquad \qquad ……(49)$$

and referring this to the normal probability scale. When we are testing hypothesis (48) we assume the admissible alternative hypotheses will be that $\rho_1 > \rho_2$ or $\rho_1 < \rho_2$, and hence we should consider both tail areas of the normal curve. If, on the other hand, we are testing the hypothesis
$$\rho_1 \geqslant \rho_2 \qquad \qquad \qquad ……(50)$$

the admissible alternative hypothesis will be that $\rho_1 < \rho_2$, and we should therefore only concern ourselves with one tail area.

E. S. Pearson has suggested the following rough test, when dealing with small samples, involving the use of Chart I, without the need for any transformation of variables. The rule to be adopted is as follows:

Using the observed values of $r_1$ and $r_2$ read off from the appropriate curves for $n_1$ and $n_2$ in Chart I the quantities $\rho_{a_1}$ and $\rho_{b_1}$, $\rho_{a_2}$ and $\rho_{b_2}$ as shown in Fig. 7 below. Here we must distinguish between the two hypotheses. If we are testing hypothesis (48) then the rule will be: Reject the hypothesis, $\rho_1 = \rho_2$, if
$$\rho_{b_2} > \rho_{a_1} \quad \text{or} \quad \rho_{b_1} > \rho_{a_2}. \qquad \qquad ……(51)$$

The risk of rejecting hypothesis (48) when it is true will be approximately ·02, provided $n_1$ and $n_2$ are not too different. If we are testing hypothesis (50) then the rule will be: Reject the hypothesis, $\rho_1 \geqslant \rho_2$, if
$$\rho_{b_2} > \rho_{a_1}. \qquad \qquad ……(52)$$

The first kind of error, i.e. the risk of rejecting hypothesis (50) when it is true, will be ·01. The basis of this rule will be discussed later.

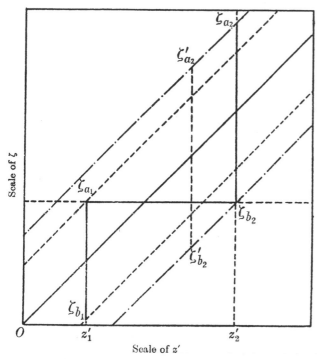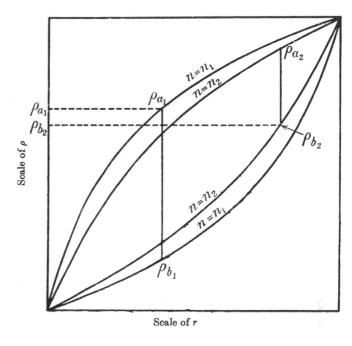

Fig. 7.

*Example I.* Suppose we consider two of the types of bacteria, i.e. Streptococcus Lactis No. 5 and Streptococcus Cremoris No. 6, given in the example on p. xxv. We may ask the question: Could the correlation between $k$, the increase per hour of the number of bacteria, and the amount of acid, $b$, produced by one cell in one hour, have a common value for both populations of bacteria?

Table V

| Type of bacteria | Number of experiments with the same culture | Correlation between $k$ and $b$ |
|---|---|---|
| Streptococcus Lactis 5 | 7 | $r_1 = +\cdot8276$ |
| Streptococcus Cremoris 6 | 6 | $r_2 = -\cdot1973$ |

We are testing the hypothesis $$\rho_1 = \rho_2,$$
and our alternative hypotheses will be that $\rho_1 > \rho_2$ or $\rho_1 < \rho_2$. Turning to Chart I we see that $\rho_{a_1} = +\cdot95$, $\rho_{b_1} = +\cdot33$; $\rho_{a_2} = +\cdot56$, $\rho_{b_2} = -\cdot755$, giving $$\rho_{b_1} < \rho_{a_2}.$$

There does not therefore seem to be any ground for rejecting the hypothesis that the correlation between the two variables is the same for both types of bacteria.

This is confirmed by an application of the $z'$-test, which gives the probability of a larger difference between the $r$'s to be $\cdot071$. It would, of course, be ridiculous to accept the hypothesis as proved on such scanty data. The results of our analysis would lead us to accept the hypothesis as proved only if it was confirmed by further experimentation.

*Example II.* A series of measurements on skulls of different Swiss races are given in "Les Crânes Valaisans".*

* *Crania Helvetica*, I. "Les Crânes Valaisans de la Vallée du Rhône", par Eugene Pittard.

Random samples of 10 were picked out from the Biel and Sierre series, and the coefficient of correlation between maximum length and maximum breadth of the skulls was worked out:

$$\text{Correlation for the Biel series} \quad = r_1 = + \cdot 777,$$

$$\text{Correlation for the Sierre series} = r_2 = - \cdot 352.$$

We ask the question: Is it possible that the correlation between maximum length and breadth of the skull can be the same for both series? We have no *a priori* knowledge which might lead us to test the hypothesis $\rho_1 \gtrless \rho_2$, and we must therefore test the hypothesis

$$\rho_1 = \rho_2,$$

and assume that if the hypothesis tested is not true, then $\rho_1$ may be $> \rho_2$ or $< \rho_2$.

Chart I shows us that the lines $r = r_1$ and $r = r_2$ cut the pair of curves $n = 10$ at the points

$$(\cdot 777, \cdot 92) \ (\cdot 777, \cdot 38) \quad \text{and} \quad (-\cdot 352, +\cdot 23) \ (-\cdot 352, -\cdot 73).$$

We see that
$$\rho_{b_1} > \rho_{a_2}.$$

We should therefore decide to reject the hypothesis $\rho_1 = \rho_2$. The $z'$-test gives the probability of a greater difference between the $r$'s to be $\cdot 0086$. This confirms the result of our approximate test.

### *Theoretical Basis of Rule given*

The position may be understood most clearly by comparing in Fig. 7 the confidence belt for $(r, \rho)$ and that for $(z', \zeta)$. For rough purposes it may be supposed that $z'$ is normally distributed about $\zeta$ with standard deviation $1/\sqrt{n-3}$. Thus the confidence belt for a given $n$ is bounded by the two parallel lines

$$z' = \zeta \pm \psi_0/\sqrt{n-3}. \qquad \qquad \ldots\ldots(53)$$

The values of $\psi_0$ for the $(z', \zeta)$ charts corresponding to Charts I, II, III and IV, respectively, would have the values shown in the table below.

### Table VI

| Chart: | I | II | III | IV |
|---|---|---|---|---|
| $\psi_0$ | 1·645 | 1·960 | 2·326 | 2·576 |
| $\psi_0'$ | 2·326 | 2·772 | 3·289 | 3·693 |
| Risk of error in using rule | ·02 | ·006 | ·001 | ·0002 |

Corresponding to the observed $r_1$ and $r_2$ we have $z_1'$ and $z_2'$ and find from the belt $\zeta_{a_1}$, $\zeta_{b_1}$ and $\zeta_{a_2}$, $\zeta_{b_2}$. Suppose we now follow the rule:

Reject the hypothesis (48) that $\zeta_1 = \zeta_2$, if either

$$\zeta_{b_2} > \zeta_{a_1} \quad \text{or} \quad \zeta_{b_1} > \zeta_{a_2}. \qquad \qquad \ldots\ldots(54)$$

This is equivalent to the rule (51) expressed above in terms of the $\rho$'s. It is also seen from the diagram to be equivalent to rejecting the hypothesis tested if

$$|z_2' - z_1'| > \psi_0/\sqrt{n_1 - 3} + \psi_0/\sqrt{n_2 - 3}. \qquad \qquad \ldots\ldots(55)$$

But provided $n_1$ and $n_2$ be of reasonable size and not too different

$$\sqrt{\frac{1}{n_1 - 3} + \frac{1}{n_2 - 3}} \simeq \frac{1}{\sqrt{2}} \left( \frac{1}{\sqrt{n_1 - 3}} + \frac{1}{\sqrt{n_2 - 3}} \right). \qquad \qquad \ldots\ldots(56)$$

Hence the rule (51) corresponds approximately to the rule of rejecting the hypothesis that $\rho_1 = \rho_2$ when

$$\frac{|z_2' - z_1'|}{\sqrt{\dfrac{1}{n_1-3} + \dfrac{1}{n_2-3}}} > \sqrt{2}\,\psi_0 = \psi_0', \qquad\qquad \ldots\ldots(57)$$

or is approximately equivalent to using the test in terms of $z'$ in equation (49) above. The relation between $\psi_0$, $\psi_0'$ and the risk of error in using the rule, i.e. the risk of rejecting the hypothesis when it is true, are shown in the table given above. It is seen that if Chart I is used the risk should be approximately ·02 or 1 in 50. Tests were carried out to see if this suggested empirical rule held good in practice. Pairs of values of $n_1$ and $n_2$ were taken and Chart I used. A fixed value for $r_2$ was chosen and $r_1$ read off from the chart so that $\rho_{b_2} = \rho_{a_1}$. Fisher's $z'$-test was then applied.

Table VII

| Case: | (i) | (ii) | (iii) | (iv) | (v) | (vi) | (vii) | (viii) | (ix) | (x) |
|---|---|---|---|---|---|---|---|---|---|---|
| $n_2$ | 10 | 10 | 15 | 10 | 10 | 20 | 25 | 20 | 25 | 25 |
| $n_1$ | 8 | 5 | 10 | 10 | 10 | 15 | 15 | 12 | 12 | 25 |
| $r_2$ | +·50 | +·80 | +·80 | +·80 | +·50 | +·50 | +·40 | +·40 | +·40 | +·50 |
| $r_1$ | −·665 | −·54 | +·0325 | −·135 | −·5975 | −·310 | −·380 | −·480 | −·440 | −·150 |
| $\dfrac{z_2' - z_1'}{\sqrt{\dfrac{1}{n_1-3} + \dfrac{1}{n_2-3}}}$ | 2·3075 | 2·1237 | 2·3783 | 2·3094 | 2·3171 | 2·3071 | 2·2953 | 2·3678 | 2·3409 | 2·3231 |
| $2\epsilon$ | ·021 | ·034 | ·017 | ·021 | ·0205 | ·021 | ·022 | ·018 | ·019 | ·020 |

We see that the results are valid precisely so long as we remember the assumptions made for the approximation of $\sqrt{\dfrac{1}{n_1-3} + \dfrac{1}{n_2-3}}$. Directly we depart from these provisions the result obtained is not even approximately equal to ·02. These empirical results do, however, suggest that, as long as we remember the limitations imposed, we have here a very useful method of testing whether $\rho_1 = \rho_2$, with a risk of the first kind of error of ·02, since a glance at Chart I is sufficient to tell us whether $\rho_{b_2} > \rho_{a_1}$.

The significance levels for $|z_2' - z_1'|$, using Charts III and IV, are so small as to be worthless from the practical point of view, although that of Chart II, giving a risk of ·006, might be used. If, however, the reader should find this approximate method useful, it will be a simple matter for him to construct confidence belts from the tables which will give a specific value to the first kind of error of the difference between the $z''$s. For example, to obtain a significance level of ·05 for the difference it would be necessary to construct a series of confidence belts in which the first kind of error, i.e. the chance of rejecting the hypothesis when true, for the samples considered separately is ·121.

## SECTION V. EFFICACY OF THE $z'$-TRANSFORMATION

We have seen ((7) and (8)) that if we write

$$z' = \tfrac{1}{2}\log_e \frac{1+r}{1-r}, \qquad \zeta = \tfrac{1}{2}\log_e \frac{1+\rho}{1-\rho}, \qquad\qquad \ldots\ldots(58)$$

then

$$\text{Mean}\,(z') = \zeta + \frac{\rho}{2(n-1)}\left\{1 + \frac{5+\rho^2}{4(n-1)} + \ldots\right\}, \qquad\qquad \ldots\ldots(59)$$

$$\sigma_{z'}^2 = \frac{1}{n-1}\left\{1 + \frac{4-\rho^2}{2(n-1)} + \frac{22-6\rho^2-3\rho^4}{6(n-1)^2} + \ldots\right\}, \qquad\qquad \ldots\ldots(60)$$

and, provided $n$ be of reasonable size, $z'$ is approximately normally distributed with standard deviation equal to $\sigma_{z'}$. It is customary to take an approximation to (59) and (60) when testing for the significance of $r$

in order to shorten the calculation needed. In this section when equations (59) and (60) are used in full this will be called Approximation I, while for a second and rougher approximation, Approximation II, we have

$$\text{Mean}\,(z') = \zeta + \frac{\rho}{2\,(n-1)}, \qquad \qquad \ldots\ldots(61)$$

$$\sigma_{z'}^2 = \frac{1}{n-3}. \qquad \qquad \ldots\ldots(62)$$

A third approximation, which is sometimes used, is to assume that $z'$ is normally distributed about $\zeta$ with standard deviation equal to $1/\sqrt{n-3}$. This last was not calculated for the small sample illustration below. Calculations for isolated values of $n$ and $\rho$ tend to show that both Approximations I and II are extraordinarily accurate even for low values of $n$, provided that $|\rho|$ is not too near unity. It was decided to investigate how accurate these approximations are when $|\rho|$ is near unity and $n$ is fairly small. To this end the probability integral of $r$ for $n = 11$, and $\rho = +\cdot9$, was calculated by Approximations I and II and compared with the true value. Over a range of $r$ incorporating nine-tenths of the probability integral the agreement is reasonably good, while the tail areas of the curve as given by both approximations are sufficiently close to the true value to render a serious mistake in tests of significance of $r$ unlikely.

It is obvious that the greatest discrepancy between the $z'$-transformation and the true values will occur when $|\rho|$ is near unity, since it is here that the slight error brought in by the transformation will be greatest. We see below how slight this discrepancy is for the case when $n = 11$ and $\rho = \cdot9$, but as a further illustration the tail areas of the distribution curve of $r$ when $n = 11$ and $\rho = \cdot5$ were also calculated, and these are appended to show that the smaller the absolute value of $\rho$, the closer becomes the agreement between the $z'$-transformation and the true values.

### Table VIII

| $n = 11$ $\rho = \cdot9$ | Probability integral | | |
|---|---|---|---|
| $r$ | Approximation II | Approximation I | Actual |
| $\cdot80$ | $\cdot11821$ | $\cdot10839$ | $\cdot10461$ |
| $\cdot81$ | $\cdot13488$ | $\cdot12454$ | $\cdot12047$ |
| $\cdot82$ | $\cdot15401^5$ | $\cdot14320$ | $\cdot13873$ |
| $\cdot83$ | $\cdot17598$ | $\cdot16477$ | $\cdot15998$ |
| $\cdot84$ | $\cdot20120$ | $\cdot18968$ | $\cdot18470$ |
| $\cdot85$ | $\cdot23014$ | $\cdot21845$ | $\cdot21342$ |
| $\cdot86$ | $\cdot26330$ | $\cdot25162$ | $\cdot24679$ |
| $\cdot87$ | $\cdot30124$ | $\cdot28980$ | $\cdot28545^-$ |
| $\cdot88$ | $\cdot34455^-$ | $\cdot33362^5$ | $\cdot33006$ |
| $\cdot89$ | $\cdot39376$ | $\cdot38367$ | $\cdot38127$ |
| $\cdot90$ | $\cdot44936$ | $\cdot44049$ | $\cdot43957$ |
| $\cdot91$ | $\cdot51163$ | $\cdot50437$ | $\cdot50517$ |
| $\cdot92$ | $\cdot58047$ | $\cdot57520$ | $\cdot57776$ |
| $\cdot93$ | $\cdot65516$ | $\cdot65215$ | $\cdot65614^5$ |
| $\cdot94$ | $\cdot73388$ | $\cdot73322$ | $\cdot73783$ |
| $\cdot95$ | $\cdot81319$ | $\cdot81457^5$ | $\cdot81851$ |

| $n = 11$ $\rho = \cdot9$ | Probability integral | | |
|---|---|---|---|
| $r$ | Approximation II | Approximation I | Actual |
| $\cdot60$ | $\cdot00988$ | $\cdot00791$ | $\cdot00881$ |
| $\cdot625$ | $\cdot01329$ | $\cdot01081$ | $\cdot01165^+$ |
| $\cdot65$ | $\cdot01793$ | $\cdot01484$ | $\cdot01553$ |
| $\cdot675$ | $\cdot02428$ | $\cdot02043$ | $\cdot02086$ |
| $\cdot70$ | $\cdot03301$ | $\cdot02825$ | $\cdot02826$ |
| $\cdot725$ | $\cdot04508$ | $\cdot03924$ | $\cdot03864$ |
| $\cdot75$ | $\cdot06155^-$ | $\cdot05478$ | $\cdot05335^-$ |

| $n = 11$ $\rho = \cdot9$ | Probability integral | | |
|---|---|---|---|
| $r$ | Approximation II | Approximation I | Actual |
| $\cdot97$ | $\cdot94808$ | $\cdot95071$ | $\cdot94987$ |
| $\cdot975$ | $\cdot97049$ | $\cdot97259$ | $\cdot97083$ |
| $\cdot98$ | $\cdot98635^-$ | $\cdot98771$ | $\cdot98571$ |
| $\cdot985$ | $\cdot99557$ | $\cdot99620$ | $\cdot99470$ |

| $n = 11$ $\rho = \cdot5$ | Probability integral | | |
|---|---|---|---|
| $r$ | Approximation II | Approximation I | Actual |
| $-\cdot25$ | $\cdot00947$ | $\cdot00857$ | $\cdot00995^+$ |
| $-\cdot20$ | $\cdot01398$ | $\cdot01275$ | $\cdot01414$ |
| $-\cdot15$ | $\cdot02009$ | $\cdot01853$ | $\cdot01974$ |
| $-\cdot10$ | $\cdot02818$ | $\cdot02620$ | $\cdot02710$ |
| $-\cdot05$ | $\cdot03870$ | $\cdot03626$ | $\cdot03666$ |
| $\cdot00$ | $\cdot05215$ | $\cdot04919$ | $\cdot04893$ |
| $\cdot05$ | $\cdot06906$ | $\cdot06558$ | $\cdot06449$ |

| $n = 11$ $\rho = \cdot5$ | Probability integral | | |
|---|---|---|---|
| $r$ | Approximation II | Approximation I | Actual |
| $\cdot80$ | $\cdot93096$ | $\cdot93200$ | $\cdot93270$ |
| $\cdot825$ | $\cdot95461$ | $\cdot95559$ | $\cdot95517$ |
| $\cdot85$ | $\cdot97311$ | $\cdot97391$ | $\cdot97267$ |
| $\cdot875$ | $\cdot98629$ | $\cdot98685$ | $\cdot98526$ |
| $\cdot90$ | $\cdot99445$ | $\cdot99477$ | $\cdot99335^+$ |

It can be seen from the above figures that for a size of sample as small as 11, the approximations are very close to the true value, although if a probability integral of $r$ is required exactly, the extra labour involved by interpolation into the tables will be worth while. As $n$ becomes larger the agreement between the transformation and the true values becomes very good indeed. We append values for $n = 160$, $\rho = \cdot 8$ to illustrate this, and have calculated out the third approximation, i.e. we have assumed that $z'$ is normally distributed about $\zeta$ with standard error equal to $1/\sqrt{n-3}$. Approximation III is sometimes in error by more than $\cdot 01$, but either Approximation I or II, and particularly the former, gives results which are very little different from the true values. The results obtained by applying Lagrange's four-point formula, equation (20), to table entries at $n = 50$, 100, 200 and 400, are reproduced here for comparison. It will be seen that around the region of the mode the Lagrangian four-point formula is exact, but that elsewhere it appears to be on a par with Fisher's $z'$-transformation, Approximation II, and is certainly not as consistently good as Approximation I.

Table IX

| $n = 160$<br>$\rho = \cdot 8$ | Probability integral | | | | |
|---|---|---|---|---|---|
| $r$ | Actual | Lagrangian four-point | Approximation I | Approximation II | Approximation III |
| $\cdot 70$ | $\cdot 00172$ | $\cdot 00168$ | $\cdot 00168$ | $\cdot 00170$ | $\cdot 00188$ |
| $\cdot 75$ | $\cdot 05388$ | $\cdot 05422$ | $\cdot 05396$ | $\cdot 05414$ | $\cdot 05769$ |
| $\cdot 80$ | $\cdot 48729$ | $\cdot 48729$ | $\cdot 48740$ | $\cdot 48743$ | $\cdot 50000$ |
| $\cdot 85$ | $\cdot 97402$ | $\cdot 97323$ | $\cdot 97407$ | $\cdot 97396$ | $\cdot 97581$ |

It is often stated that provided $n$ be large, and $|\rho|$ be not too near unity, $r$ may be assumed to be distributed normally with standard error equal to $\dfrac{1-\rho^2}{\sqrt{n-1}}$, and that tables of the normal curve may therefore be used in tests of significance. There has, however, been no attempt to specify the approximate size of sample that one would regard as large, nor to state what value of $\rho$ we should regard as being not too near unity. Some statisticians would regard a sample of 100 observations as large, while others would require as many as 500 observations before they assumed "large sample theory". In the case of the distribution of $r$, calculations tend to show that up to a size of sample as large as 400 the distribution curves of $r$ from $\rho = \cdot 0$ to $\rho = \cdot 6$ (about) are tending only very slowly to normality, while for $n = 400$ and $\rho > \cdot 6$ there is a very wide divergence from the normal distribution.

Some comparisons are shown in the following tables:

Table X

| $n = 400$<br>$\rho = 0$ | Probability integral | | $n = 400$<br>$\rho = \cdot 4$ | Probability integral | | $n = 400$<br>$\rho = \cdot 9$ | Probability integral | |
|---|---|---|---|---|---|---|---|---|
| $r$ | Actual | Normal Approximation | $r$ | Actual | Normal Approximation | $r$ | Actual | Normal Approximation |
| $-\cdot 16$ | $\cdot 00066$ | $\cdot 00070$ | $-\cdot 22$ | $\cdot 00004$ | $\cdot 00000$ | $\cdot 85$ | $\cdot 00001$ | $\cdot 00000$ |
| $-\cdot 14$ | $\cdot 00251^5$ | $\cdot 00258$ | $-\cdot 26$ | $\cdot 00083$ | $\cdot 00046$ | $\cdot 86$ | $\cdot 00017$ | $\cdot 00002$ |
| $-\cdot 12$ | $\cdot 00817$ | $\cdot 00827$ | $-\cdot 30$ | $\cdot 01121$ | $\cdot 00901$ | $\cdot 87$ | $\cdot 00261$ | $\cdot 00092$ |
| $-\cdot 10$ | $\cdot 02282$ | $\cdot 02289$ | $-\cdot 34$ | $\cdot 08132$ | $\cdot 07851$ | $\cdot 88$ | $\cdot 02588$ | $\cdot 01930$ |
| $-\cdot 08$ | $\cdot 05507$ | $\cdot 05502$ | $-\cdot 38$ | $\cdot 31550^-$ | $\cdot 32094$ | $\cdot 89$ | $\cdot 15254$ | $\cdot 15317$ |
| $-\cdot 06$ | $\cdot 11559$ | $\cdot 11536$ | $-\cdot 42$ | $\cdot 68050^-$ | $\cdot 68619$ | $\cdot 90$ | $\cdot 49100$ | $\cdot 50897$ |
| $-\cdot 04$ | $\cdot 21249$ | $\cdot 21215^-$ | $-\cdot 46$ | $\cdot 92781$ | $\cdot 92439$ | $\cdot 91$ | $\cdot 86000$ | $\cdot 85722$ |
| $-\cdot 02$ | $\cdot 34503$ | $\cdot 34476$ | $-\cdot 50$ | $\cdot 99341$ | $\cdot 99147$ | $\cdot 92$ | $\cdot 98941$ | $\cdot 97271$ |
| $\cdot 00$ | $\cdot 50000$ | $\cdot 50000$ | $-\cdot 54$ | $\cdot 99999$ | $\cdot 99957$ | $\cdot 93$ | $\cdot 99988$ | $\cdot 99921$ |

It will be noted that for $n = 400$, $\rho = 0$ the normal probability integral agrees with the probability integral of $r$ tolerably well, three and sometimes four decimal figures being in agreement. The agreement is less good for $n = 400$, $\rho = \cdot 4$, but here it is sufficiently close to render a serious mistake in tests of significance unlikely. For $n = 400$, $\rho = \cdot 9$ the tail areas of the normal curve differ from those of the $r$-curve, and in this case the replacement of the probability integral of $r$ by the normal probability integral will lead to error.

It is probably wiser never to assume normality, particularly since there are so many alternative methods, some merely involving the reading of a chart, and others, such as the $z'$-transformation, requiring a minimum of arithmetic.

## SECTION VI. CONCLUSION

The present writer would suggest that the charts at the end of this volume will serve three purposes:

(1) They will prove the quickest measure for tests of the significance of $r$ for size of sample, $n$, between 3 and 25.

(2) They will also be useful for rough determinations of significance when $n$ is $> 25$.

(3) They will provide an accurate graphical method for estimating an interval for $\rho$.

For an exact value of the probability integral for $n$ lying between 3 and 25, interpolation into the tables will be necessary, while approximations to the exact value can be obtained either by the approximate graphical method (see p. xiv) or by the $z'$-transformation (Approximations I and II). For $n$ greater than 25, it is seen from the illustration on p. xxxiii that there is little to choose between the results obtained from the adaptation of the Lagrangian four-point formula and the rougher $z'$-transformation (Approximation II). In view of the minimum of calculation involved by the latter, it is recommended that this should generally be used when the size of sample be large. When a rough approximation to several probabilities of a specified $n$ and $\rho$ be required, the diagrams of $n$ plotted on a logarithmic scale may be used.

Finally, the writer would like to make acknowledgements to the Department of Scientific and Industrial Research for a grant given in 1933–34 which enabled part of the tables presented here to be calculated; to Miss C. M. Thompson, Mrs P. C. V. Entwisle and Miss M. G. Francis for assistance in the calculation of the second part of the tables; to Mrs Ida Larmor for the setting out of the tables; and to Mr J. G. Lee for the excellent way in which he has drawn all the charts and diagrams. The writer would also like to acknowledge a real sense of indebtedness to Professor E. S. Pearson and Dr J. Neyman, both of whom have been ever ready to discuss and advise.

The major portion of this work was prepared when the writer was working under Professor Karl Pearson. It was felt that the most fitting tribute that could be paid to his memory was an attempt to bring this publication to the high level of the tables which he edited.

# APPENDIX

## THE DISTRIBUTION OF THE CORRELATION COEFFICIENT FOR ANY $n$ AND $\rho$

Assume that we have a sample of $n$ independent pairs of observations $(x_i y_i)$, $(i = 1, 2, \ldots n)$, which have been randomly drawn from the normal bivariate population of equation (1), p. vii. Then if $\rho$ be the correlation coefficient of $x$ and $y$ in the population, we may write

$$p(x_i y_i) = \left(\frac{1}{2\pi\sigma_1\sigma_2\sqrt{1-\rho^2}}\right)^n e^{-\frac{1}{2(1-\rho^2)}\sum_{i=1}^{n}\left\{\frac{(x_i-\xi_1)^2}{\sigma_1^2} - \frac{2\rho(x_i-\xi_1)(y_i-\xi_2)}{\sigma_1\sigma_2} + \frac{(y_i-\xi_2)^2}{\sigma_2^2}\right\}} . \qquad \ldots\ldots(1)$$
$$(i = 1, 2, \ldots n)$$

Following the usual notation, write

$$\bar{x} = \frac{1}{n}\sum_{i=1}^{n}x_i, \qquad\qquad \bar{y} = \frac{1}{n}\sum_{i=1}^{n}y_i$$

$$s_1^2 = \frac{1}{n}\sum_{i=1}^{n}(x_i-\bar{x})^2, \qquad s_2^2 = \frac{1}{n}\sum_{i=1}^{n}(y_i-\bar{y})^2 \left.\right\} , \qquad\qquad \ldots\ldots(2)$$

$$r = \frac{\sum_{i=1}^{n}(x_i-\bar{x})(y_i-\bar{y})}{ns_1 s_2}$$

and substitute these values in the exponent of (1):

$$\sum_{i=1}^{n}(x_i-\xi_1)^2 = \sum_{i=1}^{n}(x_i-\bar{x})^2 + n(\bar{x}-\xi_1)^2$$
$$= ns_1^2 + n(\bar{x}-\xi_1)^2. \qquad\qquad \ldots\ldots(3)$$

Similarly
$$\sum_{i=1}^{n}(y_i-\xi_2)^2 = ns_2^2 + n(\bar{y}-\xi_2)^2, \qquad\qquad \ldots\ldots(4)$$

$$\sum_{i=1}^{n}(x_i-\xi_1)(y_i-\xi_2) = \sum_{i=1}^{n}\left\{[(x_i-\bar{x})+(\bar{x}-\xi_1)][(y_i-\bar{y})+(\bar{y}-\xi_2)]\right\}$$
$$= nrs_1 s_2 + n(\bar{x}-\xi_1)(\bar{y}-\xi_2). \qquad\qquad \ldots\ldots(5)$$

Therefore by virtue of (3), (4) and (5), (1) becomes

$$p(x_i y_i) = \left(\frac{1}{2\pi\sigma_1\sigma_2\sqrt{1-\rho^2}}\right)^n e^{-\frac{n}{2(1-\rho^2)}\left[\frac{(\bar{x}-\xi_1)^2}{\sigma_1^2} - \frac{2\rho(\bar{x}-\xi_1)(\bar{y}-\xi_2)}{\sigma_1\sigma_2} + \frac{(\bar{y}-\xi_2)^2}{\sigma_2^2}\right]} \times e^{-\frac{n}{2(1-\rho^2)}\left[\frac{s_1^2}{\sigma_1^2} - \frac{2\rho rs_1 s_2}{\sigma_1\sigma_2} + \frac{s_2^2}{\sigma_2^2}\right]}.$$
$$(i = 1, 2, \ldots n)$$
$$\qquad\qquad \ldots\ldots(6)$$

Transform (6) by means of the following substitutions:

$$x_1 = \bar{x} + \frac{1}{\sqrt{2.1}}u_1 + \frac{1}{\sqrt{3.2}}u_2 + \ldots + \frac{1}{\sqrt{n(n-1)}}u_{n-1}$$

$$x_2 = \bar{x} - \frac{1}{\sqrt{2.1}}u_1 + \frac{1}{\sqrt{3.2}}u_2 + \ldots + \frac{1}{\sqrt{n(n-1)}}u_{n-1}$$

$$x_3 = \bar{x} \qquad\qquad - \frac{2}{\sqrt{3.2}}u_2 + \ldots + \frac{1}{\sqrt{n(n-1)}}u_{n-1} \qquad \left.\right\} , \qquad\qquad \ldots\ldots(7)$$

$$\vdots \qquad \vdots \qquad\qquad\qquad\qquad \vdots$$

$$x_n = \bar{x} \qquad\qquad\qquad\qquad\qquad - \frac{n-1}{\sqrt{n(n-1)}}u_{n-1}$$

$$
\left.\begin{aligned}
y_1 &= \bar{y} + \frac{1}{\sqrt{2 \cdot 1}}\, v_1 + \frac{1}{\sqrt{3 \cdot 2}}\, v_2 + \ldots + \frac{1}{\sqrt{n\,(n-1)}}\, v_{n-1} \\[2mm]
y_2 &= \bar{y} - \frac{1}{\sqrt{2 \cdot 1}}\, v_1 + \frac{1}{\sqrt{3 \cdot 2}}\, v_2 + \ldots + \frac{1}{\sqrt{n\,(n-1)}}\, v_{n-1} \\[2mm]
y_3 &= \bar{y} \qquad\qquad - \frac{2}{\sqrt{3 \cdot 2}}\, v_2 + \ldots + \frac{1}{\sqrt{n\,(n-1)}}\, v_{n-1} \\
\vdots \quad & \quad\vdots \qquad\qquad\qquad\qquad \vdots \\
y_n &= \bar{y} \qquad\qquad\qquad\qquad\quad - \frac{n-1}{\sqrt{n\,(n-1)}}\, v_{n-1}
\end{aligned}\right\} . \qquad\qquad \ldots\ldots(8)
$$

We know that
$$
p\,(\bar{x},\,\bar{y},\,u_j v_j) = p\,(x_i y_i)\left|\frac{\partial\,(x_i y_i)}{\partial\,(\bar{x},\,\bar{y},\,u_j v_j)}\right|,
$$
$$
(j=1,\,2,\ldots n-1;\qquad i=1,\,2,\ldots n)
$$

and it is easy to see that
$$
\left|\frac{\partial\,(x_i y_i)}{\partial\,(\bar{x},\,\bar{y},\,u_j v_j)}\right| = n.
$$

Therefore
$$
p\,(\bar{x},\,\bar{y},\,u_j v_j) = p\,(x_i y_i). \qquad\qquad \ldots\ldots(9)
$$
$$
(j=1,\,2,\ldots n-1;\qquad i=1,\,2,\ldots n)
$$

Using the fact that
$$
\left.\begin{aligned}
n s_1^2 &= u_1^2 + u_2^2 + \ldots + u_{n-1}^2 \\
n s_2^2 &= v_1^2 + v_2^2 + \ldots + v_{n-1}^2 \\
n r s_1 s_2 &= u_1 v_1 + u_2 v_2 + \ldots + u_{n-1} v_{n-1}
\end{aligned}\right\}, \qquad\qquad \ldots\ldots(10)
$$

we see that (9) may be written as follows:

$$
p\,(\bar{x},\,\bar{y},\,u_j v_j) = \text{Constant}\,.\,e^{-\frac{n}{2(1-\rho^2)}\left[\frac{(\bar{x}-\xi_1)^2}{\sigma_1^2} - \frac{2\rho\,(\bar{x}-\xi_1)\,(\bar{y}-\xi_2)}{\sigma_1\sigma_2} + \frac{(\bar{y}-\xi_2)^2}{\sigma_2^2}\right]} \times e^{-\frac{n}{2(1-\rho^2)}\left[\sum\limits_{j=1}^{n-1}\left(\frac{u_j^2}{\sigma_1^2} - \frac{2\rho u_j v_j}{\sigma_1\sigma_2} + \frac{v_j^2}{\sigma_2^2}\right)\right]}.
$$
$$
(j=1,\,2,\ldots n-1) \qquad\qquad\qquad\qquad\qquad \ldots\ldots(11)
$$

Transform (11) by means of the following substitutions:

$$
\left.\begin{aligned}
u_1 &= \sqrt{n}\,s_1 \cos\phi_1 \cos\phi_2 \cos\phi_3 \ldots \cos\phi_{n-2}, & w_1 &= \sqrt{n}\,s_2 \cos\psi_1 \cos\psi_2 \cos\psi_3 \ldots \cos\psi_{n-2} \\
u_2 &= \sqrt{n}\,s_1 \cos\phi_1 \cos\phi_2 \cos\phi_3 \ldots \sin\phi_{n-2}, & w_2 &= \sqrt{n}\,s_2 \cos\psi_1 \cos\psi_2 \cos\psi_3 \ldots \sin\psi_{n-2} \\
u_3 &= \sqrt{n}\,s_1 \cos\phi_1 \cos\phi_2 \cos\phi_3 \ldots \sin\phi_{n-3}, & w_3 &= \sqrt{n}\,s_2 \cos\psi_1 \cos\psi_2 \cos\psi_3 \ldots \sin\psi_{n-3} \\
\vdots \quad & & \vdots \quad & \\
u_{n-1} &= \sqrt{n}\,s_1 \sin\phi_1, & w_{n-1} &= \sqrt{n}\,s_2 \sin\psi_1 = \sqrt{n}\,s_2 r \\[2mm]
v_1 &= \quad w_1 \frac{u_2}{\Sigma_2} + w_2 \frac{u_1 u_3}{\Sigma_2 \Sigma_3} + w_3 \frac{u_1 u_4}{\Sigma_3 \Sigma_4} + \ldots + w_{n-2} \frac{u_1 u_{n-1}}{\Sigma_{n-2} \Sigma_{n-1}} + w_{n-1} \frac{u_1}{\Sigma_{n-1}} \\[2mm]
v_2 &= -w_1 \frac{u_1}{\Sigma_2} + w_2 \frac{u_2 u_3}{\Sigma_2 \Sigma_3} + w_3 \frac{u_2 u_4}{\Sigma_3 \Sigma_4} + \ldots + w_{n-2} \frac{u_2 u_{n-1}}{\Sigma_{n-2} \Sigma_{n-1}} + w_{n-1} \frac{u_2}{\Sigma_{n-1}} \\[2mm]
v_3 &= \qquad -w_2 \frac{u_1^2 + u_2^2}{\Sigma_2 \Sigma_3} + w_3 \frac{u_3 u_4}{\Sigma_3 \Sigma_4} + \ldots + w_{n-2} \frac{u_3 u_{n-1}}{\Sigma_{n-2} \Sigma_{n-1}} + w_{n-1} \frac{u_3}{\Sigma_{n-1}} \\
\vdots \quad & \qquad\qquad\qquad\qquad\qquad\qquad\qquad \vdots \qquad\qquad \vdots \\
v_{n-1} &= \qquad\qquad\qquad\qquad -w_{n-2} \frac{u_1^2 + \ldots + u_{n-2}^2}{\Sigma_{n-2} \Sigma_{n-1}} + w_{n-1} \frac{u_{n-1}}{\Sigma_{n-1}}
\end{aligned}\right\},
$$
$$
\qquad\qquad\qquad\qquad\qquad\qquad\qquad\qquad\qquad\qquad\qquad \ldots\ldots(12)
$$

where
$$
\Sigma_{n-1} = (u_1^2 + u_2^2 + \ldots + u_{n-1}^2)^{\frac{1}{2}}.
$$

# APPENDIX

Using substitutions (12) we see that (10) holds and that

$$p\left(\bar{x},\bar{y},s_1 s_2,r,\phi_k\psi_l\right)=p\left(\bar{x},\bar{y},u_i v_i\right)\times\left|\frac{\partial\left(\bar{x},\bar{y},u_i v_i\right)}{\partial\left(\bar{x},\bar{y},s_1 s_2,r,\phi_k\psi_l\right)}\right|, \qquad \ldots\ldots(13)$$

where $i=1,2,\ldots n-1;\ k=1,2,\ldots n-2;\ l=2,3,\ldots n-2$. It can be shown that

$$\left|\frac{\partial\left(\bar{x},\bar{y},u_i v_i\right)}{\partial\left(\bar{x},\bar{y},s_1 s_2,r,\phi_k\psi_l\right)}\right|=s_1^{n-2}s_2^{n-2}\left(1-r^2\right)^{\frac{n-4}{2}}f(\phi)F(\psi) \qquad \ldots\ldots(14)$$

$$(i=1,2,\ldots n-1;\ k=1,2,\ldots n-2;\ l=2,3,\ldots n-2);$$

therefore, using (14), we may write (13)

$$p\left(\bar{x},\bar{y},s_1 s_2,r,\phi_k\psi_l\right)=p\left(\bar{x},\bar{y},u_i v_i\right)s_1^{n-2}s_2^{n-2}\left(1-r^2\right)^{\frac{n-4}{2}}f(\phi)F(\psi) \qquad \ldots\ldots(15)$$

$$(i=1,2,\ldots n-1;\ k=1,2,\ldots n-2;\ l=2,3,\ldots n-2).$$

Hence

$$p\left(\bar{x},\bar{y},s_1 s_2,r\right)=\text{Constant}\int_0^{2\pi}\int_{-\frac{\pi}{2}}^{+\frac{\pi}{2}}\ldots\int_0^{2\pi}\int_{-\frac{\pi}{2}}^{+\frac{\pi}{2}}p\left(\bar{x},\bar{y},u_i v_i\right)s_1^{n-2}s_2^{n-2}\left(1-r^2\right)^{\frac{n-4}{2}}f(\phi)F(\psi)\,d\phi_k d\psi_l.$$

$$\underbrace{\qquad}_{\phi}\underbrace{\qquad}_{\psi}$$

$$\qquad\qquad\ldots\ldots(16)$$

$$(k=1,2,\ldots n-2;\ l=2,3,\ldots n-2)$$

Substituting conditions (10) and integrating with respect to $\phi_k$ and $\psi_l$ for $k=1,2,\ldots n-2;\ l=2,3,\ldots n-2$, we see that

$$p\left(\bar{x},\bar{y},s_1 s_2,r\right)=\text{Constant}\cdot s_1^{n-2}s_2^{n-2}\left(1-r^2\right)^{\frac{n-4}{2}}e^{-\frac{n}{2(1-\rho^2)}\left[\frac{(\bar{x}-\xi_1)^2}{\sigma_1^2}-\frac{2\rho(\bar{x}-\zeta_1)(\bar{y}-\xi_2)}{\sigma_1\sigma_2}+\frac{(\bar{y}-\xi_2)^2}{\sigma_2^2}\right]}$$

$$\times e^{-\frac{n}{2(1-\rho^2)}\left[\frac{s_1^2}{\sigma_1^2}-\frac{2\rho r s_1 s_2}{\sigma_1\sigma_2}+\frac{s_2^2}{\sigma_2^2}\right]}; \qquad \ldots\ldots(17)$$

therefore

$$p\left(s_1 s_2,r\right)=\text{Constant}\cdot s_1^{n-2}s_2^{n-2}\left(1-r^2\right)^{\frac{n-4}{2}}e^{-\frac{n}{2(1-\rho^2)}\left[\frac{s_1^2}{\sigma_1^2}-\frac{2\rho r s_1 s_2}{\sigma_1\sigma_2}+\frac{s_2^2}{\sigma_2^2}\right]}$$

$$\times\int_{-\infty}^{+\infty}\int_{-\infty}^{+\infty}e^{-\frac{n}{2(1-\rho^2)}\left[\frac{(\bar{x}-\xi_1)^2}{\sigma_1^2}-\frac{2\rho(\bar{x}-\xi_1)(\bar{y}-\xi_2)}{\sigma_1\sigma_2}+\frac{(\bar{y}-\xi_2)^2}{\sigma_2^2}\right]}d\bar{x}\,d\bar{y}$$

$$=\text{Constant}\cdot s_1^{n-2}s_2^{n-2}\left(1-r^2\right)^{\frac{n-4}{2}}e^{-\frac{n}{2(1-\rho^2)}\left[\frac{s_1^2}{\sigma_1^2}-\frac{2\rho r s_1 s_2}{\sigma_1\sigma_2}+\frac{s_2^2}{\sigma_2^2}\right]}. \qquad \ldots\ldots(18)$$

Make a further transformation and write

$$s=\frac{s_1 s_2}{\sigma_1\sigma_2},\quad e^t=\frac{s_1\sigma_2}{s_2\sigma_1},\quad r=r. \qquad \ldots\ldots(19)$$

Therefore

$$p\left(s,t,r\right)=p\left(s_1 s_2,r\right)\times\left|\frac{\partial\left(s_1 s_2,r\right)}{\partial\left(s,t,r\right)}\right|$$

$$=p\left(s_1 s_2,r\right)\cdot\tfrac{1}{2}\sigma_1\sigma_2. \qquad \ldots\ldots(20)$$

Hence

$$p\left(s,t,r\right)=\text{Constant}\cdot s^{n-2}\left(1-r^2\right)^{\frac{n-4}{2}}e^{-\frac{n}{2(1-\rho^2)}\left[s\cosh t-s\rho r\right]}. \qquad \ldots\ldots(21)$$

Therefore

$$p\left(r\right)=\text{Constant}\int_0^{+\infty}\int_{-\infty}^{+\infty}s^{n-2}\left(1-r^2\right)^{\frac{n-4}{2}}e^{-\frac{n}{2(1-\rho^2)}\left[s\cosh t-s\rho r\right]}ds\,dt, \qquad \ldots\ldots(22)$$

i.e.

$$p\left(r\right)=\text{Constant}\int_0^{+\infty}\left(1-r^2\right)^{\frac{n-4}{2}}\left(\cosh t-\rho r\right)^{-(n-1)}dt. \qquad \ldots\ldots(23)$$

Following the procedure of R. A. Fisher we write

$$-\rho r = \cos\theta,$$

and (22) becomes

$$p\,(r) = \text{Constant} \int_0^{+\infty} (1-r^2)^{\frac{n-4}{2}} (\cosh t + \cos\theta)^{-(n-1)}\,dt$$

$$= \text{Constant}\,\frac{(1-r^2)^{\frac{n-4}{2}}}{(n-2)!}\left[\left(\frac{d}{\sin\theta\,d\theta}\right)^{n-2}\left(\frac{\theta}{\sin\theta}\right)\right]$$

$$= \text{Constant}\,\frac{(1-r^2)^{\frac{n-4}{2}}}{(n-2)!}\,\frac{d^{n-2}}{d\,(r\rho)^{n-2}}\left(\frac{\text{arc}\cos\,(-\rho r)}{\sqrt{1-\rho^2 r^2}}\right). \qquad\qquad \ldots\ldots(24)$$

Finally, collecting the constant terms and substituting in (24) we obtain

$$p\,(r) = \frac{(1-\rho^2)^{\frac{n-1}{2}}}{\pi\,(n-3)\,!}\,(1-r^2)^{\frac{n-4}{2}}\,\frac{d^{n-2}}{d\,(r\rho)^{n-2}}\left(\frac{\text{arc}\cos\,(-\rho r)}{\sqrt{1-\rho^2 r^2}}\right). \qquad\qquad \ldots\ldots(25)$$

## REFERENCES

(1) "Student". *Biometrika*, vi (1906), 302–10.

(2) R. A. Fisher. *Biometrika*, x (1915), 507–21.

(3) H. E. Soper. *Biometrika*, ix (1913), 91–115.

(4) "A Cooperative Study." *Biometrika*, xi (1917), 328–413.

(5) R. A. Fisher. *Metron*, i, iv (1921), 1–30.

(6) F. Garwood. *Biometrika*, xxv (1933), 71–8.

(7) Whittaker and Robinson. *Calculus of Observations* (1926), p. 143.

(8) K. Pearson. *Tracts for Computers*, iii (1920).

(9) K. Pearson. *Tracts for Computers*, ii (1920), p. 31.

(10) E. S. Pearson. *Biometrika*, xxi (1929), 356–60.

(11) P. R. Rider. *Biometrika*, xxiv (1932), 382–403.

(12) J. Neyman and E. S. Pearson. *Statistical Research Memoirs*, i (1936), 1–37.

(13) F. N. David. *Biometrika*, xxix (1937), 157–60.

(14) R. A. Fisher. *Proc. Camb. Phil. Soc.* xxvi (1930), 528–35.

(15) J. Neyman. *Philos. Trans.* ccxxxvi, A (1937), 333–80, and earlier publications.

(16) R. A. Fisher. *Statistical Methods for Research Workers* (1932), pp. 97–9.

(17) K. Pearson. *Biometrika*, xxv (1933), 379–410.

(18) L. H. C. Tippett. *The Methods of Statistics* (1937), p. 179.

(19) T. Matuszewski and J. Supinska. *Travaux et Comptes Rendus de l'institut de bacteriologie à Varsovie*, xx (1935), 249–74.

(20) E. S. Pearson and S. S. Wilks. *Biometrika*, xxv (1933), 372–5.

Diagram I

ρ (Population Correlation Coefficient) = 0·0

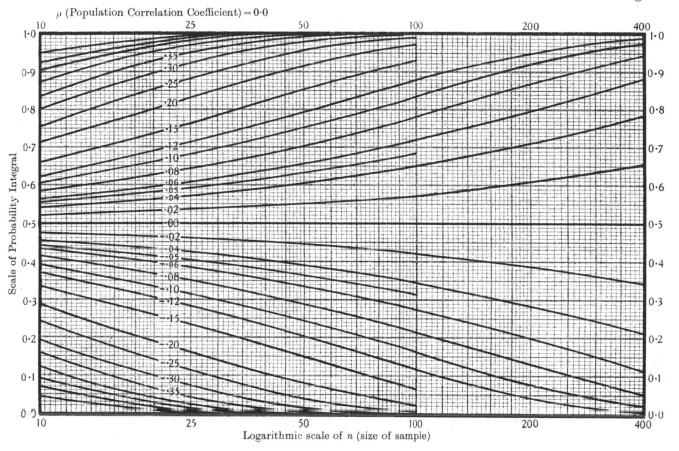

Logarithmic scale of n (size of sample)

Diagram II

ρ (Population Correlation Coefficient) = 0·1

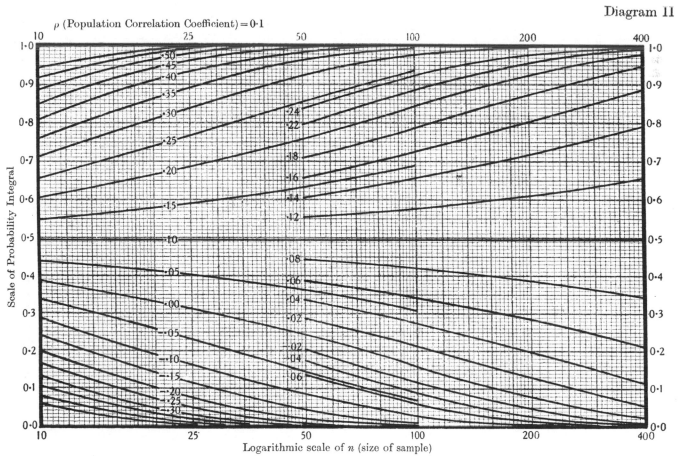

Logarithmic scale of n (size of sample)

The numbers on the curves indicate the value of r (sample correlation coefficient).

## Diagram III

$\rho$ (Population Correlation Coefficient) $= 0.2$

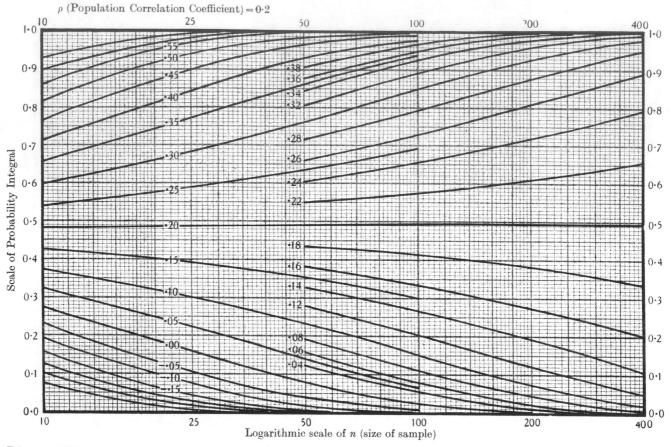

## Diagram IV

$\rho$ (Population Correlation Coefficient) $= 0.3$

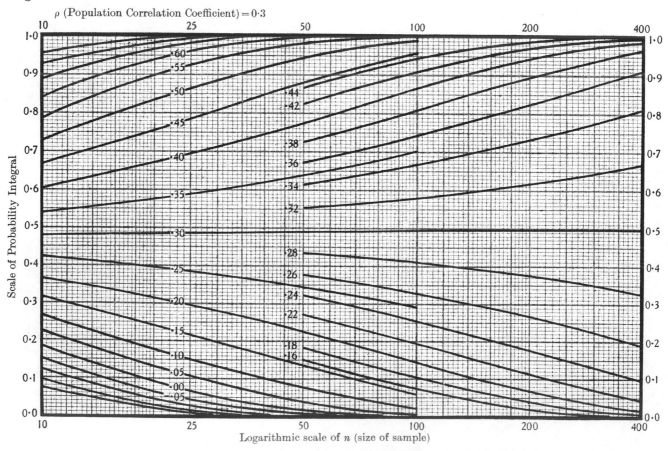

The numbers on the curves indicate the value of $r$ (sample correlation coefficient).

Diagram V

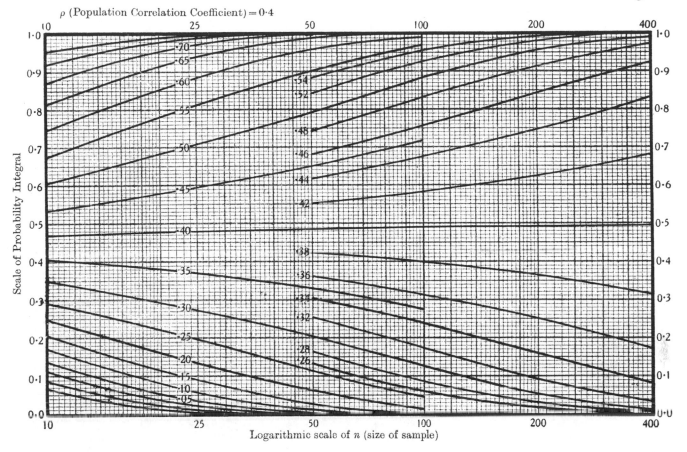

Logarithmic scale of $n$ (size of sample)

Diagram VI

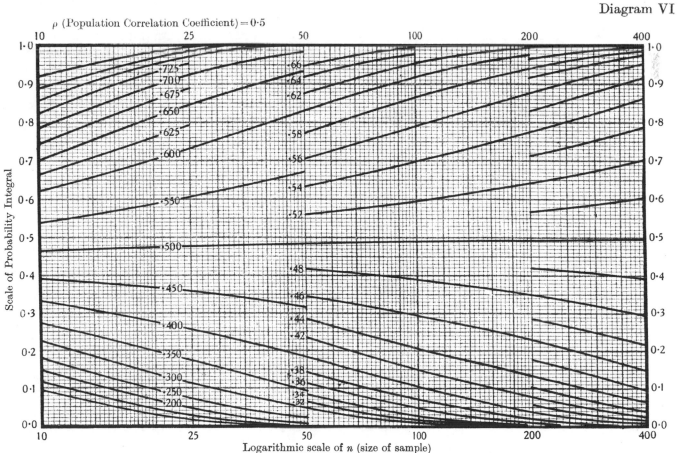

Logarithmic scale of $n$ (size of sample)

The numbers on the curves indicate the value of $r$ (sample correlation coefficient).

## Diagram VII

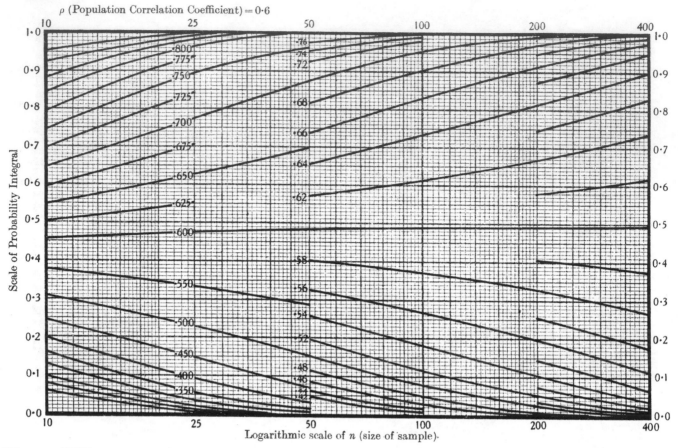

$\rho$ (Population Correlation Coefficient) $= 0.6$

Scale of Probability Integral

Logarithmic scale of $n$ (size of sample).

## Diagram VIII

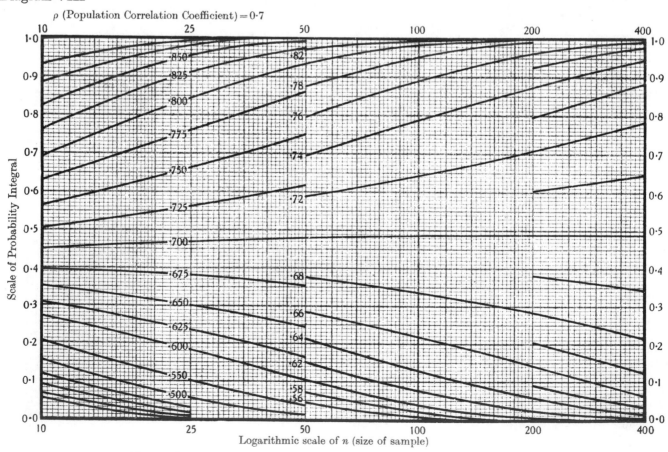

$\rho$ (Population Correlation Coefficient) $= 0.7$

Scale of Probability Integral

Logarithmic scale of $n$ (size of sample)

The numbers on the curves indicate the value of $r$ (sample correlation coefficient).

Diagram IX

ρ (Population Correlation Coefficient) = 0·8

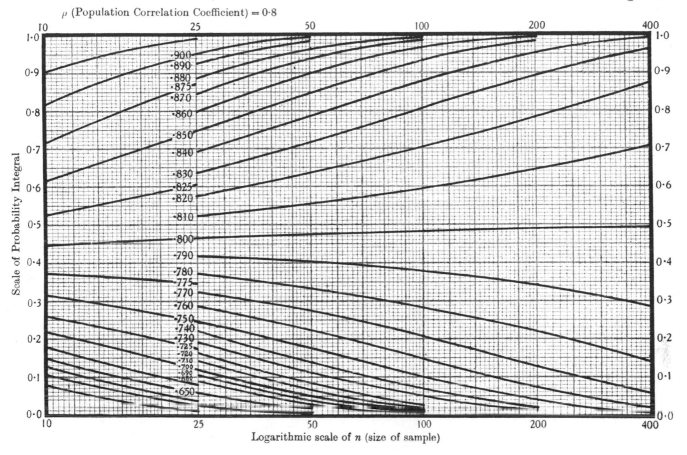

Logarithmic scale of *n* (size of sample)

Diagram X

ρ (Population Correlation Coefficient) = 0·9

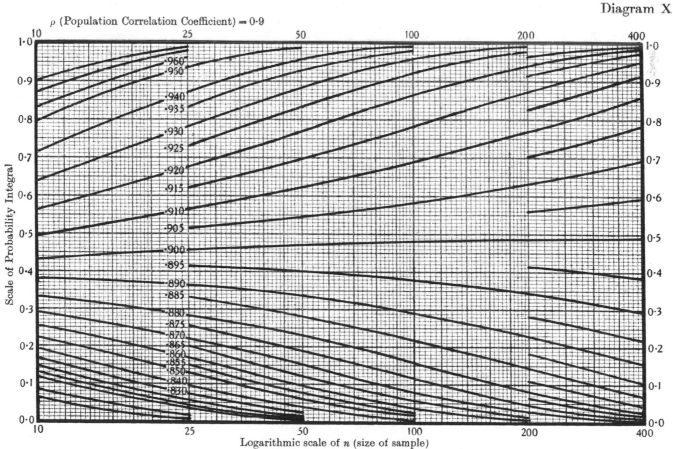

Logarithmic scale of *n* (size of sample)

The numbers on the curves indicate the value of *r* (sample correlation coefficient).

# TABLES OF THE ORDINATES AND PROBABILITY INTEGRAL
## OF THE CORRELATION COEFFICIENT

ρ=0·0——0·4   n=3

| r | ρ = 0·0 Area | ρ = 0·0 Ordinate | ρ = 0·1 Area | ρ = 0·1 Ordinate | ρ = 0·2 Area | ρ = 0·2 Ordinate | ρ = 0·3 Area | ρ = 0·3 Ordinate | ρ = 0·4 Area | ρ = 0·4 Ordinate | r |
|---|---|---|---|---|---|---|---|---|---|---|---|
| −1·00 | ·00 000 | ∞ | ·00 000 | ∞ | ·00 000 | ∞ | ·00 000 | ∞ | ·00 000 | ∞ | −1·00 |
| − ·95 | ·10 108 | 1019·41 | ·08 635 | 874·99 | ·07 315 | 744·21 | ·06 121 | 624·87 | ·05 032 | 515·25 | − ·95 |
| − ·90 | ·14 357 | 730·25 | ·12 300 | 631·33 | ·10 436 | 540·26 | ·08 748 | 456·01 | ·07 203 | 377·73 | − ·90 |
| − ·85 | ·17 660 | 604·25 | ·15 160 | 526·19 | ·12 896 | 453·08 | ·10 830 | 384·48 | ·08 931 | 319·97 | − ·85 |
| − ·80 | ·20 483 | 530·52 | ·17 627 | 465·35 | ·15 027 | 403·21 | ·12 643 | 344·03 | ·10 444 | 287·69 | − ·80 |
| − ·75 | ·23 005 | 481·24 | ·19 847 | 425·22 | ·16 957 | 370·79 | ·14 294 | 318·13 | ·11 828 | 267·34 | − ·75 |
| − ·70 | ·25 318 | 445·72 | ·21 898 | 396·74 | ·18 751 | 348·18 | ·15 838 | 300·44 | ·13 128 | 253·76 | − ·70 |
| − ·65 | ·27 477 | 418·87 | ·23 827 | 375·59 | ·20 449 | 331·78 | ·17 307 | 287·95 | ·14 372 | 244·48 | − ·65 |
| − ·60 | ·29 517 | 397·89 | ·25 663 | 359·43 | ·22 076 | 319·60 | ·18 723 | 279·03 | ·15 578 | 238·18 | − ·60 |
| − ·55 | ·31 463 | 381·13 | ·27 427 | 346·87 | ·23 650 | 310·50 | ·20 102 | 272·73 | ·16 758 | 234·08 | − ·55 |
| − ·50 | ·33 333 | 367·55 | ·29 136 | 337·02 | ·25 184 | 303·74 | ·21 454 | 268·44 | ·17 922 | 231·71 | − ·50 |
| − ·45 | ·35 142 | 356·44 | ·30 801 | 329·29 | ·26 690 | 298·82 | ·22 789 | 265·77 | ·19 077 | 230·74 | − ·45 |
| − ·40 | ·36 901 | 347·30 | ·32 432 | 323·28 | ·28 175 | 295·41 | ·24 114 | 264·44 | ·20 231 | 230·97 | − ·40 |
| − ·35 | ·38 618 | 339·80 | ·34 036 | 318·70 | ·29 646 | 293·29 | ·25 420 | 264·28 | ·21 389 | 232·26 | − ·35 |
| − ·30 | ·40 301 | 333·68 | ·35 621 | 315·35 | ·31 110 | 292·30 | ·26 758 | 265·17 | ·22 555 | 234·53 | − ·30 |
| − ·25 | ·41 957 | 328·75 | ·37 191 | 313·08 | ·32 571 | 292·30 | ·28 088 | 267·01 | ·23 736 | 237·71 | − ·25 |
| − ·20 | ·43 591 | 324·87 | ·38 753 | 311·77 | ·34 035 | 293·24 | ·29 430 | 269·76 | ·24 934 | 241·79 | − ·20 |
| − ·15 | ·45 207 | 321·95 | ·40 310 | 311·36 | ·35 505 | 295·05 | ·30 787 | 273·39 | ·26 155 | 246·77 | − ·15 |
| − ·10 | ·46 812 | 319·91 | ·41 868 | 311·80 | ·36 986 | 297·71 | ·32 165 | 277·91 | ·27 403 | 252·66 | − ·10 |
| − ·05 | ·48 408 | 318·71 | ·43 430 | 313·06 | ·38 483 | 301·21 | ·33 568 | 283·32 | ·28 683 | 259·51 | − ·05 |
| 0·00 | ·50 000 | 318·31 | ·45 000 | 315·13 | ·40 000 | 305·58 | ·35 000 | 289·66 | ·30 000 | 267·38 | 0·00 |
| + ·05 | ·51 592 | 318·71 | ·46 583 | 318·02 | ·41 541 | 310·83 | ·36 466 | 296·99 | ·31 359 | 276·35 | + ·05 |
| + ·10 | ·53 188 | 319·91 | ·48 182 | 321·75 | ·43 110 | 317·02 | ·37 972 | 305·38 | ·32 765 | 286·51 | + ·10 |
| + ·15 | ·54 793 | 321·95 | ·49 802 | 326·39 | ·44 712 | 324·22 | ·39 522 | 314·94 | ·34 226 | 298·02 | + ·15 |
| + ·20 | ·56 409 | 324·87 | ·51 447 | 331·99 | ·46 354 | 332·52 | ·41 123 | 325·79 | ·35 748 | 311·04 | + ·20 |
| + ·25 | ·58 043 | 328·75 | ·53 123 | 338·66 | ·48 040 | 342·06 | ·42 782 | 338·10 | ·37 339 | 325·78 | + ·25 |
| + ·30 | ·59 699 | 333·68 | ·54 836 | 346·53 | ·49 777 | 353·00 | ·44 507 | 352·08 | ·39 009 | 342·52 | + ·30 |
| + ·35 | ·61 382 | 339·80 | ·56 591 | 355·76 | ·51 572 | 365·56 | ·46 306 | 368·00 | ·40 769 | 361·58 | + ·35 |
| + ·40 | ·63 099 | 347·30 | ·58 396 | 366·59 | ·53 436 | 380·02 | ·48 191 | 386·21 | ·42 630 | 383·43 | + ·40 |
| + ·45 | ·64 858 | 356·44 | ·60 260 | 379·33 | ·55 376 | 396·75 | ·50 173 | 407·18 | ·44 608 | 408·63 | + ·45 |
| + ·50 | ·66 667 | 367·55 | ·62 193 | 394·39 | ·57 408 | 416·27 | ·52 268 | 431·53 | ·46 723 | 437·95 | + ·50 |
| + ·55 | ·68 537 | 381·13 | ·64 209 | 412·36 | ·59 545 | 439·27 | ·54 495 | 460·12 | ·48 996 | 472·45 | + ·55 |
| + ·60 | ·70 483 | 397·89 | ·66 323 | 434·08 | ·61 808 | 466·77 | ·56 878 | 494·15 | ·51 459 | 513·63 | + ·60 |
| + ·65 | ·72 523 | 418·87 | ·68 558 | 460·81 | ·64 223 | 500·25 | ·59 449 | 535·44 | ·54 148 | 563·68 | + ·65 |
| + ·70 | ·74 682 | 445·72 | ·70 943 | 494·50 | ·66 824 | 542·05 | ·62 249 | 586·77 | ·57 116 | 626·01 | + ·70 |
| + ·75 | ·76 995 | 481·24 | ·73 520 | 538·43 | ·69 663 | 596·05 | ·65 341 | 652·77 | ·60 437 | 706·23 | + ·75 |
| + ·80 | ·79 517 | 530·52 | ·76 354 | 598·63 | ·72 817 | 669·27 | ·68 815 | 741·91 | ·64 224 | 814·50 | + ·80 |
| + ·85 | ·82 340 | 604·25 | ·79 553 | 687·68 | ·76 413 | 776·82 | ·72 826 | 871·73 | ·68 662 | 971·83 | + ·85 |
| + ·90 | ·85 643 | 730·25 | ·83 329 | 838·35 | ·80 702 | 956·77 | ·77 671 | 1087·46 | ·74 108 | 1232·06 | + ·90 |
| + ·95 | ·89 892 | 1019·41 | ·88 228 | 1180·31 | ·86 324 | 1361·48 | ·84 106 | 1568·00 | ·81 465 | 1806·93 | + ·95 |
| +1·00 | 1·00 000 | ∞ | 1·00 000 | ∞ | 1·00 000 | ∞ | 1·00 000 | ∞ | 1·00 000 | ∞ | +1·00 |

n = 3     ρ = 0.5 —— 0.9

| r | ρ=0.5 Area | ρ=0.5 Ordinate | ρ=0.6 Area | ρ=0.6 Ordinate | ρ=0.7 Area | ρ=0.7 Ordinate | ρ=0.8 Area | ρ=0.8 Ordinate |
|---|---|---|---|---|---|---|---|---|
| −1.00 | ·00 000 | ∞ | ·00 000 | ∞ | ·00 000 | ∞ | ·00 000 | ∞ |
| − ·95 | ·04 033 | 413·97 | ·03 109 | 319·92 | ·02 252 | 232·19 | ·01 453 | 150·04 |
| − ·90 | ·05 780 | 304·70 | ·04 462 | 236·31 | ·03 235 | 172·05 | ·02 089 | 111·49 |
| − ·85 | ·07 177 | 259·17 | ·05 547 | 201·74 | ·04 026 | 147·37 | ·02 602 | 95·78 |
| − ·80 | ·08 404 | 234·02 | ·06 505 | 182·85 | ·04 727 | 134·03 | ·03 058 | 87·38 |
| − ·75 | ·09 533 | 218·43 | ·07 388 | 171·35 | ·05 376 | 126·04 | ·03 482 | 82·44 |
| − ·70 | ·10 598 | 208·27 | ·08 225 | 164·05 | ·05 993 | 121·12 | ·03 886 | 79·49 |
| − ·65 | ·11 621 | 201·60 | ·09 033 | 159·47 | ·06 590 | 118·20 | ·04 279 | 77·84 |
| − ·60 | ·12 618 | 197·36 | ·09 823 | 156·81 | ·07 177 | 116·70 | ·04 666 | 77·14 |
| − ·55 | ·13 598 | 194·94 | ·10 603 | 155·60 | ·07 759 | 116·29 | ·05 052 | 77·18 |
| − ·50 | ·14 570 | 193·97 | ·11 381 | 155·57 | ·08 342 | 116·79 | ·05 439 | 77·82 |
| − ·45 | ·15 540 | 194·20 | ·12 161 | 156·54 | ·08 928 | 118·06 | ·05 831 | 79·02 |
| − ·40 | ·16 513 | 195·48 | ·12 948 | 158·40 | ·09 523 | 120·05 | ·06 230 | 80·72 |
| − ·35 | ·17 496 | 197·72 | ·13 746 | 161·09 | ·10 130 | 122·71 | ·06 639 | 82·91 |
| − ·30 | ·18 492 | 200·85 | ·14 560 | 164·58 | ·10 752 | 126·05 | ·07 060 | 85·60 |
| − ·25 | ·19 506 | 204·86 | ·15 393 | 168·86 | ·11 392 | 130·08 | ·07 496 | 88·82 |
| − ·20 | ·20 542 | 209·74 | ·16 250 | 173·97 | ·12 054 | 134·82 | ·07 949 | 92·59 |
| − ·15 | ·21 605 | 215·51 | ·17 134 | 179·94 | ·12 741 | 140·33 | ·08 423 | 96·98 |
| − ·10 | ·22 699 | 222·22 | ·18 051 | 186·82 | ·13 458 | 146·69 | ·08 920 | 102·04 |
| − ·05 | ·23 829 | 229·93 | ·19 004 | 194·72 | ·14 210 | 153·99 | ·09 444 | 107·88 |
| 0·00 | ·25 000 | 238·73 | ·20 000 | 203·72 | ·15 000 | 162·34 | ·10 000 | 114·59 |
| + ·05 | ·26 218 | 248·73 | ·21 044 | 213·97 | ·15 835 | 171·89 | ·10 592 | 122·33 |
| + ·10 | ·27 489 | 260·05 | ·22 142 | 225·63 | ·16 721 | 182·84 | ·11 225 | 131·27 |
| + ·15 | ·28 821 | 272·89 | ·23 303 | 238·91 | ·17 666 | 195·41 | ·11 907 | 141·63 |
| + ·20 | ·30 221 | 287·45 | ·24 534 | 254·08 | ·18 678 | 209·89 | ·12 645 | 153·71 |
| + ·25 | ·31 699 | 304·00 | ·25 847 | 271·15 | ·19 769 | 226·66 | ·13 447 | 167·99 |
| + ·30 | ·33 265 | 322·88 | ·27 253 | 291·45 | ·20 950 | 246·18 | ·14 328 | 184·60 |
| + ·35 | ·34 932 | 344·53 | ·28 767 | 314·60 | ·22 236 | 269·08 | ·15 299 | 204·53 |
| + ·40 | ·36 716 | 369·48 | ·30 406 | 341·59 | ·23 647 | 296·15 | ·16 380 | 228·52 |
| + ·45 | ·38 634 | 398·48 | ·32 191 | 373·30 | ·25 207 | 328·48 | ·17 593 | 257·73 |
| + ·50 | ·40 709 | 432·48 | ·34 149 | 410·96 | ·26 943 | 367·53 | ·18 968 | 293·80 |
| + ·55 | ·42 969 | 472·82 | ·36 313 | 456·23 | ·28 896 | 415·36 | ·20 546 | 339·09 |
| + ·60 | ·45 451 | 521·35 | ·38 727 | 511·47 | ·31 116 | 474·92 | ·22 380 | 397·07 |
| + ·625 | ·46 789 | 549·49 | ·40 046 | 543·85 | ·32 347 | 510·40 | ·23 416 | 432·40 |
| + ·650 | ·48 201 | 580·81 | ·41 450 | 580·15 | ·33 672 | 550·63 | ·24 547 | 473·12 |
| + ·675 | ·49 696 | 615·88 | ·42 950 | 621·10 | ·35 105 | 596·55 | ·25 787 | 520·43 |
| + ·700 | ·51 284 | 655·44 | ·44 560 | 667·63 | ·36 661 | 649·38 | ·27 156 | 575·90 |
| + ·725 | ·52 978 | 700·44 | ·46 294 | 720·97 | ·38 359 | 710·70 | ·28 675 | 641·57 |
| + ·750 | ·54 792 | 752·17 | ·48 172 | 782·72 | ·40 223 | 782·63 | ·30 374 | 720·25 |
| + ·775 | ·56 746 | 812·36 | ·50 217 | 855·10 | ·42 283 | 868·06 | ·32 290 | 815·83 |
| + ·800 | ·58 863 | 883·48 | ·52 459 | 941·21 | ·44 578 | 971·06 | ·34 472 | 933·85 |
| + ·825 | ·61 175 | 969·14 | ·54 938 | 1045·60 | ·47 158 | 1097·60 | ·36 985 | 1082·55 |
| + ·850 | ·63 725 | 1074·99 | ·57 708 | 1175·29 | ·50 101 | 1256·88 | ·39 921 | 1274·71 |
| + ·875 | ·66 575 | 1210·28 | ·60 845 | 1341·84 | ·53 482 | 1463·96 | ·43 412 | 1531·46 |
| + ·900 | ·69 815 | 1391·88 | ·64 464 | 1566·09 | ·57 475 | 1745·85 | ·47 663 | 1890·70 |
| + ·925 | ·73 600 | 1654·64 | ·68 758 | 1890·97 | ·62 318 | 2157·80 | ·53 013 | 2429·68 |
| + ·950 | ·78 226 | 2088·24 | ·74 089 | 2426·12 | ·68 481 | 2839·55 | ·60 112 | 3341·78 |
| + ·975 | ·84 442 | 3046·02 | ·81 377 | 3601·21 | ·77 137 | 4333·00 | ·70 573 | 5363·86 |
| +1·000 | 1·00 000 | ∞ | 1·00 000 | ∞ | 1·00 000 | ∞ | 1·00 000 | ∞ |

| r | ρ=0.9 Area | ρ=0.9 Ordinate |
|---|---|---|
| −1·00 | ·00 000 | ∞ |
| − ·95 | ·00 704 | 72·82 |
| − ·90 | ·01 013 | 54·25 |
| − ·85 | ·01 263 | 46·73 |
| − ·80 | ·01 486 | 42·75 |
| − ·75 | ·01 693 | 40·45 |
| − ·70 | ·01 892 | 39·12 |
| − ·65 | ·02 086 | 38·44 |
| − ·60 | ·02 277 | 38·22 |
| − ·55 | ·02 469 | 38·38 |
| − ·50 | ·02 662 | 38·85 |
| − ·45 | ·02 829 | 39·61 |
| − ·40 | ·03 058 | 40·63 |
| − ·35 | ·03 264 | 41·93 |
| − ·30 | ·03 478 | 43·51 |
| − ·25 | ·03 700 | 45·38 |
| − ·20 | ·03 932 | 47·57 |
| − ·15 | ·04 176 | 50·12 |
| − ·10 | ·04 434 | 53·08 |
| − ·05 | ·04 708 | 56·51 |
| 0·00 | ·05 000 | 60·48 |
| + ·05 | ·05 314 | 65·09 |
| + ·10 | ·05 652 | 70·48 |
| + ·15 | ·06 020 | 76·79 |
| + ·20 | ·06 422 | 84·24 |
| + ·25 | ·06 865 | 93·11 |
| + ·30 | ·07 356 | 103·75 |
| + ·35 | ·07 906 | 116·66 |
| + ·40 | ·08 528 | 132·54 |
| + ·45 | ·09 238 | 152·34 |
| + ·50 | ·10 060 | 177·48 |
| + ·55 | ·11 025 | 210·05 |
| + ·60 | ·12 178 | 253·32 |
| + ·625 | ·12 844 | 280·52 |
| + ·650 | ·13 585 | 312·61 |
| + ·675 | ·14 413 | 350·87 |
| + ·700 | ·15 346 | 396·99 |
| + ·725 | ·16 406 | 453·34 |
| + ·750 | ·17 623 | 523·20 |
| + ·775 | ·19 037 | 611·35 |
| + ·800 | ·20 701 | 724·94 |
| + ·810 | ·21 453 | 779·78 |
| + ·820 | ·22 263 | 841·36 |
| + ·830 | ·23 138 | 910·83 |
| + ·840 | ·24 088 | 989·66 |
| + ·850 | ·25 122 | 1079·64 |
| + ·860 | ·26 252 | 1183·07 |
| + ·870 | ·27 493 | 1302·86 |
| + ·880 | ·28 864 | 1442·80 |
| + ·890 | ·30 387 | 1607·91 |
| + ·900 | ·32 091 | 1804·95 |
| + ·910 | ·34 013 | 2043·28 |
| + ·920 | ·36 195 | 2336·22 |
| + ·930 | ·38 708 | 2703·45 |
| + ·940 | ·41 637 | 3175·43 |
| + ·950 | ·45 110 | 3802·31 |
| + ·955 | ·47 107 | 4199·34 |
| + ·960 | ·49 322 | 4673·95 |
| + ·965 | ·51 799 | 5252·00 |
| + ·970 | ·54 598 | 5973·24 |
| + ·975 | ·57 805 | 6902·94 |
| + ·980 | ·61 554 | 8158·43 |
| + ·985 | ·66 057 | 9980·34 |
| + ·990 | ·71 725 | 12980·43 |
| + ·995 | ·77 670 | 19544·18 |
| +1·000 | 1·00 000 | ∞ |

ρ=0.9

| r | Area | Ordinate |
|---|---|---|
| + ·800 | ·20 701 | 724·94 |
| + ·825 | ·22 692 | 875·02 |
| + ·850 | ·25 122 | 1079·64 |
| + ·875 | ·28 161 | 1370·03 |
| + ·900 | ·32 091 | 1804·95 |
| + ·925 | ·37 406 | 2508·97 |
| + ·950 | ·45 110 | 3802·31 |
| + ·975 | ·57 805 | 6902·94 |
| +1·000 | 1·00 000 | ∞ |

ρ=0·0 — 0·4                    n=4

| r | ρ = 0·0 Area | ρ = 0·0 Ordinate | ρ = 0·1 Area | ρ = 0·1 Ordinate | ρ = 0·2 Area | ρ = 0·2 Ordinate | ρ = 0·3 Area | ρ = 0·3 Ordinate | ρ = 0·4 Area | ρ = 0·4 Ordinate | r |
|---|---|---|---|---|---|---|---|---|---|---|---|
| −1·00 | ·00 000 | 500·00 | ·00 000 | 386·33 | ·00 000 | 295·47 | ·00 000 | 222·28 | ·00 000 | 163·11 | −1·00 |
| − ·95 | ·02 500 | 500·00 | ·01 943 | 390·84 | ·01 493 | 301·85 | ·01 128 | 228·97 | ·00 831 | 169·22 | − ·95 |
| − ·90 | ·05 000 | 500·00 | ·03 909 | 395·43 | ·03 019 | 308·42 | ·02 290 | 235·94 | ·01 693 | 175·66 | − ·90 |
| − ·85 | ·07 500 | 500·00 | ·05 897 | 400·09 | ·04 578 | 315·19 | ·03 488 | 243·21 | ·02 588 | 182·44 | − ·85 |
| − ·80 | ·10 000 | 500·00 | ·07 910 | 404·83 | ·06 171 | 322·17 | ·04 723 | 250·79 | ·03 518 | 189·59 | − ·80 |
| − ·75 | ·12 500 | 500·00 | ·09 946 | 409·64 | ·07 800 | 329·36 | ·05 996 | 258·71 | ·04 484 | 197·13 | − ·75 |
| − ·70 | ·15 000 | 500·00 | ·12 006 | 414·54 | ·09 465+ | 336·78 | ·07 310 | 266·97 | ·05 490 | 205·09 | − ·70 |
| − ·65 | ·17 500 | 500·00 | ·14 091 | 419·52 | ·11 168 | 344·44 | ·08 667 | 275·61 | ·06 536 | 213·51 | − ·65 |
| − ·60 | ·20 000 | 500·00 | ·16 201 | 424·58 | ·12 910 | 352·34 | ·10 067 | 284·64 | ·07 626 | 222·42 | − ·60 |
| − ·55 | ·22 500 | 500·00 | ·18 337 | 429·73 | ·14 692 | 360·49 | ·11 514 | 294·09 | ·08 761 | 231·85 | − ·55 |
| − ·50 | ·25 000 | 500·00 | ·20 499 | 434·96 | ·16 515+ | 368·90 | ·13 009 | 303·97 | ·09 945+ | 241·84 | − ·50 |
| − ·45 | ·27 500 | 500·00 | ·22 687 | 440·28 | ·18 382 | 377·59 | ·14 554 | 314·32 | ·11 181 | 252·44 | − ·45 |
| − ·40 | ·30 000 | 500·00 | ·24 902 | 445·69 | ·20 292 | 386·57 | ·16 153 | 325·17 | ·12 471 | 263·69 | − ·40 |
| − ·35 | ·32 500 | 500·00 | ·27 144 | 451·20 | ·22 248 | 395·84 | ·17 807 | 336·54 | ·13 819 | 275·65 | − ·35 |
| − ·30 | ·35 000 | 500·00 | ·29 414 | 456·80 | ·24 251 | 405·43 | ·19 519 | 348·48 | ·15 228 | 288·38 | − ·30 |
| − ·25 | ·37 500 | 500·00 | ·31 712 | 462·50 | ·26 303 | 415·34 | ·21 293 | 361·01 | ·16 704 | 301·94 | − ·25 |
| − ·20 | ·40 000 | 500·00 | ·34 039 | 468·30 | ·28 405− | 425·59 | ·23 130 | 374·17 | ·18 249 | 316·40 | − ·20 |
| − ·15 | ·42 500 | 500·00 | ·36 395 | 474·19 | ·30 559 | 436·20 | ·25 035+ | 388·01 | ·19 869 | 331·84 | − ·15 |
| − ·10 | ·45 000 | 500·00 | ·38 781 | 480·20 | ·32 767 | 447·17 | ·27 012 | 402·57 | ·21 569 | 348·34 | − ·10 |
| − ·05 | ·47 500 | 500·00 | ·41 197 | 486·30 | ·35 031 | 458·53 | ·29 062 | 417·89 | ·23 355 | 366·00 | − ·05 |
| 0·00 | ·50 000 | 500·00 | ·43 644 | 492·52 | ·37 353 | 470·30 | ·31 192 | 434·04 | ·25 232 | 384·94 | 0·00 |
| + ·05 | ·52 500 | 500·00 | ·46 123 | 498·85 | ·39 735+ | 482·49 | ·33 404 | 451·07 | ·27 206 | 405·26 | + ·05 |
| + ·10 | ·55 000 | 500·00 | ·48 633 | 505·29 | ·42 179 | 495·13 | ·35 704 | 469·04 | ·29 287 | 427·09 | + ·10 |
| + ·15 | ·57 500 | 500·00 | ·51 176 | 511·84 | ·44 687 | 508·23 | ·38 096 | 488·02 | ·31 480 | 450·60 | + ·15 |
| + ·20 | ·60 000 | 500·00 | ·53 752 | 518·52 | ·47 262 | 521·81 | ·40 586 | 508·08 | ·33 796 | 475·94 | + ·20 |
| + ·25 | ·62 500 | 500·00 | ·56 361 | 525·31 | ·49 906 | 535·90 | ·43 179 | 529·31 | ·36 243 | 503·31 | + ·25 |
| + ·30 | ·65 000 | 500·00 | ·59 005+ | 532·23 | ·52 622 | 550·53 | ·45 881 | 551·79 | ·38 833 | 532·91 | + ·30 |
| + ·35 | ·67 500 | 500·00 | ·61 684 | 539·28 | ·55 412 | 565·72 | ·48 699 | 575·62 | ·41 576 | 564·99 | + ·35 |
| + ·40 | ·70 000 | 500·00 | ·64 398 | 546·46 | ·58 280 | 581·49 | ·51 640 | 600·89 | ·44 487 | 599·81 | + ·40 |
| + ·45 | ·72 500 | 500·00 | ·67 149 | 553·77 | ·61 228 | 597·89 | ·54 711 | 627·74 | ·47 579 | 637·68 | + ·45 |
| + ·50 | ·75 000 | 500·00 | ·69 936 | 561·22 | ·64 260 | 614·93 | ·57 920 | 656·28 | ·50 869 | 678·96 | + ·50 |
| + ·55 | ·77 500 | 500·00 | ·72 761 | 568·81 | ·67 379 | 632·65 | ·61 277 | 686·66 | ·54 375+ | 724·05 | + ·55 |
| + ·60 | ·80 000 | 500·00 | ·75 624 | 576·54 | ·70 588 | 651·09 | ·64 790 | 719·03 | ·58 117 | 773·41 | + ·60 |
| + ·65 | ·82 500 | 500·00 | ·78 527 | 584·41 | ·73 891 | 670·29 | ·68 117 | 753·55 | ·62 117 | 827·57 | + ·65 |
| + ·70 | ·85 000 | 500·00 | ·81 469 | 592·44 | ·77 292 | 690·28 | ·72 330 | 790·43 | ·66 402 | 887·15 | + ·70 |
| + ·75 | ·87 500 | 500·00 | ·84 451 | 600·62 | ·80 795+ | 711·11 | ·76 379 | 829·86 | ·70 999 | 952·87 | + ·75 |
| + ·80 | ·90 000 | 500·00 | ·87 475+ | 608·96 | ·84 405− | 732·82 | ·80 633 | 872·07 | ·75 942 | 1025·55 | + ·80 |
| + ·85 | ·92 500 | 500·00 | ·90 541 | 617·46 | ·88 125− | 755·47 | ·85 105 | 917·33 | ·81 268 | 1106·18 | + ·85 |
| + ·90 | ·95 000 | 500·00 | ·93 650+ | 626·13 | ·91 961 | 779·10 | ·89 812 | 965·92 | ·87 019 | 1195·91 | + ·90 |
| + ·95 | ·97 500 | 500·00 | ·96 803 | 634·97 | ·95 918 | 803·76 | ·94 771 | 1018·17 | ·93 244 | 1296·08 | + ·95 |
| +1·00 | 1·00 000 | 500·00 | 1·00 000 | 643·98 | 1·00 000 | 829·53 | 1·00 000 | 1074·43 | 1·00 000 | 1408·32 | +1·00 |

$n = 4$      $\rho = 0.5 — 0.9$

| r | ρ=0.5 Area | Ordinate | ρ=0.6 Area | Ordinate | ρ=0.7 Area | Ordinate | ρ=0.8 Area | Ordinate | r | ρ=0.9 Area | Ordinate |
|---|---|---|---|---|---|---|---|---|---|---|---|
| -1.00 | .00 000 | 115.34 | .00 000 | 77.06 | .00 000 | 46.93 | .00 000 | 24.03 | -1.00 | .00 000 | 8.02 |
| - .95 | .00 589 | 120.41 | .00 395 | 80.89 | .00 241 | 49.50 | .00 124 | 25.46 | - .95 | .00 041 | 8.53 |
| - .90 | .01 205 | 125.79 | .00 809 | 84.99 | .00 495 | 52.27 | .00 255 | 27.01 | - .90 | .00 085 | 9.09 |
| - .85 | .01 848 | 131.52 | .01 245 | 89.38 | .00 761 | 55.26 | .00 394 | 28.69 | - .85 | .00 132 | 9.69 |
| - .80 | .02 520 | 137.61 | .01 704 | 94.10 | .01 048 | 58.49 | .00 542 | 30.52 | - .80 | .00 182 | 10.36 |
| - .75 | .03 224 | 144.15 | .02 187 | 99.16 | .01 349 | 61.99 | .00 699 | 32.51 | - .75 | .00 236 | 11.08 |
| - .70 | .03 962 | 151.02 | .02 696 | 104.61 | .01 669 | 65.79 | .00 867 | 34.69 | - .70 | .00 293 | 11.89 |
| - .65 | .04 735 | 158.41 | .03 234 | 110.48 | .02 008 | 69.92 | .01 047 | 37.08 | - .65 | .00 355 | 12.77 |
| - .60 | .05 547 | 166.31 | .03 802 | 116.82 | .02 369 | 74.41 | .01 238 | 39.70 | - .60 | .00 421 | 13.75 |
| - .55 | .06 400 | 174.77 | .04 403 | 123.68 | .02 753 | 79.32 | .01 444 | 42.58 | - .55 | .00 493 | 14.83 |
| - .50 | .07 296 | 183.84 | .05 039 | 131.10 | .03 163 | 84.68 | .01 665 | 45.76 | - .50 | .00 570 | 16.04 |
| - .45 | .08 239 | 193.57 | .05 715 | 139.16 | .03 600 | 90.56 | .01 902 | 49.28 | - .45 | .00 653 | 17.39 |
| - .40 | .09 233 | 204.03 | .06 432 | 147.92 | .04 069 | 97.03 | .02 158 | 53.19 | - .40 | .00 744 | 18.90 |
| - .35 | .10 281 | 215.29 | .07 195 | 157.46 | .04 572 | 104.15 | .02 435 | 57.54 | - .35 | .00 843 | 20.59 |
| - .30 | .11 387 | 227.44 | .08 008 | 167.88 | .05 112 | 112.01 | .02 735 | 62.40 | - .30 | .00 950 | 22.51 |
| - .25 | .12 557 | 240.55 | .08 876 | 179.28 | .05 693 | 120.72 | .03 060 | 67.85 | - .25 | .01 068 | 24.68 |
| - .20 | .13 794 | 254.74 | .09 803 | 191.79 | .06 321 | 130.41 | .03 414 | 73.98 | - .20 | .01 198 | 27.16 |
| - .15 | .15 106 | 270.11 | .10 796 | 205.53 | .06 999 | 141.20 | .03 801 | 80.91 | - .15 | .01 340 | 29.99 |
| - .10 | .16 498 | 286.81 | .11 860 | 220.69 | .07 735 | 153.27 | .04 225 | 88.77 | - .10 | .01 498 | 33.25 |
| - .05 | .17 977 | 304.96 | .13 005 | 237.43 | .08 534 | 166.83 | .04 691 | 97.73 | - .05 | .01 674 | 37.02 |
| 0.00 | .19 550 | 324.76 | .14 238 | 256.00 | .09 406 | 182.11 | .05 204 | 108.00 | 0.00 | .01 869 | 41.41 |
| + .05 | .21 227 | 346.38 | .15 569 | 276.64 | .10 359 | 199.40 | .05 773 | 119.83 | + .05 | .02 089 | 46.56 |
| + .10 | .23 017 | 370.06 | .17 008 | 299.67 | .11 404 | 219.05 | .06 406 | 133.53 | + .10 | .02 336 | 52.64 |
| + .15 | .24 932 | 396.04 | .18 570 | 325.45 | .12 554 | 241.50 | .07 112 | 149.52 | + .15 | .02 617 | 59.89 |
| + .20 | .26 982 | 424.63 | .20 268 | 354.41 | .13 825 | 267.28 | .07 906 | 168.29 | + .20 | .02 938 | 68.60 |
| + .25 | .29 103 | 456.16 | .22 120 | 387.08 | .15 233 | 297.05 | .08 801 | 190.49 | + .25 | .03 306 | 79.17 |
| + .30 | .31 549 | 491.05 | .24 146 | 424.09 | .16 803 | 331.63 | .09 818 | 216.99 | + .30 | .03 734 | 92.15 |
| + .35 | .34 099 | 529.75 | .26 369 | 466.20 | .18 560 | 372.07 | .10 980 | 248.90 | + .35 | .04 233 | 108.29 |
| + .40 | .36 854 | 572.82 | .28 818 | 514.35 | .20 536 | 419.70 | .12 318 | 287.74 | + .40 | .04 823 | 128.65 |
| + .45 | .39 836 | 620.92 | .31 525 | 569.71 | .22 771 | 476.24 | .13 872 | 335.53 | + .45 | .05 529 | 154.74 |
| + .50 | .43 073 | 674.81 | .34 530 | 633.70 | .25 317 | 543.96 | .15 693 | 395.13 | + .50 | .06 384 | 188.83 |
| + .55 | .46 595 | 735.44 | .37 879 | 708.15 | .28 235 | 625.86 | .17 850 | 470.54 | + .55 | .07 436 | 234.34 |
| + .60 | .50 440 | 803.91 | .41 632 | 795.33 | .31 606 | 725.97 | .20 434 | 567.57 | + .60 | .08 755 | 296.69 |
| + .625 | .52 497 | 841.50 | .43 681 | 844.59 | .33 492 | 784.51 | .21 925 | 626.75 | + .625 | .09 546 | 336.76 |
| + .650 | .54 650 | 881.57 | .45 859 | 898.20 | .35 534 | 849.83 | .23 576 | 694.88 | + .650 | .10 446 | 384.78 |
| + .675 | .56 907 | 924.35 | .48 176 | 956.66 | .37 748 | 922.98 | .25 409 | 773.79 | + .675 | .11 478 | 442.91 |
| + .700 | .59 274 | 970.08 | .50 647 | 1020.55 | .40 156 | 1005.20 | .27 455 | 865.74 | + .700 | .12 671 | 514.05 |
| + .725 | .61 760 | 1019.01 | .53 284 | 1090.53 | .42 783 | 1097.97 | .29 751 | 973.66 | + .725 | .14 062 | 602.17 |
| + .750 | .64 372 | 1071.44 | .56 105 | 1167.38 | .45 656 | 1203.12 | .32 340 | 1101.28 | + .750 | .15 701 | 712.86 |
| + .775 | .67 120 | 1127.70 | .59 127 | 1251.98 | .48 810 | 1322.82 | .35 278 | 1253.46 | + .775 | .17 652 | 854.10 |
| + .800 | .70 014 | 1188.16 | .62 372 | 1345.36 | .52 285 | 1459.79 | .38 633 | 1436.61 | + .800 | .20 006 | 1037.65 |
| + .825 | .73 064 | 1253.20 | .65 862 | 1448.72 | .56 127 | 1617.34 | .42 494 | 1659.33 | + .810 | .21 088 | 1126.61 |
| + .850 | .76 284 | 1323.35 | .69 625 | 1563.48 | .60 363 | 1799.62 | .46 972 | 1933.28 | + .820 | .22 263 | 1248.44 |
| + .875 | .79 685 | 1399.04 | .73 691 | 1691.31 | .65 150 | 2011.84 | .52 216 | 2274.64 | + .830 | .23 545 | 1339.37 |
| + .900 | .83 285 | 1480.93 | .78 094 | 1834.16 | .70 482 | 2260.62 | .58 420 | 2706.28 | + .840 | .24 947 | 1467.18 |
|  |  |  |  |  |  |  |  |  | + .850 | .26 485 | 1612.67 |
| + .925 | .87 096 | 1569.62 | .82 876 | 1994.39 | .76 490 | 2554.46 | .65 850 | 3261.33 | + .860 | .28 179 | 1779.12 |
| + .950 | .91 139 | 1665.90 | .88 083 | 2174.80 | .83 301 | 2904.38 | .74 870 | 3989.04 | + .870 | .30 052 | 1970.55 |
| + .975 | .95 433 | 1770.58 | .93 770 | 2378.79 | .91 071 | 3324.90 | .86 001 | 4964.96 | + .880 | .32 130 | 2192.00 |
| +1.000 | 1.00 000 | 1884.66 | 1.00 000 | 2610.44 | 1.00 000 | 3835.42 | 1.00 000 | 6309.30 | + .890 | .34 448 | 2449.79 |
|  |  |  |  |  |  |  |  |  | + .900 | .37 045 | 2751.97 |
|  |  |  |  |  |  |  |  |  | + .910 | .39 970 | 3108.85 |
|  |  |  |  |  |  |  |  |  | + .920 | .43 285 | 3533.82 |
|  |  |  |  |  |  |  |  |  | + .930 | .47 066 | 4044.52 |
|  |  |  |  |  |  |  |  |  | + .940 | .51 410 | 4664.53 |
|  |  |  |  |  |  |  |  |  | + .950 | .56 443 | 5425.82 |
|  |  |  |  |  |  |  |  |  | + .955 | .59 265 | 5872.63 |
|  |  |  |  |  |  |  |  |  | + .960 | .62 324 | 6372.58 |
|  |  |  |  |  |  |  |  |  | + .965 | .65 648 | 6934.07 |
|  |  |  |  |  |  |  |  |  | + .970 | .69 270 | 7567.16 |
|  |  |  |  |  |  |  |  |  | + .975 | .73 229 | 8284.05 |
|  |  |  |  |  |  |  |  |  | + .980 | .77 570 | 9099.54 |
|  |  |  |  |  |  |  |  |  | + .985 | .82 348 | 10031.74 |
|  |  |  |  |  |  |  |  |  | + .990 | .87 625 | 11103.05 |
|  |  |  |  |  |  |  |  |  | + .995 | .94 538 | 12341.29 |
|  |  |  |  |  |  |  |  |  | +1.000 | 1.00 000 | 13781.45 |

| r | ρ=0.9 Area | Ordinate |
|---|---|---|
| + .800 | .20 006 | 1037.65 |
| + .825 | .22 890 | 1279.86 |
| + .850 | .26 485 | 1612.67 |
| + .875 | .31 063 | 2077.16 |
| + .900 | .37 045 | 2751.97 |
| + .925 | .45 112 | 3777.14 |
| + .950 | .56 443 | 5425.82 |
| + .975 | .73 229 | 8284.05 |
| +1.000 | 1.00 000 | 13781.45 |

ρ = 0·0 — 0·4　　　　　　　　　　n = 5

| r | ρ = 0·0 Area | Ordinate | ρ = 0·1 Area | Ordinate | ρ = 0·2 Area | Ordinate | ρ = 0·3 Area | Ordinate | ρ = 0·4 Area | Ordinate | r |
|---|---|---|---|---|---|---|---|---|---|---|---|
| −1·00 | ·00 000 | ·00 | ·00 000 | ·00 | ·00 000 | ·00 | ·00 000 | ·00 | ·00 000 | ·00 | −1·00 |
| − ·95 | ·00 666 | 198·78 | ·00 471 | 141·35 | ·00 328 | 99·03 | ·00 224 | 67·80 | ·00 147 | 44·87 | − ·95 |
| − ·90 | ·01 869 | 277·50 | ·01 334 | 200·54 | ·00 937 | 142·43 | ·00 643 | 98·66 | ·00 427 | 65·96 | − ·90 |
| − ·85 | ·03 407 | 335·36 | ·02 455 | 246·33 | ·01 739 | 177·39 | ·01 202 | 124·35 | ·00 802 | 84·01 | − ·85 |
| − ·80 | ·05 204 | 381·97 | ·03 786 | 285·19 | ·02 704 | 208·27 | ·01 882 | 147·80 | ·01 265 | 100·93 | − ·80 |
| − ·75 | ·07 215 | 421·08 | ·05 298 | 319·60 | ·03 801 | 236·74 | ·02 678 | 170·10 | ·01 811 | 117·44 | − ·75 |
| − ·70 | ·09 406 | 454·64 | ·06 976 | 350·80 | ·05 068 | 263·63 | ·03 582 | 191·85 | ·02 439 | 133·96 | − ·70 |
| − ·65 | ·11 754 | 483·79 | ·08 804 | 379·53 | ·06 452 | 289·41 | ·04 596 | 213·36 | ·03 150 | 150·72 | − ·65 |
| − ·60 | ·14 238 | 509·30 | ·10 768 | 406·24 | ·07 962 | 314·40 | ·05 716 | 234·88 | ·03 947 | 167·91 | − ·60 |
| − ·55 | ·16 842 | 531·68 | ·12 864 | 431·25 | ·09 596 | 338·80 | ·06 946 | 256·56 | ·04 831 | 185·68 | − ·55 |
| − ·50 | ·19 550 | 551·33 | ·15 078 | 454·76 | ·11 349 | 362·75 | ·08 282 | 278·52 | ·05 805 | 204·14 | − ·50 |
| − ·45 | ·22 351 | 568·52 | ·17 407 | 476·92 | ·13 222 | 386·34 | ·09 730 | 300·86 | ·06 874 | 223·41 | − ·45 |
| − ·40 | ·25 232 | 583·47 | ·19 843 | 497·83 | ·15 213 | 409·64 | ·11 292 | 323·65 | ·08 040 | 243·58 | − ·40 |
| − ·35 | ·28 182 | 596·35 | ·22 384 | 517·56 | ·17 318 | 432·68 | ·12 967 | 346·96 | ·09 311 | 264·76 | − ·35 |
| − ·30 | ·31 192 | 607·30 | ·25 018 | 536·16 | ·19 539 | 455·50 | ·14 763 | 370·82 | ·10 690 | 287·04 | − ·30 |
| − ·25 | ·34 252 | 616·40 | ·27 745 | 553·64 | ·21 873 | 478·09 | ·16 678 | 395·29 | ·12 183 | 310·51 | − ·25 |
| − ·20 | ·37 353 | 623·76 | ·30 556 | 570·02 | ·24 319 | 500·44 | ·18 716 | 420·38 | ·13 797 | 335·28 | − ·20 |
| − ·15 | ·40 487 | 629·42 | ·33 440 | 585·27 | ·26 877 | 522·52 | ·20 883 | 446·12 | ·15 539 | 361·44 | − ·15 |
| − ·10 | ·43 644 | 633·43 | ·36 407 | 599·37 | ·29 543 | 544·31 | ·23 179 | 472·50 | ·17 414 | 389·09 | − ·10 |
| − ·05 | ·46 818 | 635·82 | ·39 432 | 612·28 | ·32 318 | 565·73 | ·25 608 | 499·53 | ·19 431 | 418·31 | − ·05 |
| 0·00 | ·50 000 | 636·62 | ·42 526 | 623·95 | ·35 200 | 586·71 | ·28 175 | 527·18 | ·21 600 | 449·20 | 0·00 |
| + ·05 | ·53 182 | 635·82 | ·45 673 | 634·31 | ·38 184 | 607·16 | ·30 881 | 555·41 | ·23 927 | 481·84 | + ·05 |
| + ·10 | ·56 356 | 633·43 | ·48 868 | 643·27 | ·41 281 | 626·97 | ·33 730 | 584·15 | ·26 421 | 516·30 | + ·10 |
| + ·15 | ·59 513 | 629·42 | ·52 101 | 650·74 | ·44 454 | 645·99 | ·36 723 | 613·31 | ·29 093 | 552·65 | + ·15 |
| + ·20 | ·62 647 | 623·76 | ·55 372 | 656·59 | ·47 729 | 664·06 | ·39 863 | 642·76 | ·31 951 | 590·91 | + ·20 |
| + ·25 | ·65 748 | 616·40 | ·58 663 | 660·68 | ·51 091 | 680·97 | ·43 151 | 672·33 | ·35 005 | 631·08 | + ·25 |
| + ·30 | ·68 808 | 607·30 | ·61 972 | 662·86 | ·54 536 | 696·46 | ·46 587 | 701·77 | ·38 265 | 673·11 | + ·30 |
| + ·35 | ·71 818 | 596·35 | ·65 289 | 662·91 | ·58 054 | 710·24 | ·50 169 | 730·79 | ·41 740 | 716·87 | + ·35 |
| + ·40 | ·74 768 | 583·47 | ·68 600 | 660·60 | ·61 635 | 721·95 | ·53 893 | 758·98 | ·45 437 | 762·14 | + ·40 |
| + ·45 | ·77 649 | 568·52 | ·71 891 | 655·66 | ·65 270 | 731·14 | ·57 756 | 785·81 | ·49 362 | 808·54 | + ·45 |
| + ·50 | ·80 450 | 551·33 | ·75 152 | 647·74 | ·68 942 | 737·25 | ·61 747 | 810·60 | ·53 523 | 855·49 | + ·50 |
| + ·55 | ·83 158 | 531·68 | ·78 365 | 636·41 | ·72 635 | 739·59 | ·65 857 | 832·42 | ·57 917 | 902·10 | + ·55 |
| + ·60 | ·85 762 | 509·30 | ·81 510 | 621·15 | ·76 329 | 737·29 | ·70 065 | 850·06 | ·62 541 | 947·05 | + ·60 |
| + ·65 | ·88 246 | 483·79 | ·84 565 | 601·26 | ·79 999 | 729·21 | ·74 347 | 861·85 | ·67 381 | 988·36 | + ·65 |
| + ·70 | ·90 594 | 454·64 | ·87 511 | 575·83 | ·83 610 | 713·82 | ·78 670 | 865·50 | ·72 413 | 1023·03 | + ·70 |
| + ·75 | ·92 785 | 421·08 | ·90 313 | 543·59 | ·87 207 | 689·01 | ·82 984 | 857·73 | ·77 593 | 1046·50 | + ·75 |
| + ·80 | ·94 796 | 381·97 | ·92 933 | 502·63 | ·90 480 | 651·67 | ·87 220 | 833·60 | ·82 847 | 1051·45 | + ·80 |
| + ·85 | ·96 593 | 335·36 | ·95 320 | 449·87 | ·93 609 | 596·86 | ·91 280 | 785·20 | ·88 057 | 1025·58 | + ·85 |
| + ·90 | ·98 131 | 277·50 | ·97 403 | 379·52 | ·96 404 | 515·46 | ·95 008 | 698·04 | ·93 015 | 945·84 | + ·90 |
| + ·95 | ·99 334 | 198·78 | ·99 065 | 277·21 | ·98 685 | 385·59 | ·98 142 | 538·03 | ·97 340 | 757·76 | + ·95 |
| +1·00 | 1·00 000 | ·00 | 1·00 000 | ·00 | 1·00 000 | ·00 | 1·00 000 | ·00 | 1·00 000 | ·00 | +1·00 |

| r | ρ=0·5 Area | ρ=0·5 Ordinate | ρ=0·6 Area | ρ=0·6 Ordinate | ρ=0·7 Area | ρ=0·7 Ordinate | ρ=0·8 Area | ρ=0·8 Ordinate | r | ρ=0·9 Area | ρ=0·9 Ordinate |
|---|---|---|---|---|---|---|---|---|---|---|---|
| −1·00 | ·00 000 | ·00 | ·00 000 | ·00 | ·00 000 | ·00 | ·00 000 | ·00 | −1·00 | ·00 000 | ·00 |
| − ·95 | ·00 092 | 28·25 | ·00 054 | 16·49 | ·00 028 | 8·50 | ·00 011 | 3·48 | − ·95 | ·00 003 | ·80 |
| − ·90 | ·00 269 | 41·90 | ·00 157 | 24·65 | ·00 081 | 12·79 | ·00 033 | 5·27 | − ·90 | ·00 008 | 1·22 |
| − ·85 | ·00 509 | 53·86 | ·00 299 | 31·94 | ·00 154 | 16·70 | ·00 064 | 6·92 | − ·85 | ·00 015 | 1·62 |
| − ·80 | ·00 807 | 65·32 | ·00 476 | 39·06 | ·00 248 | 20·58 | ·00 103 | 8·59 | − ·80 | ·00 024 | 2·02 |
| − ·75 | ·01 162 | 76·75 | ·00 690 | 46·30 | ·00 361 | 24·59 | ·00 150 | 10·33 | − ·75 | ·00 035 | 2·45 |
| − ·70 | ·01 575 | 88·43 | ·00 940 | 53·84 | ·00 494 | 28·82 | ·00 206 | 12·20 | − ·70 | ·00 048 | 2·91 |
| − ·65 | ·02 047 | 100·55 | ·01 229 | 61·79 | ·00 650 | 33·37 | ·00 272 | 14·24 | − ·65 | ·00 064 | 3·42 |
| − ·60 | ·02 581 | 113·24 | ·01 559 | 70·28 | ·00 829 | 38·30 | ·00 349 | 16·48 | − ·60 | ·00 083 | 3·99 |
| − ·55 | ·03 180 | 126·63 | ·01 933 | 79·40 | ·01 033 | 43·68 | ·00 437 | 18·96 | − ·55 | ·00 104 | 4·62 |
| − ·50 | ·03 849 | 140·84 | ·02 354 | 89·26 | ·01 266 | 49·59 | ·00 539 | 21·72 | − ·50 | ·00 125 | 5·34 |
| − ·45 | ·04 590 | 156·00 | ·02 827 | 99·98 | ·01 530 | 56·12 | ·00 655 | 24·82 | − ·45 | ·00 156 | 6·16 |
| − ·40 | ·05 410 | 172·23 | ·03 355 | 111·67 | ·01 829 | 63·37 | ·00 788 | 28·32 | − ·40 | ·00 191 | 7·10 |
| − ·35 | ·06 314 | 189·65 | ·03 945 | 124·47 | ·02 166 | 71·45 | ·00 939 | 32·27 | − ·35 | ·00 229 | 8·17 |
| − ·30 | ·07 309 | 208·40 | ·04 602 | 138·53 | ·02 545 | 80·49 | ·01 111 | 36·78 | − ·30 | ·00 273 | 9·41 |
| − ·25 | ·08 401 | 228·62 | ·05 333 | 154·02 | ·02 972 | 90·65 | ·01 308 | 41·92 | − ·25 | ·00 323 | 10·86 |
| − ·20 | ·09 598 | 250·49 | ·06 145 | 171·14 | ·03 454 | 102·08 | ·01 532 | 47·83 | − ·20 | ·00 382 | 12·54 |
| − ·15 | ·10 909 | 274·17 | ·07 047 | 190·10 | ·03 995 | 115·02 | ·01 788 | 54·64 | − ·15 | ·00 449 | 14·52 |
| − ·10 | ·12 344 | 299·86 | ·08 050 | 211·15 | ·04 606 | 129·71 | ·02 080 | 62·54 | − ·10 | ·00 528 | 16·86 |
| − ·05 | ·13 911 | 327·76 | ·09 163 | 234·60 | ·05 296 | 146·44 | ·02 415 | 71·74 | − ·05 | ·00 619 | 19·65 |
| 0·00 | ·15 625 | 358·10 | ·10 400 | 260·76 | ·06 075 | 165·58 | ·02 800 | 82·51 | 0·00 | ·00 725 | 22·98 |
| + ·05 | ·17 497 | 391·13 | ·11 775 | 290·03 | ·06 957 | 187·57 | ·03 243 | 95·19 | + ·05 | ·00 850 | 27·01 |
| + ·10 | ·19 541 | 427·11 | ·13 306 | 322·86 | ·07 956 | 212·91 | ·03 756 | 110·22 | + ·10 | ·00 996 | 31·91 |
| + ·15 | ·21 773 | 466·33 | ·15 011 | 359·76 | ·09 092 | 242·26 | ·04 383 | 129·14 | + ·15 | ·01 171 | 37·93 |
| + ·20 | ·24 211 | 509·09 | ·16 912 | 401·33 | ·10 387 | 276·40 | ·05 044 | 149·65 | + ·20 | ·01 378 | 45·39 |
| + ·25 | ·26 871 | 555·72 | ·19 033 | 448·27 | ·11 866 | 316·29 | ·05 854 | 175·68 | + ·25 | ·01 628 | 54·74 |
| + ·30 | ·29 775 | 606·51 | ·21 405 | 501·37 | ·13 561 | 363·13 | ·06 809 | 207·42 | + ·30 | ·01 930 | 66·60 |
| + ·35 | ·32 944 | 661·77 | ·24 059 | 561·55 | ·15 511 | 418·41 | ·07 942 | 246·48 | + ·35 | ·02 299 | 81·84 |
| + ·40 | ·36 400 | 721·75 | ·27 033 | 629·83 | ·17 763 | 483·98 | ·09 290 | 295·00 | + ·40 | ·02 756 | 101·75 |
| + ·45 | ·40 169 | 786·62 | ·30 373 | 707·37 | ·20 372 | 562·17 | ·10 911 | 355·92 | + ·45 | ·03 327 | 128·18 |
| + ·50 | ·44 274 | 856·39 | ·34 125 | 795·39 | ·23 410 | 655·88 | ·12 877 | 433·30 | + ·50 | ·04 007 | 163·98 |
| + ·55 | ·48 741 | 930·77 | ·38 346 | 895·14 | ·26 963 | 768·74 | ·15 281 | 532·81 | + ·55 | ·04 990 | 213·57 |
| + ·60 | ·53 589 | 1009·03 | ·43 097 | 1007·71 | ·31 137 | 905·20 | ·18 255 | 662·54 | + ·60 | ·06 224 | 284·13 |
| + ·625 | ·56 161 | 1049·17 | ·45 693 | 1069·02 | ·33 496 | 983·92 | ·20 008 | 742·21 | + ·625 | ·06 990 | 330·74 |
| + ·650 | ·58 835 | 1089·62 | ·48 445 | 1133·68 | ·36 063 | 1070·61 | ·21 976 | 834·18 | + ·650 | ·07 886 | 387·69 |
| + ·675 | ·61 610 | 1129·97 | ·51 364 | 1201·60 | ·38 857 | 1166·04 | ·24 191 | 940·75 | + ·675 | ·08 910 | 457·91 |
| + ·700 | ·64 484 | 1169·66 | ·54 456 | 1272·51 | ·41 901 | 1270·96 | ·26 694 | 1064·70 | + ·700 | ·10 190 | 545·45 |
| + ·725 | ·67 457 | 1207·97 | ·57 728 | 1345·91 | ·45 220 | 1386·07 | ·29 532 | 1209·40 | + ·725 | ·11 686 | 655·85 |
| + ·750 | ·70 522 | 1243·92 | ·61 187 | 1421·02 | ·48 840 | 1511·86 | ·32 762 | 1378·85 | + ·750 | ·13 495 | 796·89 |
| + ·775 | ·73 673 | 1276·18 | ·64 834 | 1496·53 | ·52 789 | 1648·45 | ·36 451 | 1577·77 | + ·775 | ·15 705 | 979·65 |
| + ·800 | ·76 898 | 1302·93 | ·68 668 | 1570·47 | ·57 091 | 1795·27 | ·40 680 | 1811·55 | + ·800 | ·18 441 | 1220·21 |
| + ·825 | ·80 181 | 1321·68 | ·72 682 | 1639·76 | ·61 772 | 1950·42 | ·45 542 | 2085·92 | + ·810 | ·19 719 | 1337·61 |
| + ·850 | ·83 497 | 1328·96 | ·76 859 | 1699·62 | ·63 483 | 2109·71 | ·51 148 | 2406·01 | + ·820 | ·21 121 | 1469·92 |
| + ·875 | ·86 812 | 1319·63 | ·81 166 | 1742·59 | ·68 648 | 2264·65 | ·57 613 | 2773·92 | + ·830 | ·22 664 | 1619·48 |
| + ·900 | ·90 075 | 1286·02 | ·85 548 | 1756·53 | ·78 153 | 2398·74 | ·65 052 | 3182·44 | + ·840 | ·24 367 | 1789·05 |
|  |  |  |  |  |  |  |  |  | + ·850 | ·26 250 | 1981·91 |
| + ·925 | ·93 211 | 1215·55 | ·89 907 | 1720·60 | ·84 265 | 2479·25 | ·73 533 | 3598·75 | + ·860 | ·28 339 | 2201·94 |
| + ·950 | ·96 104 | 1085·21 | ·94 078 | 1595·22 | ·90 449 | 2436·67 | ·82 969 | 3917·02 | + ·870 | ·30 664 | 2453·69 |
| + ·975 | ·98 569 | 840·63 | ·96 304 | 1286·20 | ·94 172 | 2093·74 | ·92 757 | 3783·13 | + ·880 | ·33 259 | 2742·58 |
| +1·000 | 1·00 000 | ·00 | 1·00 000 | ·00 | 1·00 000 | ·00 | 1·00 000 | ·00 | + ·890 | ·36 164 | 3074·87 |
|  |  |  |  |  |  |  |  |  | + ·900 | ·39 426 | 3457·80 |
|  |  |  |  |  |  |  |  |  | + ·910 | ·43 099 | 3899·42 |
|  |  |  |  |  |  |  |  |  | + ·920 | ·47 247 | 4408·19 |
|  |  |  |  |  |  |  |  |  | + ·930 | ·51 941 | 4991·87 |
|  |  |  |  |  |  |  |  |  | + ·940 | ·57 257 | 5654·90 |
|  |  |  |  |  |  |  |  |  | + ·950 | ·63 275 | 6392·40 |
|  |  |  |  |  |  |  |  |  | + ·955 | ·66 568 | 6781·92 |
|  |  |  |  |  |  |  |  |  | + ·960 | ·70 058 | 7176·32 |
|  |  |  |  |  |  |  |  |  | + ·965 | ·73 743 | 7563·03 |
|  |  |  |  |  |  |  |  |  | + ·970 | ·77 616 | 7921·21 |
|  |  |  |  |  |  |  |  |  | + ·975 | ·81 654 | 8216·35 |
|  |  |  |  |  |  |  |  |  | + ·980 | ·85 812 | 8389·83 |
|  |  |  |  |  |  |  |  |  | + ·985 | ·90 005 | 8337·28 |
|  |  |  |  |  |  |  |  |  | + ·990 | ·93 963 | 7854·33 |
|  |  |  |  |  |  |  |  |  | + ·995 | ·97 593 | 6446·39 |
|  |  |  |  |  |  |  |  |  | +1·000 | 1·00 000 | ·00 |

| | ρ=0·9 | |
|---|---|---|
| r | Area | Ordinate |
| + ·800 | ·18 441 | 1220·21 |
| + ·825 | ·21 873 | 1540·97 |
| + ·850 | ·26 250 | 1981·91 |
| + ·875 | ·31 929 | 2593·12 |
| + ·900 | ·39 426 | 3457·80 |
| + ·925 | ·49 523 | 4690·26 |
| + ·950 | ·63 275 | 6392·40 |
| + ·975 | ·81 654 | 8216·35 |
| +1·000 | 1·00 000 | ·00 |

ρ=0·0 — 0·4        n=6

| r | ρ=0·0 Area | ρ=0·0 Ordinate | ρ=0·1 Area | ρ=0·1 Ordinate | ρ=0·2 Area | ρ=0·2 Ordinate | ρ=0·3 Area | ρ=0·3 Ordinate | ρ=0·4 Area | ρ=0·4 Ordinate | r |
|---|---|---|---|---|---|---|---|---|---|---|---|
| -1·00 | ·00 000 | ·000 | ·00 000 | ·00 | ·00 000 | ·00 | ·00 000 | ·00 | ·00 000 | ·00 | -1·00 |
| - ·95 | ·00 184 | 73·125 | ·00 118 | 47·28 | ·00 075 | 30·03 | ·00 046 | 18·55 | ·00 027 | 10·99 | - ·95 |
| - ·90 | ·00 725 | 142·500 | ·00 472 | 94·07 | ·00 302 | 60·81 | ·00 187 | 38·13 | ·00 111 | 22·88 | - ·90 |
| - ·85 | ·01 603 | 208·125 | ·01 058 | 140·28 | ·00 684 | 92·30 | ·00 429 | 58·76 | ·00 257 | 35·74 | - ·85 |
| - ·80 | ·02 800 | 270·000 | ·01 874 | 185·83 | ·01 226 | 124·49 | ·00 777 | 80·50 | ·00 470 | 49·64 | - ·80 |
| - ·75 | ·04 297 | 328·125 | ·02 915 | 230·63 | ·01 930 | 157·34 | ·01 236 | 103·38 | ·00 757 | 64·65 | - ·75 |
| - ·70 | ·06 075 | 382·500 | ·04 179 | 274·58 | ·02 800 | 190·82 | ·01 812 | 127·43 | ·01 119 | 80·85 | - ·70 |
| - ·65 | ·08 116 | 433·125 | ·05 659 | 317·59 | ·03 839 | 224·87 | ·02 512 | 152·69 | ·01 566 | 98·33 | - ·65 |
| - ·60 | ·10 400 | 480·000 | ·07 353 | 359·54 | ·05 050 | 259·44 | ·03 341 | 179·18 | ·02 103 | 117·16 | - ·60 |
| - ·55 | ·12 909 | 523·125 | ·09 253 | 400·33 | ·06 434 | 294·48 | ·04 306 | 206·94 | ·02 740 | 137·45 | - ·55 |
| - ·50 | ·15 625 | 562·500 | ·11 354 | 439·82 | ·07 995 | 329·89 | ·05 413 | 235·97 | ·03 481 | 159·29 | - ·50 |
| - ·45 | ·18 528 | 598·125 | ·13 649 | 477·89 | ·09 734 | 365·60 | ·06 668 | 266·29 | ·04 336 | 182·79 | - ·45 |
| - ·40 | ·21 600 | 630·000 | ·16 130 | 514·41 | ·11 651 | 401·49 | ·08 078 | 297·90 | ·05 312 | 208·03 | - ·40 |
| - ·35 | ·24 822 | 658·125 | ·18 790 | 549·23 | ·13 749 | 437·46 | ·09 649 | 330·80 | ·06 420 | 235·13 | - ·35 |
| - ·30 | ·28 175 | 682·500 | ·21 619 | 582·20 | ·16 026 | 473·37 | ·11 388 | 364·95 | ·07 666 | 264·20 | - ·30 |
| - ·25 | ·31 640 | 703·125 | ·24 609 | 613·16 | ·18 482 | 509·07 | ·13 301 | 400·33 | ·09 065 | 295·32 | - ·25 |
| - ·20 | ·35 200 | 720·000 | ·27 747 | 641·93 | ·21 116 | 544·37 | ·15 393 | 436·88 | ·10 625 | 328·62 | - ·20 |
| - ·15 | ·38 828 | 733·125 | ·31 024 | 668·33 | ·23 925 | 579·08 | ·17 671 | 474·49 | ·12 355 | 364·17 | - ·15 |
| - ·10 | ·42 525 | 742·500 | ·34 426 | 692·18 | ·26 905 | 612·97 | ·20 140 | 513·07 | ·14 271 | 402·05 | - ·10 |
| - ·05 | ·43 290 | 748·125 | ·37 941 | 713·27 | ·30 053 | 645·79 | ·22 803 | 552·46 | ·16 381 | 442·32 | - ·05 |
| 0·00 | ·50 000 | 750·000 | ·41 554 | 731·39 | ·33 361 | 677·23 | ·25 665 | 592·47 | ·18 697 | 485·02 | 0·00 |
| + ·05 | ·56 710 | 748·125 | ·45 250 | 746·31 | ·36 822 | 706·99 | ·28 729 | 632·84 | ·21 232 | 530·14 | + ·05 |
| + ·10 | ·57 475 | 742·500 | ·49 011 | 757·78 | ·40 427 | 734·67 | ·31 994 | 673·26 | ·24 003 | 577·63 | + ·10 |
| + ·15 | ·61 172 | 733·125 | ·52 821 | 765·56 | ·44 165 | 759·87 | ·35 461 | 713·34 | ·27 015 | 627·35 | + ·15 |
| + ·20 | ·64 800 | 720·000 | ·56 660 | 769·38 | ·48 021 | 782·10 | ·39 126 | 752·61 | ·30 279 | 679·10 | + ·20 |
| + ·25 | ·68 360 | 703·125 | ·60 508 | 768·94 | ·51 980 | 800·84 | ·42 984 | 790·47 | ·33 809 | 732·53 | + ·25 |
| + ·30 | ·71 825 | 682·500 | ·64 342 | 763·96 | ·56 023 | 815·48 | ·47 027 | 826·21 | ·39 852 | 787·15 | + ·30 |
| + ·35 | ·75 178 | 658·125 | ·68 140 | 754·11 | ·60 127 | 825·36 | ·51 241 | 858·93 | ·41 682 | 842·24 | + ·35 |
| + ·40 | ·78 400 | 630·000 | ·71 875 | 739·06 | ·64 267 | 829·70 | ·55 610 | 887·57 | ·46 027 | 896·80 | + ·40 |
| + ·45 | ·81 472 | 598·125 | ·75 521 | 718·45 | ·68 409 | 827·66 | ·60 108 | 910·82 | ·50 643 | 949·48 | + ·45 |
| + ·50 | ·84 375 | 562·500 | ·79 049 | 691·90 | ·72 531 | 818·27 | ·64 706 | 927·11 | ·55 516 | 998·43 | + ·50 |
| + ·55 | ·87 091 | 523·125 | ·82 429 | 659·02 | ·76 582 | 800·46 | ·69 365 | 934·53 | ·60 619 | 1041·18 | + ·55 |
| + ·60 | ·89 600 | 480·000 | ·85 628 | 619·39 | ·80 520 | 772·99 | ·74 033 | 930·76 | ·65 913 | 1074·42 | + ·60 |
| + ·65 | ·91 884 | 433·125 | ·88 611 | 572·56 | ·84 294 | 734·52 | ·78 649 | 913·01 | ·71 341 | 1093·74 | + ·65 |
| + ·70 | ·93 925 | 382·500 | ·91 341 | 518·05 | ·87 844 | 683·50 | ·83 134 | 877·89 | ·76 817 | 1093·72 | + ·70 |
| + ·75 | ·95 703 | 328·125 | ·93 778 | 455·38 | ·91 105 | 618·19 | ·87 392 | 821·29 | ·82 226 | 1065·22 | + ·75 |
| + ·80 | ·97 200 | 270·000 | ·95 880 | 384·02 | ·93 999 | 536·66 | ·91 303 | 738·25 | ·87 405 | 999·24 | + ·80 |
| + ·85 | ·98 397 | 208·125 | ·97 603 | 303·41 | ·96 441 | 436·70 | ·94 720 | 622·74 | ·92 132 | 881·50 | + ·85 |
| + ·90 | ·99 275 | 142·500 | ·98 898 | 212·95 | ·98 331 | 315·86 | ·97 464 | 467·45 | ·96 105 | 693·60 | + ·90 |
| + ·95 | ·99 816 | 73·125 | ·99 715 | 112·04 | ·99 560 | 171·33 | ·99 314 | 263·48 | ·98 910 | 410·83 | + ·95 |
| +1·00 | 1·00 000 | ·000 | 1·00 000 | ·00 | 1·00 000 | ·00 | 1·00 000 | ·00 | 1·00 000 | ·00 | +1·00 |

n = 6

ρ = 0.5 —— 0.9

| r | ρ=0.5 Area | Ordinate | ρ=0.6 Area | Ordinate | ρ=0.7 Area | Ordinate | ρ=0.8 Area | Ordinate |
|---|---|---|---|---|---|---|---|---|
| -1.00 | .00 000 | .00 | .00 000 | .00 | .00 000 | .00 | .00 000 | .00 |
| - .95 | .00 015 | 6.12 | .00 008 | 3.10 | .00 003 | 1.35 | .00 001 | .44 |
| - .90 | .00 076 | 12.89 | .00 032 | 6.60 | .00 014 | 2.89 | .00 005 | .95 |
| - .85 | .00 145 | 20.37 | .00 075 | 10.54 | .00 033 | 4.66 | .00 011 | 1.54 |
| - .80 | .00 267 | 28.64 | .00 138 | 14.97 | .00 061 | 6.68 | .00 020 | 2.23 |
| - .75 | .00 432 | 37.76 | .00 225 | 19.96 | .00 100 | 9.00 | .00 033 | 3.03 |
| - .70 | .00 646 | 47.84 | .00 339 | 25.59 | .00 151 | 11.66 | .00 050 | 3.96 |
| - .65 | .00 913 | 58.96 | .00 482 | 31.92 | .00 217 | 14.71 | .00 073 | 5.05 |
| - .60 | .01 238 | 71.24 | .00 659 | 39.06 | .00 299 | 18.20 | .00 101 | 6.32 |
| - .55 | .01 628 | 84.78 | .00 874 | 47.09 | .00 400 | 22.22 | .00 137 | 7.79 |
| - .50 | .02 088 | 99.73 | .01 132 | 56.16 | .00 522 | 26.83 | .00 180 | 9.52 |
| - .45 | .02 628 | 116.21 | .01 438 | 66.38 | .00 670 | 32.13 | .00 232 | 11.55 |
| - .40 | .03 254 | 134.39 | .01 798 | 77.92 | .00 845 | 38.25 | .00 296 | 13.93 |
| - .35 | .03 970 | 154.44 | .02 219 | 90.95 | .01 054 | 45.31 | .00 372 | 16.73 |
| - .30 | .04 802 | 176.55 | .02 710 | 105.68 | .01 300 | 53.47 | .00 464 | 20.03 |
| - .25 | .05 744 | 200.93 | .03 279 | 122.35 | .01 591 | 62.92 | .00 574 | 23.94 |
| - .20 | .06 814 | 227.80 | .03 937 | 141.22 | .01 932 | 73.89 | .00 705 | 28.59 |
| - .15 | .08 027 | 257.41 | .04 696 | 162.61 | .02 332 | 86.65 | .00 861 | 34.12 |
| - .10 | .09 394 | 290.01 | .05 568 | 186.88 | .02 802 | 101.53 | .01 048 | 40.75 |
| - .05 | .10 933 | 325.89 | .06 570 | 214.44 | .03 351 | 118.92 | .01 271 | 48.71 |
| 0.00 | .12 659 | 365.35 | .07 719 | 245.76 | .03 996 | 139.31 | .01 538 | 58.32 |
| + .05 | .14 591 | 408.70 | .09 034 | 281.39 | .04 751 | 163.28 | .01 857 | 69.98 |
| + .10 | .16 753 | 456.23 | .10 541 | 321.94 | .05 636 | 191.55 | .02 242 | 84.21 |
| + .15 | .19 162 | 508.26 | .12 264 | 368.14 | .06 675 | 224.98 | .02 705 | 101.67 |
| + .20 | .21 843 | 565.04 | .14 233 | 420.76 | .07 896 | 264.66 | .03 265 | 123.24 |
| + .25 | .24 821 | 626.81 | .16 484 | 480.72 | .09 334 | 311.99 | .03 946 | 150.07 |
| + .30 | .28 120 | 693.69 | .19 054 | 548.97 | .11 030 | 368.33 | .04 777 | 183.69 |
| + .35 | .31 765 | 765.62 | .21 989 | 626.56 | .13 036 | 435.94 | .05 798 | 226.18 |
| + .40 | .35 784 | 842.35 | .25 337 | 714.54 | .15 413 | 517.20 | .07 058 | 280.35 |
| + .45 | .40 196 | 923.20 | .29 154 | 813.87 | .18 236 | 615.10 | .08 627 | 350.06 |
| + .50 | .45 021 | 1006.98 | .33 496 | 925.29 | .21 597 | 733.19 | .10 594 | 440.67 |
| + .55 | .50 267 | 1091.61 | .38 427 | 1048.92 | .25 608 | 875.62 | .13 081 | 559.69 |
| + .60 | .55 933 | 1173.81 | .44 004 | 1183.82 | .30 401 | 1046.92 | .16 255 | 717.69 |
| + .625 | .58 916 | 1212.47 | .47 052 | 1254.67 | .33 139 | 1144.78 | .18 169 | 815.75 |
| + .650 | .61 993 | 1248.41 | .50 278 | 1326.99 | .36 132 | 1251.36 | .20 347 | 929.54 |
| + .675 | .65 155 | 1280.54 | .53 687 | 1399.79 | .39 403 | 1366.92 | .22 832 | 1061.83 |
| + .700 | .68 391 | 1307.53 | .57 277 | 1471.73 | .42 975 | 1491.35 | .25 674 | 1215.82 |
| + .725 | .71 687 | 1327.73 | .61 043 | 1540.93 | .46 867 | 1624.05 | .28 932 | 1395.10 |
| + .750 | .75 023 | 1339.14 | .64 977 | 1604.80 | .51 101 | 1763.58 | .32 674 | 1603.56 |
| + .775 | .78 373 | 1339.29 | .69 060 | 1659.81 | .55 689 | 1907.18 | .36 977 | 1845.03 |
| + .800 | .81 707 | 1325.11 | .73 265 | 1701.20 | .60 636 | 2050.08 | .41 929 | 2122.58 |
| + .825 | .84 984 | 1292.86 | .77 549 | 1722.48 | .65 932 | 2184.35 | .47 621 | 2436.93 |
| + .850 | .88 153 | 1237.97 | .81 853 | 1714.92 | .71 540 | 2297.17 | .54 141 | 2783.31 |
| + .875 | .91 150 | 1154.70 | .86 090 | 1666.68 | .77 383 | 2368.12 | .61 551 | 3144.98 |
| + .900 | .93 897 | 1036.09 | .90 140 | 1561.73 | .83 319 | 2364.83 | .69 848 | 3479.97 |
| + .925 | .96 294 | 873.43 | .93 833 | 1378.27 | .89 103 | 2235.99 | .78 856 | 3693.41 |
| + .950 | .98 219 | 655.98 | .96 941 | 1086.58 | .94 328 | 1899.93 | .88 072 | 3578.10 |
| + .975 | .99 517 | 370.38 | .99 143 | 645.89 | .98 323 | 1225.57 | .96 133 | 2682.19 |
| +1.000 | 1.00 000 | .00 | 1.00 000 | .00 | 1.00 000 | .00 | 1.00 000 | .00 |

ρ = 0.9

| r | Area | Ordinate |
|---|---|---|
| -1.00 | .00 000 | .00 |
| - .95 | .00 000 | .07 |
| - .90 | .00 001 | .15 |
| - .85 | .00 002 | .25 |
| - .80 | .00 003 | .36 |
| - .75 | .00 005 | .50 |
| - .70 | .00 008 | .66 |
| - .65 | .00 012 | .85 |
| - .60 | .00 017 | 1.07 |
| - .55 | .00 023 | 1.33 |
| - .50 | .00 030 | 1.64 |
| - .45 | .00 039 | 2.02 |
| - .40 | .00 051 | 2.46 |
| - .35 | .00 064 | 3.00 |
| - .30 | .00 081 | 3.64 |
| - .25 | .00 101 | 4.41 |
| - .20 | .00 125 | 5.35 |
| - .15 | .00 155 | 6.50 |
| - .10 | .00 190 | 7.91 |
| - .05 | .00 234 | 9.65 |
| 0.00 | .00 288 | 11.80 |
| + .05 | .00 353 | 14.50 |
| + .10 | .00 434 | 17.90 |
| + .15 | .00 534 | 22.24 |
| + .20 | .00 658 | 27.81 |
| + .25 | .00 815 | 35.06 |
| + .30 | .01 013 | 44.60 |
| + .35 | .01 266 | 57.33 |
| + .40 | .01 593 | 74.61 |
| + .45 | .02 023 | 98.48 |
| + .50 | .02 594 | 132.11 |
| + .55 | .03 369 | 180.65 |
| + .60 | .04 440 | 252.62 |
| + .625 | .05 131 | 301.63 |
| + .650 | .05 958 | 362.77 |
| + .675 | .06 957 | 439.77 |
| + .700 | .08 174 | 537.74 |
| + .725 | .09 669 | 663.81 |
| + .750 | .11 525 | 828.01 |
| + .775 | .13 853 | 1044.63 |
| + .800 | .16 809 | 1334.30 |
| + .810 | .18 213 | 1476.92 |
| + .820 | .19 769 | 1638.33 |
| + .830 | .21 497 | 1821.38 |
| + .840 | .23 420 | 2029.33 |
| + .850 | .25 565 | 2265.98 |
| + .860 | .27 963 | 2535.59 |
| + .870 | .30 649 | 2842.96 |
| + .880 | .33 663 | 3193.29 |
| + .890 | .37 051 | 3591.95 |
| + .900 | .40 865 | 4043.96 |
| + .910 | .45 158 | 4552.99 |
| + .920 | .49 990 | 5119.39 |
| + .930 | .55 414 | 5736.51 |
| + .940 | .61 473 | 6383.78 |
| + .950 | .68 176 | 7013.72 |
| + .955 | .71 755 | 7294.27 |
| + .960 | .75 463 | 7527.03 |
| + .965 | .79 269 | 7683.58 |
| + .970 | .83 127 | 7723.89 |
| + .975 | .86 964 | 7591.48 |
| + .980 | .90 676 | 7206.50 |
| + .985 | .94 110 | 6455.59 |
| + .990 | .97 044 | 5176.85 |
| + .995 | .99 161 | 3137.54 |
| +1.000 | 1.00 000 | .00 |

ρ = 0.9

| r | Area | Ordinate |
|---|---|---|
| + .800 | .16 809 | 1334.30 |
| + .825 | .20 609 | 1725.33 |
| + .850 | .25 565 | 2265.98 |
| + .875 | .32 116 | 3012.42 |
| + .900 | .40 865 | 4043.96 |
| + .925 | .52 626 | 5422.46 |
| + .950 | .68 176 | 7013.72 |
| + .975 | .86 964 | 7591.48 |
| +1.000 | 1.00 000 | .00 |

ρ = 0·0 — 0·4                    n = 7

| r | ρ = 0·0 Area | Ordinate | ρ = 0·1 Area | Ordinate | ρ = 0·2 Area | Ordinate | ρ = 0·3 Area | Ordinate | ρ = 0·4 Area | Ordinate | r |
|---|---|---|---|---|---|---|---|---|---|---|---|
| −1·00 | ·00 000 | ·00 | ·00 000 | ·00 | ·00 000 | ·00 | ·00 000 | ·00 | ·00 000 | ·00 | −1·00 |
| − ·95 | ·00 052 | 25·84 | ·00 038 | 15·19 | ·00 015⁻ | 8·75 | ·00 008 | 4·87 | ·00 006 | 2·58 | − ·95 |
| − ·90 | ·00 288 | 70·30 | ·00 179 | 42·38 | ·00 098 | 24·93 | ·00 055⁻ | 14·14 | ·00 031 | 7·62 | − ·90 |
| − ·85 | ·00 771 | 124·08 | ·00 754 | 76·72 | ·00 274 | 46·12 | ·00 156 | 26·66 | ·00 085⁺ | 14·60 | − ·85 |
| − ·80 | ·01 537 | 183·35 | ·00 949 | 116·30 | ·00 568 | 71·46 | ·00 327 | 42·10 | ·00 179 | 23·44 | − ·80 |
| − ·75 | ·02 609 | 245·63 | ·01 627 | 159·85 | ·00 961 | 100·42 | ·00 582 | 60·33 | ·00 322 | 34·17 | − ·75 |
| − ·70 | ·03 996 | 309·15 | ·02 567 | 206·44 | ·01 576 | 132·64 | ·00 934 | 81·28 | ·00 524 | 46·85 | − ·70 |
| − ·65 | ·05 700 | 372·52 | ·03 696 | 255·27 | ·02 330 | 167·80 | ·01 399 | 104·93 | ·00 795 | 61·59 | − ·65 |
| − ·60 | ·07 719 | 434·60 | ·05 102 | 305·66 | ·03 259 | 205·61 | ·01 988 | 131·26 | ·01 143 | 78·49 | − ·60 |
| − ·55 | ·10 043 | 494·46 | ·06 761 | 356·96 | ·05 011 | 245·82 | ·02 716 | 160·28 | ·01 582 | 97·70 | − ·55 |
| − ·50 | ·12 658 | 551·33 | ·08 678 | 408·59 | ·05 723 | 288·14 | ·03 595⁺ | 191·98 | ·02 124 | 119·36 | − ·50 |
| − ·45 | ·15 550⁻ | 604·53 | ·10 847 | 459·99 | ·07 273 | 332·29 | ·04 640 | 226·34 | ·02 780 | 143·61 | − ·45 |
| − ·40 | ·18 697 | 653·49 | ·13 276 | 510·59 | ·09 044 | 377·96 | ·05 863 | 263·34 | ·03 565⁻ | 170·62 | − ·40 |
| − ·35 | ·22 077⁺ | 697·73 | ·15 953 | 559·87 | ·11 052 | 424·83 | ·07 278 | 302·91 | ·04 491 | 200·54 | − ·35 |
| − ·30 | ·25 665⁻ | 736·85 | ·18 872 | 607·28 | ·13 297 | 472·53 | ·08 896 | 344·98 | ·05 575⁻ | 233·54 | − ·30 |
| − ·25 | ·29 439 | 770·51 | ·22 029 | 652·32 | ·15 778 | 520·67 | ·10 732 | 389·43 | ·06 832 | 269·77 | − ·25 |
| − ·20 | ·33 361 | 798·41 | ·25 386 | 694·44 | ·18 504 | 568·81 | ·12 794 | 436·10 | ·08 279 | 309·36 | − ·20 |
| − ·15 | ·37 410 | 820·34 | ·28 955⁻ | 733·15 | ·21 468 | 616·47 | ·15 096 | 484·78 | ·09 932 | 352·43 | − ·15 |
| − ·10 | ·41 554 | 836·13 | ·32 711 | 767·91 | ·24 666 | 663·12 | ·17 645⁺ | 535·18 | ·11 809 | 399·06 | − ·10 |
| − ·05 | ·45 761 | 845·65 | ·36 630 | 798·24 | ·28 095⁺ | 708·17 | ·20 450⁻ | 586·95 | ·13 928 | 449·29 | − ·05 |
| 0·00 | ·50 000ᵉ | 848·83 | ·40 687 | 823·62 | ·31 743 | 750·99 | ·23 516 | 639·65 | ·16 308 | 503·10 | 0·00 |
| + ·05 | ·54 239 | 845·65 | ·44 858 | 843·56 | ·35 600 | 790·87 | ·26 847 | 692·72 | ·18 965⁻ | 560·37 | + ·05 |
| + ·10 | ·58 446 | 836·13 | ·49 114 | 857·59 | ·39 646 | 827·06 | ·30 443 | 745·49 | ·21 917 | 620·88 | + ·10 |
| + ·15 | ·62 590 | 820·34 | ·53 423 | 865·26 | ·43 862 | 858·73 | ·34 300 | 797·15 | ·25 179 | 684·25 | + ·15 |
| + ·20 | ·66 639 | 798·41 | ·57 758 | 866·13 | ·48 225⁺ | 884·99 | ·38 411 | 846·70 | ·28 763 | 749·91 | + ·20 |
| + ·25 | ·70 561 | 770·51 | ·62 071 | 859·80 | ·52 704 | 904·89 | ·42 762 | 893·00 | ·32 680 | 817·07 | + ·25 |
| + ·30 | ·74 335⁻ | 736·85 | ·66 338 | 845·93 | ·57 262 | 917·44 | ·47 333 | 934·67 | ·36 934 | 884·59 | + ·30 |
| + ·35 | ·77 923 | 697·73 | ·70 519 | 824·21 | ·61 863 | 921·57 | ·52 098 | 970·10 | ·41 524 | 950·96 | + ·35 |
| + ·40 | ·81 303⁺ | 653·49 | ·74 566 | 794·40 | ·66 461 | 916·22 | ·57 021 | 997·43 | ·46 439 | 1014·18 | + ·40 |
| + ·45 | ·84 450⁻ | 604·53 | ·78 443 | 756·38 | ·71 009 | 900·29 | ·62 055⁺ | 1014·55 | ·51 656 | 1071·64 | + ·45 |
| + ·50 | ·87 342 | 551·33 | ·82 121 | 710·11 | ·75 445⁺ | 872·71 | ·67 145⁻ | 1019·07 | ·57 140 | 1120·03 | + ·50 |
| + ·55 | ·89 957 | 494·46 | ·85 540 | 655·69 | ·79 716 | 832·49 | ·72 221 | 1008·34 | ·62 835⁻ | 1155·12 | + ·55 |
| + ·60 | ·92 281 | 434·60 | ·88 677 | 593·44 | ·83 747 | 778·80 | ·77 199 | 979·51 | ·68 661 | 1171·74 | + ·60 |
| + ·65 | ·94 300 | 372·52 | ·91 458 | 523·87 | ·87 480 | 711·02 | ·81 981 | 929·65 | ·74 511 | 1163·57 | + ·65 |
| + ·70 | ·96 004 | 309·15 | ·93 877 | 447·82 | ·90 836 | 628·96 | ·86 455⁻ | 855·90 | ·80 243 | 1123·23 | + ·70 |
| + ·75 | ·97 391 | 245·63 | ·95 931 | 366·56 | ·93 806 | 533·05 | ·90 496 | 755·92 | ·85 676 | 1042·49 | + ·75 |
| + ·80 | ·98 463 | 183·35 | ·97 551 | 281·92 | ·96 176 | 424·74 | ·93 969 | 628·49 | ·90 586 | 913·07 | + ·80 |
| + ·85 | ·99 229 | 124·08 | ·98 748 | 196·62 | ·97 977 | 307·09 | ·96 737 | 474·80 | ·94 714 | 728·55 | + ·85 |
| + ·90 | ·99 712 | 70·30 | ·99 513 | 114·82 | ·99 207 | 186·02 | ·98 682 | 300·94 | ·97 780 | 489·11 | + ·90 |
| + ·95 | ·99 948 | 25·84 | ·99 903 | 43·51 | ·99 847 | 73·17 | ·99 740 | 124·05 | ·99 544 | 214·20 | + ·95 |
| +1·00 | 1·00 000 | ·00 | 1·00 000 | ·00 | 1·00 000 | ·00 | 1·00 000 | ·00 | 1·00 000 | ·00 | +1·00 |

n=7

ρ=0.5 — 0.9

| r | ρ=0.5 Area | ρ=0.5 Ordinate | ρ=0.6 Area | ρ=0.6 Ordinate | ρ=0.7 Area | ρ=0.7 Ordinate | ρ=0.8 Area | ρ=0.8 Ordinate |
|---|---|---|---|---|---|---|---|---|
| −1·00 | ·00 000 | ·00 | ·00 000 | ·00 | ·00 000 | ·00 | ·00 000 | ·00 |
| − ·95 | ·00 003 | 1·27 | ·00 001 | ·56 | ·00 000 | ·20 | ·00 000 | ·05 |
| − ·90 | ·00 015⁵ | 3·81 | ·00 007 | 1·69 | ·00 002 | ·63 | ·00 001 | ·16 |
| − ·85 | ·00 043 | 7·39 | ·00 019 | 3·34 | ·00 007 | 1·25 | ·00 002 | ·33 |
| − ·80 | ·00 092 | 12·05 | ·00 041 | 5·51 | ·00 015⁵ | 2·08 | ·00 004 | ·56 |
| − ·75 | ·00 166 | 17·84 | ·00 076 | 8·26 | ·00 028 | 3·16 | ·00 007 | ·85 |
| − ·70 | ·00 272 | 24·84 | ·00 125 | 11·67 | ·00 047 | 4·53 | ·00 012 | 1·24 |
| − ·65 | ·00 416 | 33·20 | ·00 193 | 15·83 | ·00 074 | 6·22 | ·00 020 | 1·72 |
| − ·60 | ·00 606 | 43·03 | ·00 284 | 20·83 | ·00 110 | 8·30 | ·00 030 | 2·32 |
| − ·55 | ·00 847 | 54·50 | ·00 403 | 26·81 | ·00 158 | 10·85 | ·00 043 | 3·08 |
| − ·50 | ·01 153 | 67·80 | ·00 554 | 33·92 | ·00 220 | 13·93 | ·00 061 | 4·01 |
| − ·45 | ·01 530 | 83·12 | ·00 744 | 42·31 | ·00 298 | 17·66 | ·00 084 | 5·16 |
| − ·40 | ·01 988 | 100·70 | ·00 980 | 52·20 | ·00 397 | 22·16 | ·00 113 | 6·58 |
| − ·35 | ·02 541 | 120·78 | ·01 269 | 63·81 | ·00 521 | 27·58 | ·00 150 | 8·32 |
| − ·30 | ·03 201 | 143·64 | ·01 621 | 77·42 | ·00 675⁵ | 34·10 | ·00 197 | 10·48 |
| − ·25 | ·03 982 | 169·60 | ·02 047 | 93·33 | ·00 865⁵ | 41·94 | ·00 256 | 13·13 |
| − ·20 | ·04 902 | 198·97 | ·02 559 | 111·92 | ·01 097 | 51·36 | ·00 329 | 16·41 |
| − ·15 | ·05 978 | 232·12 | ·03 171 | 133·60 | ·01 381 | 62·69 | ·00 421 | 20·47 |
| − ·10 | ·07 230 | 269·42 | ·03 901 | 158·87 | ·01 728 | 76·33 | ·00 536 | 25·50 |
| − ·05 | ·08 680 | 311·28 | ·04 767 | 188·29 | ·02 149 | 92·77 | ·00 678 | 31·77 |
| 0·00 | ·10 351 | 358·10 | ·05 792 | 222·51 | ·02 661 | 112·60 | ·00 856 | 39·60 |
| + ·05 | ·12 270 | 410·29 | ·07 001 | 262·29 | ·03 282 | 136·56 | ·01 078 | 49·43 |
| + ·10 | ·14 464 | 468·23 | ·08 425 | 308·46 | ·04 035 | 165·58 | ·01 355 | 61·82 |
| + ·15 | ·16 963 | 532·27 | ·10 098 | 361·98 | ·04 948 | 200·77 | ·01 701 | 77·52 |
| + ·20 | ·19 797 | 602·63 | ·12 059 | 423·92 | ·06 055 | 243·54 | ·02 137 | 97·54 |
| + ·25 | ·23 000 | 679·42 | ·14 353 | 495·43 | ·07 399 | 295·61 | ·02 686 | 123·21 |
| + ·30 | ·26 602 | 762·49 | ·17 031 | 577·72 | ·09 031 | 359·10 | ·03 382 | 156·38 |
| + ·35 | ·30 634 | 851·33 | ·20 150 | 671·98 | ·11 013 | 436·64 | ·04 267 | 199·53 |
| + ·40 | ·35 123 | 944·93 | ·23 773 | 779·26 | ·13 426 | 531·38 | ·05 399 | 256·18 |
| + ·45 | ·40 089 | 1041·52 | ·27 966 | 900·25 | ·16 362 | 647·10 | ·06 859 | 331·09 |
| + ·50 | ·45 539 | 1138·25 | ·32 798 | 1034·91 | ·19 939 | 788·14 | ·08 752 | 431·03 |
| + ·55 | ·51 464 | 1230·82 | ·38 336 | 1181·85 | ·24 294 | 959·18 | ·11 226 | 565·53 |
| + ·60 | ·57 829 | 1312·87 | ·44 631 | 1337·35 | ·29 588 | 1164·60 | ·14 487 | 747·90 |
| + ·625 | ·61 156 | 1347·24 | ·48 073 | 1416·15 | ·32 643 | 1281·15 | ·16 496 | 862·58 |
| + ·650 | ·64 560 | 1375·33 | ·51 720 | 1493·80 | ·36 001 | 1406·95 | ·18 816 | 996·60 |
| + ·675 | ·68 026 | 1395·41 | ·55 540 | 1568·34 | ·39 685 | 1541·49 | ·21 498 | 1153·22 |
| + ·700 | ·71 529 | 1405·52 | ·59 548 | 1637·17 | ·43 715 | 1683·53 | ·24 604 | 1336·03 |
| + ·725 | ·75 043 | 1403·39 | ·63 718 | 1696·94 | ·48 107 | 1830·79 | ·28 203 | 1548·74 |
| + ·750 | ·78 534 | 1386·41 | ·68 022 | 1743·33 | ·52 870 | 1979·38 | ·32 375⁺ | 1794·82 |
| + ·775 | ·81 960 | 1351·71 | ·72 419 | 1770·89 | ·58 000 | 2123·17 | ·37 207 | 2076·66 |
| + ·800 | ·85 275 | 1296·13 | ·76 855 | 1772·80 | ·63 474 | 2252·77 | ·42 788 | 2393·94 |
| + ·825 | ·88 421 | 1216·35 | ·81 255 | 1740·74 | ·69 241 | 2354·22 | ·49 202 | 2740·69 |
| + ·850 | ·91 334 | 1109·20 | ·85 522 | 1664·80 | ·75 205 | 2407·27 | ·56 503 | 3099·80 |
| + ·875 | ·93 942 | 971·87 | ·89 533 | 1533·76 | ·81 213 | 2383·40 | ·64 680 | 3433·10 |
| + ·900 | ·96 167 | 802·95 | ·93 135 | 1336·07 | ·87 026 | 2244·07 | ·73 585 | 3664·14 |
| + ·925 | ·97 931 | 603·72 | ·96 150 | 1062·40 | ·92 298 | 1941·20 | ·82 800 | 3650·23 |
| + ·950 | ·99 167 | 381·46 | ·98 383 | 712·23 | ·96 556 | 1426·14 | ·91 433 | 3147·79 |
| + ·975 | ·99 837 | 157·00 | ·99 669 | 312·14 | ·99 244 | 690·66 | ·97 861 | 1831·57 |
| +1·000 | 1·00 000 | ·00 | 1·00 000 | ·00 | 1·00 000 | ·00 | 1·00 000 | ·00 |

ρ=0.9

| r | Area | Ordinate |
|---|---|---|
| −1·00 | ·00 000 | ·00 |
| − ·95 | ·00 000 | ·01 |
| − ·90 | ·00 000 | ·02 |
| − ·85 | ·00 000 | ·04 |
| − ·80 | ·00 000 | ·06 |
| − ·75 | ·00 001 | ·10 |
| − ·70 | ·00 001 | ·14 |
| − ·65 | ·00 002 | ·20 |
| − ·60 | ·00 003⁺ | ·27 |
| − ·55 | ·00 005⁺ | ·37 |
| − ·50 | ·00 007 | ·49 |
| − ·45 | ·00 010 | ·63 |
| − ·40 | ·00 014 | ·82 |
| − ·35 | ·00 018 | 1·05 |
| − ·30 | ·00 024 | 1·35 |
| − ·25 | ·00 032 | 1·72 |
| − ·20 | ·00 042 | 2·19 |
| − ·15 | ·00 054 | 2·80 |
| − ·10 | ·00 070 | 3·56 |
| − ·05 | ·00 090 | 4·55 |
| 0·00 | ·00 116 | 5·82 |
| + ·05 | ·00 149 | 7·48 |
| + ·10 | ·00 192 | 9·65 |
| + ·15 | ·00 246 | 12·53 |
| + ·20 | ·00 318 | 16·38 |
| + ·25 | ·00 413 | 21·58 |
| + ·30 | ·00 537 | 28·71 |
| + ·35 | ·00 704 | 38·62 |
| + ·40 | ·00 930 | 52·62 |
| + ·45 | ·01 241 | 72·76 |
| + ·50 | ·01 674 | 102·38 |
| + ·55 | ·02 289 | 147·00 |
| + ·60 | ·03 184 | 216·12 |
| + ·625 | ·03 783 | 264·71 |
| + ·650 | ·04 519 | 326·69 |
| + ·675 | ·05 431 | 406·49 |
| + ·700 | ·06 571 | 510·28 |
| + ·725 | ·08 009⁵ | 646·76 |
| + ·750 | ·09 842 | 828·27 |
| + ·775 | ·12 203 | 1072·49 |
| + ·800 | ·15 278 | 1404·91 |
| + ·810 | ·16 764 | 1570·30 |
| + ·820 | ·18 426 | 1758·43 |
| + ·830 | ·20 289 | 1972·67 |
| + ·840 | ·22 382 | 2216·83 |
| + ·850 | ·24 734 | 2495·12 |
| + ·860 | ·27 385 | 2812·15 |
| + ·870 | ·30 373 | 3172·67 |
| + ·880 | ·33 746 | 3581·28 |
| + ·890 | ·37 553 | 4041·76 |
| + ·900 | ·41 847 | 4555·85 |
| + ·910 | ·46 682 | 5121·16 |
| + ·920 | ·52 103 | 5727·58 |
| + ·930 | ·58 143 | 6351·05 |
| + ·940 | ·64 795⁺ | 6943·26 |
| + ·950 | ·71 990 | 7414·56 |
| + ·955 | ·75 736 | 7559·16 |
| + ·960 | ·79 533 | 7607·06 |
| + ·965 | ·83 321 | 7521·65 |
| + ·970 | ·87 025 | 7257·25 |
| + ·975 | ·90 540 | 6758·90 |
| + ·980 | ·93 735⁵ | 5964·97 |
| + ·985 | ·96 446 | 4816·92 |
| + ·990 | ·98 488 | 3288·18 |
| + ·995 | ·99 686 | 1471·65 |
| +1·000 | 1·00 000 | ·00 |

ρ=0.9

| r | Area | Ordinate |
|---|---|---|
| + ·800 | ·15 278 | 1404·91 |
| + ·825 | ·19 330 | 1860·32 |
| + ·850 | ·24 734 | 2495·12 |
| + ·875 | ·32 008 | 3370·70 |
| + ·900 | ·41 847 | 4555·85 |
| + ·925 | ·55 045⁵ | 6039·47 |
| + ·950 | ·71 990 | 7414·56 |
| + ·975 | ·90 540 | 6758·90 |
| +1·000 | 1·00 000 | ·00 |

ρ=0·0——0·4        n=8

| r | ρ=0·0 Area | Ordinate | ρ=0·1 Area | Ordinate | ρ=0·2 Area | Ordinate | ρ=0·3 Area | Ordinate | ρ=0·4 Area | Ordinate | r |
|---|---|---|---|---|---|---|---|---|---|---|---|
| −1·00 | ·00 000 | ·00 | ·00 000 | ·00 | ·00 000 | ·00 | ·00 000 | ·00 | ·00 000 | ·00 | −1·00 |
| − ·95 | ·00 015 | 8·91 | ·00 008 | 4·76 | ·00 004 | 2·49 | ·00 002 | 1·25 | ·00 001 | ·59 | − ·95 |
| − ·90 | ·00 116 | 33·84 | ·00 063 | 18·63 | ·00 033 | 9·97 | ·00 018 | 5·12 | ·00 009 | 2·47 | − ·90 |
| − ·85 | ·00 376 | 72·19 | ·00 208 | 40·95 | ·00 112 | 22·48 | ·00 058 | 11·80 | ·00 029 | 5·82 | − ·85 |
| − ·80 | ·00 856 | 121·50 | ·00 485 | 71·02 | ·00 266 | 40·02 | ·00 140 | 21·48 | ·00 071 | 10·80 | − ·80 |
| − ·75 | ·01 605 | 179·44 | ·00 930 | 108·11 | ·00 521 | 62·53 | ·00 278 | 34·34 | ·00 139 | 17·61 | − ·75 |
| − ·70 | ·02 661 | 243·84 | ·01 577 | 151·45 | ·00 900 | 89·96 | ·00 489 | 50·58 | ·00 249 | 26·48 | − ·70 |
| − ·65 | ·04 051 | 312·66 | ·02 454 | 200·21 | ·01 428 | 122·17 | ·00 790 | 70·35 | ·00 408 | 37·63 | − ·65 |
| − ·60 | ·05 792 | 384·00 | ·03 586 | 253·56 | ·02 129 | 158·99 | ·01 199 | 93·82 | ·00 629 | 51·30 | − ·60 |
| − ·55 | ·07 892 | 456·10 | ·04 996 | 310·59 | ·03 026 | 200·22 | ·01 734 | 121·13 | ·00 925 | 67·75 | − ·55 |
| − ·50 | ·10 352 | 527·34 | ·06 697 | 370·40 | ·04 138 | 245·57 | ·02 416 | 152·40 | ·01 312 | 87·25 | − ·50 |
| − ·45 | ·13 162 | 596·26 | ·08 703 | 432·05 | ·05 488 | 294·69 | ·03 265 | 187·72 | ·01 803 | 110·08 | − ·45 |
| − ·40 | ·16 308 | 661·50 | ·11 019 | 494·55 | ·07 091 | 347·19 | ·04 300 | 227·13 | ·02 418 | 136·53 | − ·40 |
| − ·35 | ·19 769 | 721·88 | ·13 648 | 556·91 | ·08 964 | 402·57 | ·05 543 | 270·64 | ·03 175 | 166·89 | − ·35 |
| − ·30 | ·23 517 | 776·34 | ·16 586 | 618·14 | ·11 120 | 460·28 | ·07 014 | 318·19 | ·04 094 | 201·43 | − ·30 |
| − ·25 | ·27 521 | 823·97 | ·19 826 | 677·22 | ·13 570 | 519·66 | ·08 732 | 369·65 | ·05 197 | 240·45 | − ·25 |
| − ·20 | ·31 744 | 864·00 | ·23 353 | 733·11 | ·16 319 | 579·99 | ·10 716 | 424·79 | ·06 507 | 284·17 | − ·20 |
| − ·15 | ·36 147 | 895·79 | ·27 150 | 784·83 | ·19 370 | 640·42 | ·12 985 | 483·31 | ·08 047 | 332·81 | − ·15 |
| − ·10 | ·40 687 | 918·84 | ·31 193 | 831·37 | ·22 722 | 700·04 | ·15 554 | 544·75 | ·09 843 | 386·52 | − ·10 |
| − ·05 | ·45 320 | 932·82 | ·35 453 | 871·77 | ·26 367 | 757·84 | ·18 437 | 608·54 | ·11 921 | 445·36 | − ·05 |
| 0·00 | ·50 000 | 937·50 | ·39 899 | 905·10 | ·30 295 | 812·68 | ·21 642 | 673·93 | ·14 305 | 509·27 | 0·00 |
| + ·05 | ·54 680 | 932·82 | ·44 491 | 930·49 | ·34 487 | 863·38 | ·25 177 | 740·00 | ·17 022 | 578·06 | + ·05 |
| + ·10 | ·59 313 | 918·84 | ·49 189 | 947·15 | ·38 920 | 908·63 | ·29 042 | 805·60 | ·20 093 | 651·31 | + ·10 |
| + ·15 | ·63 853 | 895·79 | ·53 947 | 954·37 | ·43 562 | 947·08 | ·33 230 | 869·36 | ·23 541 | 728·36 | + ·15 |
| + ·20 | ·68 256 | 864·00 | ·58 716 | 951·55 | ·48 377 | 977·31 | ·37 729 | 929·66 | ·27 382 | 808·23 | + ·20 |
| + ·25 | ·72 479 | 823·97 | ·63 445 | 938·24 | ·53 319 | 997·87 | ·42 518 | 984·60 | ·31 626 | 889·50 | + ·25 |
| + ·30 | ·76 483 | 776·34 | ·68 080 | 914·13 | ·58 337 | 1007·32 | ·47 563 | 1031·99 | ·36 276 | 970·27 | + ·30 |
| + ·35 | ·80 231 | 721·88 | ·72 568 | 879·12 | ·63 371 | 1004·28 | ·52 821 | 1069·38 | ·41 323 | 1048·02 | + ·35 |
| + ·40 | ·83 692 | 661·50 | ·76 854 | 833·34 | ·68 357 | 987·47 | ·58 235 | 1094·05 | ·46 746 | 1119·51 | + ·40 |
| + ·45 | ·86 838 | 596·26 | ·80 884 | 777·15 | ·73 221 | 955·78 | ·63 735 | 1103·06 | ·52 501 | 1180·66 | + ·45 |
| + ·50 | ·89 648 | 527·34 | ·84 609 | 711·26 | ·77 889 | 908·44 | ·69 235 | 1093·37 | ·58 527 | 1226·48 | + ·50 |
| + ·55 | ·92 108 | 456·10 | ·87 982 | 636·70 | ·82 279 | 845·06 | ·74 633 | 1061·99 | ·64 731 | 1251·03 | + ·55 |
| + ·60 | ·94 208 | 384·00 | ·90 964 | 554·91 | ·86 313 | 765·85 | ·79 814 | 1006·21 | ·70 990 | 1247·49 | + ·60 |
| + ·65 | ·95 949 | 312·66 | ·93 522 | 467·80 | ·89 913 | 671·79 | ·84 651 | 924·01 | ·77 146 | 1208·46 | + ·65 |
| + ·70 | ·97 339 | 243·84 | ·95 637 | 377·81 | ·93 009 | 564·92 | ·89 009 | 814·59 | ·83 003 | 1126·64 | + ·70 |
| + ·75 | ·98 395 | 179·44 | ·97 301 | 287·97 | ·95 546 | 448·65 | ·92 753 | 679·20 | ·88 331 | 996·09 | + ·75 |
| + ·80 | ·99 144 | 121·50 | ·98 523 | 202·00 | ·97 488 | 328·13 | ·95 765 | 522·33 | ·92 879 | 814·61 | + ·80 |
| + ·85 | ·99 624 | 72·19 | ·99 334 | 124·36 | ·98 833 | 210·79 | ·97 956 | 353·40 | ·96 401 | 587·92 | + ·85 |
| + ·90 | ·99 884 | 33·84 | ·99 789 | 60·42 | ·99 619 | 106·94 | ·99 306 | 189·15 | ·98 716 | 336·77 | + ·90 |
| + ·95 | ·99 985 | 8·91 | ·99 972 | 16·49 | ·99 947 | 30·50 | ·99 900 | 57·02 | ·99 806 | 109·05 | + ·95 |
| +1·00 | 1·00 000 | ·00 | 1·00 000 | ·00 | 1·00 000 | ·00 | 1·00 000 | ·00 | 1·00 000 | ·00 | +1·00 |

n = 8      ρ = 0.5 —— 0.9

| r | ρ=0.5 Area | ρ=0.5 Ordinate | ρ=0.6 Area | ρ=0.6 Ordinate | ρ=0.7 Area | ρ=0.7 Ordinate | ρ=0.8 Area | ρ=0.8 Ordinate | r | ρ=0.9 Area | ρ=0.9 Ordinate |
|---|---|---|---|---|---|---|---|---|---|---|---|
| −1.00 | ·00 000 | ·00 | ·00 000 | ·00 | ·00 000 | ·00 | ·00 000 | ·00 | −1.00 | ·00 000 | ·00 |
| −·95 | ·00 001 | ·26 | ·00 000 | ·10 | ·00 000 | ·03 | ·00 000 | ·01 | −·95 | ·00 000 | ·00 |
| −·90 | ·00 004 | 1·10 | ·00 001 | ·42 | ·00 000 | ·13 | ·00 000 | ·03 | −·90 | ·00 000 | ·00 |
| −·85 | ·00 013 | 2·62 | ·00 005 | 1·03 | ·00 002 | ·33 | ·00 000 | ·07 | −·85 | ·00 000 | ·01 |
| −·80 | ·00 032 | 4·95 | ·00 012 | 1·98 | ·00 004 | ·63 | ·00 001 | ·13 | −·80 | ·00 000 | ·01 |
| −·75 | ·00 065 | 8·22 | ·00 025 | 3·33 | ·00 008 | 1·08 | ·00 002 | ·23 | −·75 | ·00 000 | ·02 |
| −·70 | ·00 117 | 12·58 | ·00 047 | 5·19 | ·00 015 | 1·71 | ·00 003 | ·38 | −·70 | ·00 000 | ·03 |
| −·65 | ·00 193 | 18·23 | ·00 079 | 7·66 | ·00 025 | 2·57 | ·00 006 | ·57 | −·65 | ·00 000 | ·05 |
| −·60 | ·00 301 | 25·35 | ·00 125 | 10·84 | ·00 041 | 3·69 | ·00 009 | ·83 | −·60 | ·00 001 | ·07 |
| −·55 | ·00 448 | 34·18 | ·00 189 | 14·89 | ·00 063 | 5·16 | ·00 014 | 1·18 | −·55 | ·00 001 | ·10 |
| −·50 | ·00 645 | 44·96 | ·00 276 | 19·98 | ·00 093 | 7·06 | ·00 021 | 1·64 | −·50 | ·00 002 | ·14 |
| −·45 | ·00 901 | 58·00 | ·00 391 | 26·31 | ·00 134 | 9·47 | ·00 031 | 2·25 | −·45 | ·00 003 | ·19 |
| −·40 | ·01 229 | 73·61 | ·00 541 | 34·12 | ·00 189 | 12·53 | ·00 044 | 3·03 | −·40 | ·00 004 | ·27 |
| −·35 | ·01 642 | 92·15 | ·00 734 | 43·68 | ·00 261 | 16·38 | ·00 062 | 4·04 | −·35 | ·00 005 | ·36 |
| −·30 | ·02 156 | 114·03 | ·00 981 | 55·33 | ·00 355 | 21·22 | ·00 085 | 5·34 | −·30 | ·00 007 | ·49 |
| −·25 | ·02 789 | 139·67 | ·01 292 | 69·47 | ·00 475 | 27·27 | ·00 116 | 7·03 | −·25 | ·00 010 | ·66 |
| −·20 | ·03 560 | 169·57 | ·01 680 | 86·55 | ·00 630 | 34·83 | ·00 156 | 9·19 | −·20 | ·00 014 | ·88 |
| −·15 | ·04 493 | 204·24 | ·02 163 | 107·11 | ·00 827 | 44·26 | ·00 209 | 11·98 | −·15 | ·00 019 | 1·17 |
| −·10 | ·05 611 | 244·24 | ·02 758 | 131·79 | ·01 076 | 56·00 | ·00 277 | 15·57 | −·10 | ·00 026 | 1·57 |
| −·05 | ·06 945 | 290·13 | ·03 489 | 161·34 | ·01 392 | 70·61 | ·00 366 | 20·22 | −·05 | ·00 035 | 2·09 |
| 0·00 | ·08 523 | 342·52 | ·04 381 | 196·61 | ·01 788 | 88·81 | ·00 481 | 26·24 | 0·00 | ·00 047 | 2·80 |
| +·05 | ·10 382 | 401·96 | ·05 466 | 238·59 | ·02 287 | 111·47 | ·00 631 | 34·07 | +·05 | ·00 063 | 3·77 |
| +·10 | ·12 556 | 468·99 | ·06 780 | 288·43 | ·02 912 | 139·69 | ·00 826 | 44·29 | +·10 | ·00 086 | 5·08 |
| +·15 | ·15 085 | 544·03 | ·08 366 | 347·39 | ·03 696 | 174·87 | ·01 079 | 57·69 | +·15 | ·00 116 | 6·89 |
| +·20 | ·18 010 | 627·31 | ·10 272 | 416·87 | ·04 675 | 218·74 | ·01 410 | 75·35 | +·20 | ·00 156 | 9·42 |
| +·25 | ·21 372 | 719·81 | ·12 554 | 498·39 | ·05 901 | 273·48 | ·01 842 | 98·76 | +·25 | ·00 211 | 12·97 |
| +·30 | ·25 211 | 818·08 | ·15 278 | 593·47 | ·07 433 | 341·77 | ·02 411 | 129·97 | +·30 | ·00 288 | 18·05 |
| +·35 | ·29 564 | 924·04 | ·18 514 | 703·52 | ·09 347 | 426·94 | ·03 160 | 171·88 | +·35 | ·00 395 | 25·40 |
| +·40 | ·34 459 | 1034·75 | ·22 340 | 829·63 | ·11 737 | 532·99 | ·04 154 | 228·55 | +·40 | ·00 548 | 36·23 |
| +·45 | ·39 914 | 1147·03 | ·26 838 | 972·16 | ·14 719 | 664·66 | ·05 480 | 305·76 | +·45 | ·00 767 | 52·50 |
| +·50 | ·45 924 | 1256·06 | ·32 087 | 1130·10 | ·18 435 | 827·21 | ·07 259 | 411·69 | +·50 | ·01 087 | 77·48 |
| +·55 | ·52 458 | 1354·86 | ·38 159 | 1300·15 | ·23 052 | 1025·98 | ·09 664 | 558·02 | +·55 | ·01 565 | 116·83 |
| +·60 | ·59 440 | 1433·64 | ·45 098 | 1475·18 | ·28 762 | 1265·10 | ·12 934 | 761·16 | +·60 | ·02 296 | 180·59 |
| +·625 | ·63 061 | 1461·57 | ·48 893 | 1560·75 | ·32 092 | 1400·14 | ·14 994 | 890·81 | +·625 | ·02 802 | 226·92 |
| +·650 | ·66 739 | 1479·32 | ·52 898 | 1642·00 | ·35 771 | 1544·83 | ·17 407 | 1043·60 | +·650 | ·03 442 | 287·38 |
| +·675 | ·70 447 | 1484·67 | ·57 097 | 1715·87 | ·39 823 | 1697·70 | ·20 235 | 1223·34 | +·675 | ·04 255 | 367·03 |
| +·700 | ·74 150 | 1475·21 | ·61 468 | 1778·44 | ·44 264 | 1856·10 | ·23 549 | 1434·02 | +·700 | ·05 299 | 473·04 |
| +·725 | ·77 808 | 1448·38 | ·65 976 | 1824·91 | ·49 104 | 2015·70 | ·27 434 | 1679·43 | +·725 | ·06 651 | 615·63 |
| +·750 | ·81 375 | 1401·54 | ·70 574 | 1849·45 | ·54 338 | 2169·84 | ·31 978 | 1962·40 | +·750 | ·08 420 | 809·47 |
| +·775 | ·84 797 | 1332·14 | ·75 199 | 1845·19 | ·59 941 | 2308·64 | ·37 277 | 2283·34 | +·775 | ·10 758 | 1075·82 |
| +·800 | ·88 015 | 1237·97 | ·79 769 | 1804·24 | ·65 857 | 2418·01 | ·43 422 | 2637·69 | +·800 | ·13 884 | 1445·38 |
| +·825 | ·90 965 | 1117·49 | ·84 183 | 1718·12 | ·71 990 | 2478·46 | ·50 482 | 3011·33 | +·810 | ·15 421 | 1631·37 |
| +·850 | ·93 581 | 970·50 | ·88 316 | 1578·45 | ·78 187 | 2464·22 | ·58 470 | 3372·91 | +·820 | ·17 156 | 1844·17 |
| +·875 | ·95 797 | 798·80 | ·92 025 | 1378·57 | ·84 223 | 2343·30 | ·67 287 | 3661·62 | +·830 | ·19 119 | 2087·72 |
| +·900 | ·97 559 | 607·69 | ·95 156 | 1116·43 | ·89 786 | 2080·31 | ·76 629 | 3769·68 | +·840 | ·21 343 | 2366·36 |
| | | | | | | | | | +·850 | ·23 865 | 2684·78 |
| +·925 | ·98 828 | 407·53 | ·97 562 | 799·89 | ·94 482 | 1646·42 | ·85 844 | 3525·08 | +·860 | ·26 728 | 3047·80 |
| +·950 | ·99 604 | 216·63 | ·99 133 | 456·02 | ·97 878 | 1045·85 | ·93 776 | 2706·04 | +·870 | ·29 977 | 3459·99 |
| +·975 | ·99 943 | 64·99 | ·99 870 | 147·35 | ·99 654 | 380·27 | ·98 808 | 1222·23 | +·880 | ·33 665 | 3925·05 |
| +1·000 | 1·00 000 | ·00 | 1·00 000 | ·00 | 1·00 000 | ·00 | 1·00 000 | ·00 | +·890 | ·37 845 | 4444·55 |
| | | | | | | | | | +·900 | ·42 572 | 5016·00 |
| | | | | | | | | | +·910 | ·47 892 | 5629·57 |
| | | | | | | | | | +·920 | ·53 838 | 6262·78 |
| | | | | | | | | | +·930 | ·60 410 | 6872·21 |
| | | | | | | | | | +·940 | ·67 549 | 7380·96 |
| | | | | | | | | | +·950 | ·75 096 | 7661·16 |
| | | | | | | | | | +·955 | ·78 931 | 7656·72 |
| | | | | | | | | | +·960 | ·82 730 | 7514·37 |
| | | | | | | | | | +·965 | ·86 416 | 7196·98 |
| | | | | | | | | | +·970 | ·89 891 | 6665·01 |
| | | | | | | | | | +·975 | ·93 039 | 5881·97 |
| | | | | | | | | | +·980 | ·95 728 | 4826·07 |
| | | | | | | | | | +·985 | ·97 822 | 3513·26 |
| | | | | | | | | | +·990 | ·99 214 | 2041·54 |
| | | | | | | | | | +·995 | ·99 880 | 674·74 |
| | | | | | | | | | +1·000 | 1·00 000 | ·00 |

| ρ=0.9 r | Area | Ordinate |
|---|---|---|
| +·800 | ·13 884 | 1445·38 |
| +·825 | ·18 107 | 1960·00 |
| +·850 | ·23 865 | 2684·78 |
| +·875 | ·31 763 | 3685·75 |
| +·900 | ·42 572 | 5016·00 |
| +·925 | ·57 048 | 6574·31 |
| +·950 | ·75 096 | 7661·16 |
| +·975 | ·93 039 | 5881·97 |
| +1·000 | 1·00 000 | ·00 |

ρ=0·0——0·4                                        n=9

| r | ρ=0·0 Area | Ordinate | ρ=0·1 Area | Ordinate | ρ=0·2 Area | Ordinate | ρ=0·3 Area | Ordinate | ρ=0·4 Area | Ordinate | r |
|---|---|---|---|---|---|---|---|---|---|---|---|
| −1·00 | ·00 000 | ·00 | ·00 000 | ·00 | ·00 000 | ·00 | ·00 000 | ·00 | ·00 000 | ·00 | −1·00 |
| − ·95 | ·00 005 | 3·02 | ·00 003 | 1·47 | ·00 001 | ·69 | ·00 001 | ·31 | ·00 000 | ·13 | − ·95 |
| − ·90 | ·00 047 | 16·03 | ·00 023 | 8·06 | ·00 011 | 3·92 | ·00 005 | 1·82 | ·00 002 | ·79 | − ·90 |
| − ·85 | ·00 188 | 41·32 | ·00 094 | 21·50 | ·00 046 | 10·78 | ·00 022 | 5·14 | ·00 009 | 2·28 | − ·85 |
| − ·80 | ·00 481 | 79·21 | ·00 251 | 42·66 | ·00 126 | 22·04 | ·00 060 | 10·78 | ·00 026 | 4·89 | − ·80 |
| − ·75 | ·01 003 | 128·96 | ·00 534 | 71·93 | ·00 275 | 38·30 | ·00 134 | 19·23 | ·00 061 | 8·93 | − ·75 |
| − ·70 | ·01 788 | 189·20 | ·00 984 | 109·29 | ·00 518 | 60·01 | ·00 258 | 30·95 | ·00 119 | 14·72 | − ·70 |
| − ·65 | ·02 911 | 258·15 | ·01 640 | 154·47 | ·00 885 | 87·49 | ·00 450 | 46·39 | ·00 211 | 22·61 | − ·65 |
| − ·60 | ·04 381 | 333·77 | ·02 541 | 206·90 | ·01 403 | 120·93 | ·00 729 | 65·95 | ·00 349 | 32·98 | − ·60 |
| − ·55 | ·06 257 | 413·87 | ·03 720 | 265·84 | ·02 104 | 160·41 | ·01 117 | 90·03 | ·00 546 | 46·21 | − ·55 |
| − ·50 | ·08 524 | 496·20 | ·05 209 | 330·31 | ·03 017 | 205·86 | ·01 637 | 118·99 | ·00 817 | 62·73 | − ·50 |
| − ·45 | ·11 219 | 578·53 | ·07 031 | 399·18 | ·04 172 | 257·08 | ·02 315 | 153·13 | ·01 180 | 83·00 | − ·45 |
| − ·40 | ·14 305 | 658·72 | ·09 206 | 471·20 | ·05 597 | 313·71 | ·03 178 | 192·70 | ·01 654 | 107·46 | − ·40 |
| − ·35 | ·17 798 | 734·71 | ·11 746 | 544·95 | ·07 318 | 375·25 | ·04 252 | 237·86 | ·02 262 | 136·60 | − ·35 |
| − ·30 | ·21 642 | 804·64 | ·14 655 | 618·94 | ·09 357 | 441·03 | ·05 566 | 288·69 | ·03 029 | 170·90 | − ·30 |
| − ·25 | ·25 830 | 866·82 | ·17 933 | 691·61 | ·11 734 | 510·20 | ·07 148 | 345·14 | ·03 980 | 210·81 | − ·25 |
| − ·20 | ·30 295 | 919·77 | ·21 567 | 761·33 | ·14 463 | 581·74 | ·09 026 | 407·03 | ·05 147 | 256·77 | − ·20 |
| − ·15 | ·35 008 | 962·26 | ·25 538 | 826·48 | ·17 553 | 654·46 | ·11 227 | 473·98 | ·06 559 | 309·16 | − ·15 |
| − ·10 | ·39 899 | 993·32 | ·29 821 | 885·42 | ·21 007 | 726·99 | ·13 774 | 545·46 | ·08 249 | 368·27 | − ·10 |
| − ·05 | ·44 919 | 1012·24 | ·34 380 | 936·57 | ·24 820 | 797·78 | ·16 688 | 620·65 | ·10 253 | 434·27 | − ·05 |
| 0·00 | ·50 000 | 1018·59 | ·39 171 | 978·46 | ·28 979 | 865·14 | ·19 985 | 698·50 | ·12 604 | 507·13 | 0·00 |
| + ·05 | ·55 081 | 1012·24 | ·44 146 | 1009·68 | ·33 463 | 927·20 | ·23 675 | 777·65 | ·15 335 | 586·61 | + ·05 |
| + ·10 | ·60 101 | 993·32 | ·49 248 | 1029·05 | ·38 239 | 982·02 | ·27 760 | 856·41 | ·18 480 | 672·14 | + ·10 |
| + ·15 | ·64 992 | 962·26 | ·54 416 | 1035·54 | ·43 267 | 1027·55 | ·32 235 | 932·73 | ·22 065 | 762·75 | + ·15 |
| + ·20 | ·69 705 | 919·77 | ·59 581 | 1028·41 | ·48 496 | 1061·74 | ·37 080 | 1004·19 | ·26 113 | 856·97 | + ·20 |
| + ·25 | ·74 170 | 866·82 | ·64 676 | 1007·20 | ·53 862 | 1082·54 | ·42 264 | 1068·00 | ·30 637 | 952·68 | + ·25 |
| + ·30 | ·78 358 | 804·64 | ·69 629 | 971·79 | ·59 296 | 1088·08 | ·47 742 | 1121·01 | ·35 638 | 1047·05 | + ·30 |
| + ·35 | ·82 202 | 734·71 | ·74 371 | 922·48 | ·64 715 | 1076·68 | ·53 450 | 1159·76 | ·41 100 | 1136·34 | + ·35 |
| + ·40 | ·85 695 | 658·72 | ·78 832 | 859·99 | ·70 032 | 1047·01 | ·59 309 | 1180·62 | ·46 985 | 1215·85 | + ·40 |
| + ·45 | ·88 781 | 578·53 | ·82 951 | 785·54 | ·75 153 | 998·27 | ·65 221 | 1179·91 | ·53 232 | 1279·81 | + ·45 |
| + ·50 | ·91 476 | 496·20 | ·86 671 | 700·85 | ·79 983 | 930·34 | ·71 067 | 1154·15 | ·59 746 | 1321·44 | + ·50 |
| + ·55 | ·93 743 | 413·87 | ·89 946 | 608·22 | ·84 426 | 843·95 | ·76 715 | 1100·46 | ·66 397 | 1333·12 | + ·55 |
| + ·60 | ·95 619 | 333·77 | ·92 744 | 510·46 | ·88 394 | 740·94 | ·82 022 | 1016·99 | ·73 014 | 1306·81 | + ·60 |
| + ·65 | ·97 089 | 258·15 | ·95 048 | 410·96 | ·91 813 | 624·47 | ·86 835 | 903·63 | ·79 389 | 1234·97 | + ·65 |
| + ·70 | ·98 212 | 189·20 | ·96 857 | 313·58 | ·94 624 | 499·21 | ·91 012 | 762·81 | ·85 278 | 1111·97 | + ·70 |
| + ·75 | ·98 997 | 128·96 | ·98 194 | 222·57 | ·96 801 | 371·51 | ·94 427 | 600·46 | ·90 421 | 936·53 | + ·75 |
| + ·80 | ·99 519 | 79·21 | ·99 101 | 142·39 | ·98 349 | 249·41 | ·96 998 | 427·14 | ·94 566 | 715·16 | + ·80 |
| + ·85 | ·99 812 | 41·32 | ·99 643 | 77·38 | ·99 320 | 142·36 | ·98 707 | 258·83 | ·97 526 | 466·86 | + ·85 |
| + ·90 | ·99 953 | 16·03 | ·99 907 | 31·28 | ·99 814 | 60·49 | ·99 630 | 116·98 | ·99 250 | 228·19 | + ·90 |
| + ·95 | ·99 995 | 3·02 | ·99 991 | 6·15 | ·99 981 | 12·51 | ·99 963 | 25·79 | ·99 916 | 54·64 | + ·95 |
| +1·00 | 1·00 000 | ·00 | 1·00 000 | ·00 | 1·00 000 | ·00 | 1·00 000 | ·00 | 1·00 000 | ·00 | +1·00 |

n = 9     ρ = 0.5 —— 0.9

| r | ρ=0.5 Area | ρ=0.5 Ordinate | ρ=0.6 Area | ρ=0.6 Ordinate | ρ=0.7 Area | ρ=0.7 Ordinate | ρ=0.8 Area | ρ=0.8 Ordinate | r | ρ=0.9 Area | ρ=0.9 Ordinate |
|---|---|---|---|---|---|---|---|---|---|---|---|
| -1.00 | .00 000 | .00 | .00 000 | .00 | .00 000 | .00 | .00 000 | .00 | -1.00 | .00 000 | .00 |
| - .95 | .00 000 | .05 | .00 000 | .02 | .00 000 | .00 | .00 000 | .00 | - .95 | .00 000 | .00 |
| - .90 | .00 001 | .31 | .00 000 | .10 | .00 000 | .03 | .00 000 | .00 | - .90 | .00 000 | .00 |
| - .85 | .00 004 | .91 | .00 001 | .31 | .00 000 | .08 | .00 000 | .01 | - .85 | .00 000 | .00 |
| - .80 | .00 011 | 2.00 | .00 004 | .70 | .00 001 | .19 | .00 000 | .03 | - .80 | .00 000 | .00 |
| - .75 | .00 025 | 3.72 | .00 009 | 1.32 | .00 002 | .37 | .00 000 | .06 | - .75 | .00 000 | .00 |
| - .70 | .00 050 | 6.27 | .00 017 | 2.27 | .00 005̄ | .64 | .00 001 | .11 | - .70 | .00 000 | .01 |
| - .65 | .00 090 | 9.85 | .00 032 | 3.64 | .00 009 | 1.04 | .00 002 | .19 | - .65 | .00 000 | .01 |
| - .60 | .00 150 | 14.69 | .00 055 | 5.55 | .00 015 | 1.62 | .00 003 | .29 | - .60 | .00 000 | .02 |
| - .55 | .00 239 | 21.08 | .00 089 | 8.14 | .00 025 | 2.42 | .00 005̄ | .45 | - .55 | .00 000 | .03 |
| - .50 | .00 364 | 29.33 | .00 138 | 11.58 | .00 040 | 3.52 | .00 007 | .66 | - .50 | .00 000 | .04 |
| - .45 | .00 536 | 39.81 | .00 207 | 16.09 | .00 061 | 4.99 | .00 011 | .96 | - .45 | .00 001 | .06 |
| - .40 | .00 766 | 52.93 | .00 301 | 21.93 | .00 091 | 6.97 | .00 017 | 1.37 | - .40 | .00 001 | .09 |
| - .35 | .01 070 | 69.16 | .00 428 | 29.41 | .00 132 | 9.57 | .00 025 | 1.93 | - .35 | .00 002 | .12 |
| - .30 | .01 464 | 89.03 | .00 598 | 38.90 | .00 188 | 12.99 | .00 037 | 2.68 | - .30 | .00 002 | .17 |
| - .25 | .01 968 | 113.15 | .00 821 | 50.86 | .00 264 | 17.45 | .00 052 | 3.70 | - .25 | .00 003 | .25 |
| - .20 | .02 604 | 142.16 | .01 112 | 65.83 | .00 365 | 23.24 | .00 074 | 5.06 | - .20 | .00 005̄ | .35 |
| - .15 | .03 398 | 176.78 | .01 486 | 84.46 | .00 499 | 30.73 | .00 104 | 6.89 | - .15 | .00 007 | .48 |
| - .10 | .04 382 | 217.80 | .01 964 | 107.54 | .00 676 | 40.41 | .00 144 | 9.36 | - .10 | .00 010 | .68 |
| - .05 | .05 589 | 266.02 | .02 570 | 135.99 | .00 908 | 52.88 | .00 199 | 12.66 | - .05 | .00 014 | .95 |
| 0.00 | .07 056 | 322.29 | .03 334 | 170.89 | .01 210 | 68.91 | .00 273 | 17.11 | 0.00 | .00 019 | 1.33 |
| + .05 | .08 826 | 387.41 | .04 292 | 213.52 | .01 604 | 89.51 | .00 373 | 23.11 | + .05 | .00 027 | 1.86 |
| + .10 | .10 946 | 462.13 | .05 485̄ | 265.33 | .02 115̄ | 115.94 | .00 507 | 31.22 | + .10 | .00 038 | 2.63 |
| + .15 | .13 464 | 547.03 | .06 963 | 327.98 | .02 776 | 149.85 | .00 690 | 42.24 | + .15 | .00 054 | 3.73 |
| + .20 | .16 434 | 642.43 | .08 786 | 403.32 | .03 630 | 193.30 | .00 936 | 57.27 | + .20 | .00 077 | 5.33 |
| + .25 | .19 906 | 748.20 | .11 021 | 493.28 | .04 729 | 248.93 | .01 271 | 77.88 | + .25 | .00 109 | 7.67 |
| + .30 | .23 932 | 863.56 | .13 746 | 599.83 | .06 145̄ | 320.04 | .01 728 | 106.28 | + .30 | .00 155 | 11.16 |
| + .35 | .28 555 | 986.80 | .17 050̄ | 724.70 | .07 962 | 410.75 | .02 353 | 145.67 | + .35 | .00 223 | 16.43 |
| + .40 | .33 808 | 1114.86 | .21 026 | 869.09 | .10 293 | 526.05 | .03 211 | 200.65 | + .40 | .00 324 | 24.55 |
| + .45 | .39 704 | 1242.93 | .25 773 | 1032.99 | .13 274 | 671.79 | .04 396 | 277.86 | + .45 | .00 476 | 37.28 |
| + .50 | .46 225 | 1363.83 | .31 385 | 1214.32 | .17 072 | 854.38 | .06 042 | 386.97 | + .50 | .00 710 | 57.71 |
| + .55 | .53 313 | 1467.50 | .37 936 | 1407.47 | .21 889 | 1079.97 | .08 342 | 541.89 | + .55 | .01 075̄ | 91.38 |
| + .60 | .60 849 | 1540.47 | .45 461 | 1601.27 | .27 951 | 1352.44 | .11 570 | 762.41 | + .60 | .01 662 | 148.52 |
| + .625 | .64 728 | 1560.26 | .49 580 | 1692.73 | .31 521 | 1505.93 | .13 650̄ | 905.43 | + .625 | .02 085̄ | 191.46 |
| + .650 | .68 639 | 1565.77 | .53 918 | 1776.21 | .35 488 | 1669.36 | .16 120 | 1075.58 | + .650 | .02 631 | 248.83 |
| + .675 | .72 543 | 1554.43 | .58 451 | 1847.45 | .39 874 | 1840.16 | .19 054 | 1277.29 | + .675 | .03 345̄ | 326.22 |
| + .700 | .76 395̄ | 1523.66 | .63 141 | 1901.22 | .44 692 | 2014.01 | .22 536 | 1515.00 | + .700 | .04 286 | 431.66 |
| + .725 | .80 143 | 1470.99 | .67 937 | 1931.40 | .49 941 | 2184.26 | .26 662 | 1792.55 | + .725 | .05 537 | 576.84 |
| + .750 | .83 730 | 1394.27 | .72 773 | 1930.95 | .55 602 | 2341.13 | .31 534 | 2111.97 | + .750 | .07 217 | 778.77 |
| + .775 | .87 093 | 1291.97 | .77 560 | 1892.17 | .61 625 | 2470.79 | .37 255̄ | 2471.30 | + .775 | .09 498 | 1062.37 |
| + .800 | .90 168 | 1163.62 | .82 195 | 1807.20 | .67 918 | 2554.55 | .43 915 | 2860.84 | + .800 | .12 627 | 1463.91 |
| + .825 | .92 890 | 1010.35 | .86 552 | 1669.00 | .74 339 | 2568.27 | .51 565̄ | 3257.05 | + .810 | .14 191 | 1668.50 |
| + .850 | .95 202 | 835.66 | .90 492 | 1472.97 | .80 267 | 2482.96 | .60 168 | 3612.89 | + .820 | .15 974 | 1904.09 |
| + .875 | .97 056 | 646.15 | .93 869 | 1219.54 | .86 646 | 2267.79 | .69 527 | 3844.58 | + .830 | .18 011 | 2175.23 |
| + .900 | .98 429 | 452.62 | .96 549 | 918.20 | .91 887 | 1898.32 | .79 177 | 3818.02 | + .840 | .20 338 | 2486.86 |
|  |  |  |  |  |  |  |  |  | + .850 | .23 000 | 2844.14 |
| + .925 | .99 329 | 270.74 | .98 439 | 592.77 | .96 008 | 1374.59 | .88 253 | 3351.42 | + .860 | .26 043 | 3252.10 |
| + .950 | .99 809 | 121.08 | .99 529 | 287.39 | .98 679 | 755.00 | .95 440 | 2290.26 | + .870 | .29 522 | 3715.02 |
| + .975 | .99 979 | 26.48 | .99 947 | 68.47 | .99 839 | 206.11 | .99 342 | 803.00 | + .880 | .33 493 | 4235.38 |
| +1.000 | 1.00 000 | .00 | 1.00 000 | .00 | 1.00 000 | .00 | 1.00 000 | .00 | + .890 | .38 012 | 4812.05 |
|  |  |  |  |  |  |  |  |  | + .900 | .43 133 | 5437.47 |
|  |  |  |  |  |  |  |  |  | + .910 | .48 897 | 6093.09 |
|  |  |  |  |  |  |  |  |  | + .920 | .55 317 | 6742.56 |
|  |  |  |  |  |  |  |  |  | + .930 | .62 359 | 7321.74 |
|  |  |  |  |  |  |  |  |  | + .940 | .69 903 | 7725.62 |
|  |  |  |  |  |  |  |  |  | + .950 | .77 699 | 7794.37 |
|  |  |  |  |  |  |  |  |  | + .955 | .81 563 | 7636.46 |
|  |  |  |  |  |  |  |  |  | + .960 | .85 308 | 7308.89 |
|  |  |  |  |  |  |  |  |  | + .965 | .88 839 | 6780.67 |
|  |  |  |  |  |  |  |  |  | + .970 | .92 051 | 6027.22 |
|  |  |  |  |  |  |  |  |  | + .975 | .94 827 | 5040.34 |
|  |  |  |  |  |  |  |  |  | + .980 | .97 056 | 3844.79 |
|  |  |  |  |  |  |  |  |  | + .985 | .98 651 | 2523.17 |
|  |  |  |  |  |  |  |  |  | + .990 | .99 586 | 1248.12 |
|  |  |  |  |  |  |  |  |  | + .995 | .99 953 | 304.63 |
|  |  |  |  |  |  |  |  |  | +1.000 | 1.00 000 | .00 |

| r | ρ=0.9 Area | ρ=0.9 Ordinate |
|---|---|---|
| + .800 | .12 627 | 1463.91 |
| + .825 | .16 959 | 2033.01 |
| + .850 | .23 000 | 2844.14 |
| + .875 | .31 442 | 3967.98 |
| + .900 | .43 133 | 5437.47 |
| + .925 | .58 765̄ | 7046.37 |
| + .950 | .77 699 | 7794.37 |
| + .975 | .94 827 | 5040.34 |
| +1.000 | 1.00 000 | .00 |

$\rho = 0.0 — 0.4$      $n = 10$

| r | $\rho=0.0$ Area | Ordinate | $\rho=0.1$ Area | Ordinate | $\rho=0.2$ Area | Ordinate | $\rho=0.3$ Area | Ordinate | $\rho=0.4$ Area | Ordinate | r |
|---|---|---|---|---|---|---|---|---|---|---|---|
| −1·00 | ·00 000 | ·00 | ·00 000 | ·00 | ·00 000 | ·00 | ·00 000 | ·00 | ·00 000 | ·00 | −1·00 |
| − ·95 | ·00 001 | 1·01 | ·00 001 | ·45 | ·00 000 | ·19 | ·00 000 | ·08 | ·00 000 | ·03 | − ·95 |
| − ·90 | ·00 019 | 7·50 | ·00 009 | 3·44 | ·00 004 | 1·53 | ·00 002 | ·64 | ·00 000 | ·25 | − ·90 |
| − ·85 | ·00 092 | 23·37 | ·00 043 | 11·15 | ·00 019 | 5·11 | ·00 008 | 2·21 | ·00 003 | ·88 | − ·85 |
| − ·80 | ·00 273 | 51·03 | ·00 131 | 25·33 | ·00 060 | 12·00 | ·00 026 | 5·34 | ·00 011 | 2·19 | − ·80 |
| − ·75 | ·00 624 | 91·59 | ·00 309 | 47·29 | ·00 146 | 23·18 | ·00 065⁺ | 10·64 | ·00 027 | 4·47 | − ·75 |
| − ·70 | ·01 210 | 145·09 | ·00 619 | 77·94 | ·00 301 | 39·56 | ·00 137 | 18·72 | ·00 058 | 8·09 | − ·70 |
| − ·65 | ·02 095 | 210·66 | ·01 104 | 117·77 | ·00 552 | 61·92 | ·00 258 | 30·23 | ·00 111 | 13·43 | − ·65 |
| − ·60 | ·03 334 | 286·72 | ·01 812 | 166·85 | ·00 931 | 90·90 | ·00 446 | 45·82 | ·00 195⁺ | 20·95 | − ·60 |
| − ·55 | ·04 976 | 371·15 | ·02 787 | 224·86 | ·01 473 | 127·01 | ·00 724 | 66·13 | ·00 324 | 31·14 | − ·55 |
| − ·50 | ·07 056 | 461·43 | ·04 074 | 291·10 | ·02 214 | 170·55 | ·01 117 | 91·82 | ·00 512 | 44·57 | − ·50 |
| − ·45 | ·09 595⁺ | 554·77 | ·05 710 | 364·50 | ·03 191 | 221·63 | ·01 652 | 123·45 | ·00 777 | 61·84 | − ·45 |
| − ·40 | ·12 604 | 648·27 | ·07 729 | 443·69 | ·04 442 | 280·14 | ·02 362 | 161·56 | ·01 138 | 83·58 | − ·40 |
| − ·35 | ·16 074 | 739·03 | ·10 154 | 526·99 | ·06 004 | 345·68 | ·03 279 | 206·59 | ·01 621 | 110·50 | − ·35 |
| − ·30 | ·19 985⁻ | 824·22 | ·13 002 | 612·48 | ·07 910 | 417·62 | ·04 440 | 258·84 | ·02 253 | 143·28 | − ·30 |
| − ·25 | ·24 302 | 901·22 | ·16 279 | 698·04 | ·10 189 | 495·03 | ·05 880 | 318·47 | ·03 065⁻ | 182·65 | − ·25 |
| − ·20 | ·28 979 | 967·68 | ·19 979 | 781·38 | ·12 867 | 576·65 | ·07 637 | 385·42 | ·04 092 | 229·29 | − ·20 |
| − ·15 | ·33 958 | 1021·57 | ·24 085 | 860·14 | ·15 960 | 660·97 | ·09 746 | 459·39 | ·05 371 | 283·82 | − ·15 |
| − ·10 | ·39 171 | 1061·26 | ·28 569 | 931·94 | ·19 478 | 746·13 | ·12 241 | 539·76 | ·06 944 | 346·77 | − ·10 |
| − ·05 | ·44 545⁻ | 1085·57 | ·33 389 | 994·42 | ·23 420 | 830·01 | ·15 153 | 625·59 | ·08 853 | 418·49 | − ·05 |
| 0·00 | ·50 000⁺ | 1093·75 | ·38 494 | 1045·39 | ·27 772 | 910·20 | ·18 504 | 715·49 | ·11 143 | 499·09 | 0·00 |
| + ·05 | ·55 455⁺ | 1085·57 | ·43 820 | 1082·81 | ·32 511 | 984·11 | ·22 311 | 807·66 | ·13 858 | 588·33 | + ·05 |
| + ·10 | ·60 829 | 1061·26 | ·49 296 | 1104·96 | ·37 598 | 1048·94 | ·26 581 | 899·79 | ·17 040 | 685·53 | + ·10 |
| + ·15 | ·66 042 | 1021·57 | ·54 842 | 1110·49 | ·42 981 | 1101·85 | ·31 305 | 989·04 | ·20 725 | 789·44 | + ·15 |
| + ·20 | ·71 021 | 967·68 | ·60 372 | 1098·49 | ·48 592 | 1139·99 | ·36 461 | 1072·04 | ·24 942 | 898·05 | + ·20 |
| + ·25 | ·75 698 | 901·22 | ·65 797 | 1068·59 | ·54 352 | 1160·70 | ·42 008 | 1144·96 | ·29 709 | 1008·47 | + ·25 |
| + ·30 | ·80 015⁻ | 824·22 | ·71 028 | 1021·01 | ·60 166 | 1161·60 | ·47 886 | 1203·51 | ·35 024 | 1116·77 | + ·30 |
| + ·35 | ·83 926 | 739·03 | ·75 979 | 956·66 | ·65 931 | 1140·83 | ·54 012 | 1243·14 | ·40 864 | 1217·79 | + ·35 |
| + ·40 | ·87 396 | 648·27 | ·80 570 | 877·13 | ·71 536 | 1097·21 | ·60 279 | 1259·23 | ·47 179 | 1305·16 | + ·40 |
| + ·45 | ·90 405⁻ | 554·77 | ·84 729 | 784·74 | ·76 865⁻ | 1030·51 | ·66 558 | 1247·45 | ·53 880 | 1371·20 | + ·45 |
| + ·50 | ·92 944 | 461·43 | ·88 401 | 682·54 | ·81 804 | 941·67 | ·72 700 | 1204·17 | ·60 841 | 1407·27 | + ·50 |
| + ·55 | ·95 024 | 371·15 | ·91 544 | 574·23 | ·86 249 | 833·02 | ·78 543 | 1127·09 | ·67 888 | 1404·16 | + ·55 |
| + ·60 | ·96 666 | 286·72 | ·94 140 | 464·10 | ·90 108 | 708·50 | ·83 914 | 1015·97 | ·74 802 | 1353·14 | + ·60 |
| + ·65 | ·97 905⁺ | 210·66 | ·96 190 | 356·82 | ·93 316 | 573·73 | ·88 650⁻ | 873·46 | ·81 328 | 1247·48 | + ·65 |
| + ·70 | ·98 790 | 145·09 | ·97 721 | 257·23 | ·95 840 | 436·02 | ·92 607 | 706·04 | ·87 182 | 1084·83 | + ·70 |
| + ·75 | ·99 376 | 91·59 | ·98 783 | 170·01 | ·97 686 | 304·07 | ·95 687 | 524·71 | ·92 089 | 870·38 | + ·75 |
| + ·80 | ·99 727 | 51·03 | ·99 448 | 99·20 | ·98 906 | 187·37 | ·97 857 | 345·25⁻ | ·95 826 | 620·62 | + ·80 |
| + ·85 | ·99 908 | 23·37 | ·99 807 | 47·59 | ·99 601 | 95·03 | ·99 175⁻ | 187·37 | ·98 287 | 366·47 | + ·85 |
| + ·90 | ·99 981 | 7·50 | ·99 958 | 16·01 | ·99 909 | 33·82 | ·99 801 | 71·51 | ·99 558 | 152·84 | + ·90 |
| + ·95 | ·99 999 | 1·01 | ·99 997 | 2·27 | ·99 993 | 5·07 | ·99 985⁻ | 11·53 | ·99 963 | 27·06 | + ·95 |
| +1·00 | 1·00 000 | ·00 | 1·00 000 | ·00 | 1·00 000 | ·00 | 1·00 000 | ·00 | 1·00 000 | ·00 | +1·00 |

n=10

ρ=0.5 — 0.9

| r | ρ=0.5 Area | ρ=0.5 Ordinate | ρ=0.6 Area | ρ=0.6 Ordinate | ρ=0.7 Area | ρ=0.7 Ordinate | ρ=0.8 Area | ρ=0.8 Ordinate | r | ρ=0.9 Area | ρ=0.9 Ordinate |
|---|---|---|---|---|---|---|---|---|---|---|---|
| −1·00 | ·00 000 | ·00 | ·00 000 | ·00 | ·00 000 | ·00 | ·00 000 | ·00 | −1·00 | ·00 000 | ·00 |
| − ·95 | ·00 000 | ·01 | ·00 000 | ·00 | ·00 000 | ·00 | ·00 000 | ·00 | − ·95 | ·00 000 | ·00 |
| − ·90 | ·00 000 | ·09 | ·00 000 | ·03 | ·00 000 | ·01 | ·00 000 | ·00 | − ·90 | ·00 000 | ·00 |
| − ·85 | ·00 001 | ·31 | ·00 000 | ·09 | ·00 000 | ·02 | ·00 000 | ·00 | − ·85 | ·00 000 | ·00 |
| − ·80 | ·00 004 | ·80 | ·00 001 | ·24 | ·00 000 | ·06 | ·00 000 | ·01 | − ·80 | ·00 000 | ·00 |
| − ·75 | ·00 010 | 1·67 | ·00 003 | ·52 | ·00 001 | ·12 | ·00 000 | ·02 | − ·75 | ·00 000 | ·00 |
| − ·70 | ·00 021 | 3·09 | ·00 007 | ·98 | ·00 002 | ·23 | ·00 000 | ·03 | − ·70 | ·00 000 | ·00 |
| − ·65 | ·00 042 | 5·25 | ·00 014 | 1·71 | ·00 003 | ·42 | ·00 000 | ·06 | − ·65 | ·00 000 | ·00 |
| − ·60 | ·00 076 | 8·41 | ·00 025 | 2·81 | ·00 006 | ·70 | ·00 001 | ·10 | − ·60 | ·00 000 | ·00 |
| − ·55 | ·00 128 | 12·85 | ·00 042 | 4·39 | ·00 010 | 1·12 | ·00 002 | ·17 | − ·55 | ·00 000 | ·01 |
| − ·50 | ·00 207 | 18·91 | ·00 070 | 6·63 | ·00 017 | 1·73 | ·00 003 | ·26 | − ·50 | ·00 000 | ·01 |
| − ·45 | ·00 321 | 27·00 | ·00 111 | 9·73 | ·00 028 | 2·60 | ·00 004 | ·41 | − ·45 | ·00 000 | ·02 |
| − ·40 | ·00 481 | 37·60 | ·00 169 | 13·93 | ·00 044 | 3·83 | ·00 007 | ·61 | − ·40 | ·00 000 | ·03 |
| − ·35 | ·00 702 | 51·29 | ·00 252 | 19·56 | ·00 068 | 5·52 | ·00 010 | ·91 | − ·35 | ·00 000 | ·04 |
| − ·30 | ·01 000 | 68·70 | ·00 367 | 27·02 | ·00 101 | 7·85 | ·00 016 | 1·33 | − ·30 | ·00 001 | ·06 |
| − ·25 | ·01 396 | 90·58 | ·00 526 | 36·79 | ·00 148 | 11·03 | ·00 024 | 1·92 | − ·25 | ·00 001 | ·09 |
| − ·20 | ·01 915 | 117·77 | ·00 740 | 49·48 | ·00 213 | 15·32 | ·00 036 | 2·76 | − ·20 | ·00 002 | ·13 |
| − ·15 | ·02 584 | 151·22 | ·01 026 | 65·82 | ·00 303 | 21·09 | ·00 052 | 3·92 | − ·15 | ·00 002 | ·20 |
| − ·10 | ·03 439 | 191·94 | ·01 406 | 86·73 | ·00 427 | 28·82 | ·00 076 | 5·55 | − ·10 | ·00 004 | ·29 |
| − ·05 | ·04 518 | 241·06 | ·01 903 | 113·28 | ·00 595 | 39·13 | ·00 109 | 7·83 | − ·05 | ·00 005 | ·42 |
| 0·00 | ·05 865 | 299·70 | ·02 550 | 146·80 | ·00 824 | 52·84 | ·00 156 | 11·02 | 0·00 | ·00 008 | ·62 |
| + ·05 | ·07 532 | 369·02 | ·03 385 | 188·84 | ·01 131 | 71·03 | ·00 222 | 15·49 | + ·05 | ·00 012 | ·91 |
| + ·10 | ·09 575 | 450·06 | ·04 456 | 241·23 | ·01 544 | 95·11 | ·00 313 | 21·75 | + ·10 | ·00 017 | 1·35 |
| + ·15 | ·12 054 | 543·65 | ·05 818 | 306·06 | ·02 096 | 126·91 | ·00 443 | 30·57 | + ·15 | ·00 025 | 1·99 |
| + ·20 | ·15 033 | 650·25 | ·07 541 | 385·66 | ·02 830 | 168·83 | ·00 625 | 43·03 | + ·20 | ·00 038 | 2·98 |
| + ·25 | ·18 579 | 769·73 | ·09 704 | 482·55 | ·03 805 | 223·95 | ·00 882 | 60·70 | + ·25 | ·00 056 | 4·48 |
| + ·30 | ·22 750 | 900·04 | ·12 399 | 599·22 | ·05 098 | 296·31 | ·01 245 | 85·91 | + ·30 | ·00 084 | 6·82 |
| + ·35 | ·27 604 | 1041·59 | ·15 733 | 737·88 | ·06 804 | 390·62 | ·01 759 | 122·04 | + ·35 | ·00 127 | 10·51 |
| + ·40 | ·33 175 | 1187·27 | ·19 817 | 899·90 | ·09 051 | 513·21 | ·02 491 | 174·12 | + ·40 | ·00 193 | 16·44 |
| + ·45 | ·39 474 | 1331·27 | ·24 770 | 1084·96 | ·11 995 | 671·10 | ·03 539 | 247·62 | + ·45 | ·00 297 | 26·17 |
| + ·50 | ·46 469 | 1463·74 | ·30 700 | 1289·77 | ·15 835 | 872·31 | ·05 045 | 359·56 | + ·50 | ·00 465 | 42·50 |
| + ·55 | ·54 070 | 1571·18 | ·37 688 | 1506·11 | ·20 803 | 1123·76 | ·07 219 | 520·21 | + ·55 | ·00 742 | 70·66 |
| + ·60 | ·62 110 | 1636·20 | ·45 755 | 1718·18 | ·27 163 | 1429·27 | ·10 370 | 754·96 | + ·60 | ·01 208 | 120·77 |
| + ·625 | ·66 217 | 1646·44 | ·50 173 | 1814·81 | ·30 948 | 1601·19 | ·12 445 | 909·82 | + ·625 | ·01 556 | 159·71 |
| + ·650 | ·70 327 | 1638·20 | ·54 819 | 1899·37 | ·35 177 | 1783·34 | ·14 946 | 1095·94 | + ·650 | ·02 019 | 213·01 |
| + ·675 | ·74 390 | 1608·76 | ·59 655 | 1966·34 | ·39 870 | 1971·83 | ·17 955 | 1318·47 | + ·675 | ·02 638 | 286·66 |
| + ·700 | ·78 351 | 1555·63 | ·64 630 | 2009·23 | ·45 037 | 2160·46 | ·21 572 | 1582·39 | + ·700 | ·03 476 | 389·45 |
| + ·725 | ·82 147 | 1476·81 | ·69 675 | 2020·75 | ·50 665 | 2339·99 | ·25 905 | 1891·62 | + ·725 | ·04 620 | 534·40 |
| + ·750 | ·85 712 | 1371·13 | ·74 701 | 1993·01 | ·56 718 | 2497·22 | ·31 069 | 2247·23 | + ·750 | ·06 199 | 740·79 |
| + ·775 | ·88 980 | 1238·65 | ·79 601 | 1918·20 | ·63 118 | 2614·27 | ·37 176 | 2644·46 | + ·775 | ·08 399 | 1037·29 |
| + ·800 | ·91 885 | 1081·22 | ·84 247 | 1789·53 | ·69 737 | 2668·18 | ·44 314 | 3067·85 | + ·800 | ·11 495 | 1466·04 |
| + ·825 | ·94 370 | 903·03 | ·88 500 | 1602·83 | ·76 383 | 2631·18 | ·52 510 | 3483·12 | + ·810 | ·13 069 | 1687·33 |
| + ·850 | ·96 389 | 711·33 | ·92 214 | 1358·89 | ·82 792 | 2473·52 | ·61 670 | 3826·37 | + ·820 | ·14 881 | 1943·91 |
| + ·875 | ·97 923 | 516·69 | ·95 254 | 1066·60 | ·88 629 | 2169·89 | ·71 494 | 3991·30 | + ·830 | ·16 970 | 2241·09 |
| + ·900 | ·98 981 | 333·27 | ·97 522 | 746·59 | ·93 512 | 1712·69 | ·81 350 | 3823·56 | + ·840 | ·19 379 | 2584·23 |
|  |  |  |  |  |  |  |  |  | + ·850 | ·22 156 | 2979·24 |
| + ·925 | ·99 612 | 177·81 | ·98 992 | 434·29 | ·97 089 | 1134·69 | ·90 187 | 3150·60 | + ·860 | ·25 356 | 3431·29 |
| + ·950 | ·99 907 | 66·90 | ·99 738 | 179·06 | ·99 168 | 538·90 | ·96 611 | 1916·67 | + ·870 | ·29 039 | 3944·27 |
| + ·975 | ·99 992 | 10·67 | ·99 974 | 31·45 | ·99 922 | 110·46 | ·99 612 | 521·68 | + ·880 | ·33 266 | 4519·20 |
| +1·000 | 1·00 000 | ·00 | 1·00 000 | ·00 | 1·00 000 | ·00 | 1·00 000 | ·00 | + ·890 | ·38 097 | 5151·78 |
|  |  |  |  |  |  |  |  |  | + ·900 | ·43 584 | 5828·60 |
|  |  |  |  |  |  |  |  |  | + ·910 | ·49 759 | 6521·26 |
|  |  |  |  |  |  |  |  |  | + ·920 | ·56 615 | 7178·23 |
|  |  |  |  |  |  |  |  |  | + ·930 | ·64 076 | 7713·82 |
|  |  |  |  |  |  |  |  |  | + ·940 | ·71 959 | 7996·42 |
|  |  |  |  |  |  |  |  |  | + ·950 | ·79 924 | 7841·74 |
|  |  |  |  |  |  |  |  |  | + ·955 | ·83 774 | 7531·59 |
|  |  |  |  |  |  |  |  |  | + ·960 | ·87 423 | 7030·03 |
|  |  |  |  |  |  |  |  |  | + ·965 | ·90 769 | 6317·49 |
|  |  |  |  |  |  |  |  |  | + ·970 | ·93 705 | 5389·95 |
|  |  |  |  |  |  |  |  |  | + ·975 | ·96 127 | 4271·20 |
|  |  |  |  |  |  |  |  |  | + ·980 | ·97 955 | 3029·05 |
|  |  |  |  |  |  |  |  |  | + ·985 | ·99 156 | 1792·01 |
|  |  |  |  |  |  |  |  |  | + ·990 | ·99 780 | 754·60 |
|  |  |  |  |  |  |  |  |  | + ·995 | ·99 980 | 136·01 |
|  |  |  |  |  |  |  |  |  | +1·000 | 1·00 000 | ·00 |

ρ=0.9

| r | Area | Ordinate |
|---|---|---|
| + ·800 | ·11 495 | 1466·04 |
| + ·825 | ·15 888 | 2085·10 |
| + ·850 | ·22 156 | 2979·24 |
| + ·875 | ·31 081 | 4224·10 |
| + ·900 | ·43 584 | 5828·60 |
| + ·925 | ·60 278 | 7468·23 |
| + ·950 | ·79 924 | 7841·74 |
| + ·975 | ·96 127 | 4271·20 |
| +1·000 | 1·00 000 | ·00 |

ρ=0·0 —— 0·4        n = 11

| r | ρ=0·0 Area | ρ=0·0 Ordinate | ρ=0·1 Area | ρ=0·1 Ordinate | ρ=0·2 Area | ρ=0·2 Ordinate | ρ=0·3 Area | ρ=0·3 Ordinate | ρ=0·4 Area | ρ=0·4 Ordinate | r |
|---|---|---|---|---|---|---|---|---|---|---|---|
| −1·00 | ·00 000 | ·00 | ·00 000 | ·00 | ·00 000 | ·00 | ·00 000 | ·00 | ·00 000 | ·00 | −1·00 |
| − ·95 | ·00 000 | ·34 | ·00 000 | ·14 | ·00 000 | ·05 | ·00 000 | ·02 | ·00 000 | ·01 | − ·95 |
| − ·90 | ·00 007 | 3·48 | ·00 003 | 1·46 | ·00 000 | ·59 | ·00 000 | ·22 | ·00 000 | ·08 | − ·90 |
| − ·85 | ·00 045 | 13·10 | ·00 020 | 5·73 | ·00 008 | 2·40 | ·00 003 | ·94 | ·00 000 | ·34 | − ·85 |
| − ·80 | ·00 155 | 32·59 | ·00 069 | 14·90 | ·00 029 | 6·47 | ·00 011 | 2·63 | ·00 004 | ·97 | − ·80 |
| − ·75 | ·00 392 | 64·48 | ·00 180 | 30·82 | ·00 078 | 13·91 | ·00 031 | 5·84 | ·00 011 | 2·22 | − ·75 |
| − ·70 | ·00 823 | 110·28 | ·00 391 | 55·10 | ·00 176 | 25·85 | ·00 074 | 11·22 | ·00 028 | 4·40 | − ·70 |
| − ·65 | ·01 518 | 170·38 | ·00 747 | 89·00 | ·00 346 | 43·43 | ·00 149 | 19·52 | ·00 058 | 7·90 | − ·65 |
| − ·60 | ·02 549 | 244·13 | ·01 298 | 133·37 | ·00 621 | 67·72 | ·00 275 | 31·55 | ·00 110 | 13·19 | − ·60 |
| − ·55 | ·03 980 | 329·91 | ·02 099 | 188·52 | ·01 036 | 99·67 | ·00 472 | 48·15 | ·00 194 | 20·80 | − ·55 |
| − ·50 | ·05 865 | 425·31 | ·03 201 | 254·27 | ·01 632 | 140·04 | ·00 766 | 70·22 | ·00 323 | 31·39 | − ·50 |
| − ·45 | ·08 244 | 527·29 | ·04 658 | 329·89 | ·02 452 | 189·38 | ·01 185 | 98·64 | ·00 514 | 45·66 | − ·45 |
| − ·40 | ·11 143 | 632·37 | ·06 515 | 414·10 | ·03 541 | 247·94 | ·01 764 | 134·25 | ·00 787 | 64·43 | − ·40 |
| − ·35 | ·14 567 | 736·81 | ·08 811 | 505·13 | ·04 946 | 315·62 | ·02 541 | 177·84 | ·01 167 | 88·59 | − ·35 |
| − ·30 | ·18 503 | 836·83 | ·11 574 | 600·74 | ·06 712 | 391·97 | ·03 557 | 230·03 | ·01 683 | 119·07 | − ·30 |
| − ·25 | ·22 921 | 928·73 | ·14 821 | 698·31 | ·08 879 | 476·06 | ·04 856 | 291·27 | ·02 370 | 156·85 | − ·25 |
| − ·20 | ·27 771 | 1009·12 | ·18 555 | 794·88 | ·11 483 | 566·57 | ·06 485 | 361·75 | ·03 266 | 202·94 | − ·20 |
| − ·15 | ·32 988 | 1074·98 | ·22 763 | 887·29 | ·14 552 | 661·66 | ·08 489 | 441·31 | ·04 415 | 258·26 | − ·15 |
| − ·10 | ·38 493 | 1123·87 | ·27 416 | 972·26 | ·18 103 | 759·03 | ·10 912 | 529·42 | ·05 865 | 323·64 | − ·10 |
| − ·05 | ·44 196 | 1153·95 | ·32 468 | 1046·54 | ·22 142 | 855·92 | ·13 795 | 625·00 | ·07 673 | 399·73 | − ·05 |
| 0·00 | ·50 000 | 1164·10 | ·37 858 | 1107·05 | ·26 657 | 949·18 | ·17 172 | 726·44 | ·09 881 | 486·84 | 0·00 |
| + ·05 | ·55 804 | 1153·95 | ·43 511 | 1150·99 | ·31 622 | 1035·30 | ·21 066 | 831·45 | ·12 556 | 584·86 | + ·05 |
| + ·10 | ·61 507 | 1123·87 | ·49 337 | 1176·02 | ·36 992 | 1110·55 | ·25 487 | 937·05 | ·15 746 | 693·05 | + ·10 |
| + ·15 | ·67 012 | 1074·98 | ·55 237 | 1180·37 | ·42 703 | 1171·11 | ·30 431 | 1039·52 | ·19 501 | 809·89 | + ·15 |
| + ·20 | ·72 229 | 1009·12 | ·61 104 | 1163·01 | ·48 674 | 1213·24 | ·35 870 | 1134·42 | ·23 856 | 932·85 | + ·20 |
| + ·25 | ·77 079 | 928·73 | ·66 830 | 1123·74 | ·54 799 | 1233·55 | ·41 754 | 1216·69 | ·28 833 | 1058·17 | + ·25 |
| + ·30 | ·81 497 | 836·83 | ·72 306 | 1063·29 | ·60 966 | 1229·19 | ·48 007 | 1280·75 | ·34 433 | 1180·68 | + ·30 |
| + ·35 | ·85 433 | 736·81 | ·77 431 | 983·38 | ·67 046 | 1198·19 | ·54 522 | 1320·82 | ·40 624 | 1293·66 | + ·35 |
| + ·40 | ·88 857 | 632·37 | ·82 112 | 886·74 | ·72 902 | 1139·72 | ·61 166 | 1331·30 | ·47 340 | 1388·77 | + ·40 |
| + ·45 | ·91 756 | 527·29 | ·86 276 | 777·05 | ·78 398 | 1054·45 | ·67 777 | 1307·30 | ·54 466 | 1456·28 | + ·45 |
| + ·50 | ·94 135 | 425·31 | ·89 868 | 658·86 | ·83 406 | 944·78 | ·74 175 | 1245·35 | ·61 839 | 1485·58 | + ·50 |
| + ·55 | ·96 020 | 329·91 | ·92 859 | 537·38 | ·87 813 | 815·03 | ·80 165 | 1144·27 | ·69 240 | 1466·08 | + ·55 |
| + ·60 | ·97 451 | 244·13 | ·95 246 | 418·24 | ·91 533 | 671·55 | ·85 556 | 1006·07 | ·76 403 | 1388·90 | + ·60 |
| + ·65 | ·98 482 | 170·38 | ·97 055 | 307·09 | ·94 519 | 522·51 | ·90 174 | 836·93 | ·83 025 | 1249·15 | + ·65 |
| + ·70 | ·99 177 | 110·28 | ·98 339 | 209·16 | ·96 765 | 377·49 | ·93 891 | 647·80 | ·88 794 | 1049·15 | + ·70 |
| + ·75 | ·99 608 | 64·48 | ·99 176 | 128·73 | ·98 318 | 246·69 | ·96 645 | 454·52 | ·93 437 | 801·88 | + ·75 |
| + ·80 | ·99 845 | 32·59 | ·99 660 | 68·50 | ·99 272 | 139·53 | ·98 462 | 276·64 | ·96 777 | 533·90 | + ·80 |
| + ·85 | ·99 955 | 13·10 | ·99 896 | 29·01 | ·99 764 | 62·88 | ·99 471 | 134·46 | ·98 807 | 285·17 | + ·85 |
| + ·90 | ·99 993 | 3·48 | ·99 981 | 8·12 | ·99 955 | 18·74 | ·99 893 | 43·33 | ·99 738 | 101·48 | + ·90 |
| + ·95 | 1·00 000 | ·34 | 1·00 000 | ·83 | ·99 998 | 2·04 | ·99 988 | 5·11 | ·99 984 | 13·29 | + ·95 |
| +1·00 | 1·00 000 | ·00 | 1·00 000 | ·00 | 1·00 000 | ·00 | 1·00 000 | ·00 | 1·00 000 | ·00 | +1·00 |

$n_2 = 11$                    $\rho = 0.5 \text{---} 0.9$

| r | $\rho=0.5$ Area | Ordinate | $\rho=0.6$ Area | Ordinate | $\rho=0.7$ Area | Ordinate | $\rho=0.8$ Area | Ordinate |
|---|---|---|---|---|---|---|---|---|
| −1.00 | .00 000 | .00 | .00 000 | .00 | .00 000 | .00 | .00 000 | .00 |
| − .95 | .00 000 | .00 | .00 000 | .00 | .00 000 | .00 | .00 000 | .00 |
| − .90 | .00 000 | .02 | .00 000 | .01 | .00 000 | .00 | .00 000 | .00 |
| − .85 | .00 000 | .11 | .00 000 | .03 | .00 000 | .01 | .00 000 | .00 |
| − .80 | .00 000 | .31 | .00 000 | .08 | .00 000 | .02 | .00 000 | .00 |
| − .75 | .00 002 | .74 | .00 001 | .20 | .00 000 | .04 | .00 000 | .00 |
| − .70 | .00 009 | 1.51 | .00 003 | .42 | .00 001 | .09 | .00 000 | .01 |
| − .65 | .00 020 | 2.78 | .00 006 | .80 | .00 001 | .17 | .00 000 | .02 |
| − .60 | .00 038 | 4.77 | .00 011 | 1.41 | .00 002 | .30 | .00 000 | .04 |
| − .55 | .00 069 | 7.76 | .00 020 | 2.35 | .00 004 | .51 | .00 000 | .06 |
| − .50 | .00 118 | 12.08 | .00 035 | 3.76 | .00 008 | .84 | .00 001 | .10 |
| − .45 | .00 193 | 18.15 | .00 059 | 5.83 | .00 013 | 1.34 | .00 002 | .17 |
| − .40 | .00 303 | 26.48 | .00 095 | 8.77 | .00 021 | 2.08 | .00 003 | .27 |
| − .35 | .00 462 | 37.70 | .00 149 | 12.90 | .00 034 | 3.16 | .00 004 | .43 |
| − .30 | .00 686 | 52.54 | .00 227 | 18.61 | .00 054 | 4.71 | .00 007 | .65 |
| − .25 | .00 995 | 71.87 | .00 338 | 26.38 | .00 082 | 6.91 | .00 011 | .99 |
| − .20 | .01 414 | 96.71 | .00 495 | 36.86 | .00 124 | 10.01 | .00 017 | 1.49 |
| − .15 | .01 974 | 128.20 | .00 713 | 50.84 | .00 185 | 14.35 | .00 026 | 2.21 |
| − .10 | .02 710 | 167.66 | .01 011 | 69.32 | .00 271 | 20.37 | .00 040 | 3.27 |
| − .05 | .03 666 | 216.51 | .01 415 | 93.53 | .00 392 | 28.70 | .00 060 | 4.80 |
| 0.00 | .04 893 | 276.25 | .01 958 | 124.99 | .00 563 | 40.16 | .00 089 | 7.04 |
| + .05 | .06 449 | 348.45 | .02 680 | 165.55 | .00 801 | 55.88 | .00 132 | 10.29 |
| + .10 | .08 400 | 434.44 | .03 633 | 217.39 | .01 131 | 77.34 | .00 194 | 15.02 |
| + .15 | .10 818 | 535.54 | .04 877 | 283.09 | .01 587 | 106.54 | .00 286 | 21.93 |
| + .20 | .13 782 | 652.40 | .06 491 | 365.55 | .02 214 | 146.17 | .00 419 | 32.04 |
| + .25 | .17 367 | 781.91 | .08 566 | 467.92 | .03 072 | 199.73 | .00 614 | 46.90 |
| + .30 | .21 655 | 931.80 | .11 209 | 593.39 | .04 242 | 271.79 | .00 900 | 68.83 |
| + .35 | .26 706 | 1089.82 | .14 543 | 744.74 | .05 830 | 368.23 | .01 320 | 101.35 |
| + .40 | .32 563 | 1253.34 | .18 702 | 923.68 | .07 977 | 496.34 | .01 940 | 149.79 |
| + .45 | .39 234 | 1413.46 | .23 825 | 1124.63 | .10 861 | 664.75 | .02 858 | 222.30 |
| + .50 | .46 671 | 1557.29 | .30 037 | 1358.02 | .14 707 | 882.89 | .04 223 | 331.21 |
| + .55 | .54 751 | 1667.54 | .37 425 | 1597.69 | .19 787 | 1159.22 | .06 261 | 495.09 |
| + .60 | .63 254 | 1722.77 | .45 998 | 1827.66 | .26 403 | 1497.43 | .09 311 | 741.14 |
| + .625 | .67 565 | 1722.29 | .50 697 | 1928.86 | .30 392 | 1687.79 | .11 363 | 906.37 |
| + .650 | .71 847 | 1699.11 | .55 629 | 2013.49 | .34 851 | 1888.67 | .13 872 | 1107.09 |
| + .675 | .76 039 | 1650.55 | .60 745 | 2074.79 | .39 833 | 2094.72 | .16 933 | 1349.30 |
| + .700 | .80 077 | 1574.50 | .65 977 | 2105.04 | .45 322 | 2297.63 | .20 658 | 1638.62 |
| + .725 | .83 888 | 1469.81 | .71 237 | 2095.98 | .51 306 | 2485.26 | .25 169 | 1979.08 |
| + .750 | .87 402 | 1336.70 | .76 417 | 2039.33 | .57 722 | 2640.85 | .30 596 | 2370.71 |
| + .775 | .90 550 | 1177.25 | .81 388 | 1927.85 | .64 465 | 2742.43 | .37 059 | 2805.67 |
| + .800 | .93 270 | 995.96 | .86 006 | 1756.78 | .71 367 | 2762.99 | .44 643 | 3261.77 |
| + .825 | .95 517 | 800.14 | .90 122 | 1526.04 | .78 187 | 2672.56 | .53 348 | 3693.13 |
| + .850 | .97 267 | 600.28 | .93 593 | 1242.87 | .84 613 | 2443.04 | .63 020 | 4017.99 |
| + .875 | .98 526 | 409.60 | .96 307 | 924.83 | .90 273 | 2058.49 | .73 243 | 4108.41 |
| + .900 | .99 335 | 243.28 | .98 212 | 601.85 | .94 786 | 1532.03 | .83 226 | 3796.61 |
| + .925 | .99 775 | 115.78 | .99 347 | 315.46 | .97 867 | 928.67 | .91 764 | 2936.70 |
| + .950 | .99 955 | 36.65 | .99 858 | 110.61 | .99 474 | 381.37 | .97 489 | 1590.43 |
| + .975 | .99 997 | 4.26 | .99 991 | 14.33 | .99 963 | 58.69 | .99 776 | 336.04 |
| +1.000 | 1.00 000 | .00 | 1.00 000 | .00 | 1.00 000 | .00 | 1.00 000 | .00 |

| r | $\rho=0.9$ Area | Ordinate |
|---|---|---|
| −1.00 | .00 000 | .00 |
| − .95 | .00 000 | .00 |
| − .90 | .00 000 | .00 |
| − .85 | .00 000 | .00 |
| − .80 | .00 000 | .00 |
| − .75 | .00 000 | .00 |
| − .70 | .00 000 | .00 |
| − .65 | .00 000 | .00 |
| − .60 | .00 000 | .00 |
| − .55 | .00 000 | .00 |
| − .50 | .00 000 | .00 |
| − .45 | .00 000 | .01 |
| − .40 | .00 000 | .01 |
| − .35 | .00 000 | .01 |
| − .30 | .00 000 | .02 |
| − .25 | .00 000 | .03 |
| − .20 | .00 001 | .05 |
| − .15 | .00 001 | .08 |
| − .10 | .00 001 | .12 |
| − .05 | .00 002 | .19 |
| 0.00 | .00 003 | .29 |
| + .05 | .00 005 | .44 |
| + .10 | .00 008 | .68 |
| + .15 | .00 012 | 1.06 |
| + .20 | .00 019 | 1.65 |
| + .25 | .00 030 | 2.60 |
| + .30 | .00 046 | 4.13 |
| + .35 | .00 072 | 6.67 |
| + .40 | .00 116 | 10.92 |
| + .45 | .00 186 | 18.21 |
| + .50 | .00 306 | 31.02 |
| + .55 | .00 514 | 54.17 |
| + .60 | .00 881 | 97.35 |
| + .625 | .01 165 | 132.09 |
| + .650 | .01 553 | 180.79 |
| + .675 | .02 086 | 249.74 |
| + .700 | .02 826 | 348.37 |
| + .725 | .03 864 | 490.86 |
| + .750 | .05 335 | 698.67 |
| + .775 | .07 439 | 1004.20 |
| + .800 | .10 461 | 1455.70 |
| + .810 | .12 047 | 1691.90 |
| + .820 | .13 873 | 1967.74 |
| + .830 | .15 998 | 2289.22 |
| + .840 | .18 470 | 2662.66 |
| + .850 | .21 342 | 3094.34 |
| + .860 | .24 679 | 3589.73 |
| + .870 | .28 545 | 4152.26 |
| + .880 | .33 006 | 4781.30 |
| + .890 | .38 127 | 5468.94 |
| + .900 | .43 957 | 6195.13 |
| + .910 | .50 517 | 6920.66 |
| + .920 | .57 776 | 7577.64 |
| + .930 | .65 614 | 8059.47 |
| + .940 | .73 783 | 8207.03 |
| + .950 | .81 851 | 7823.00 |
| + .955 | .85 656 | 7365.70 |
| + .960 | .89 182 | 6704.97 |
| + .965 | .92 326 | 5836.48 |
| + .970 | .94 987 | 4779.55 |
| + .975 | .97 083 | 3589.03 |
| + .980 | .98 571 | 2366.34 |
| + .985 | .99 470 | 1262.04 |
| + .990 | .99 882 | 452.39 |
| + .995 | .99 992 | 60.21 |
| +1.000 | 1.00 000 | .00 |

| r | $\rho=0.9$ Area | Ordinate |
|---|---|---|
| + .800 | .10 461 | 1455.70 |
| + .825 | .14 895 | 2120.39 |
| + .850 | .21 342 | 3094.34 |
| + .875 | .30 697 | 4458.74 |
| + .900 | .43 957 | 6195.13 |
| + .925 | .61 635 | 7848.63 |
| + .950 | .81 851 | 7823.00 |
| + .975 | .97 083 | 3589.03 |
| +1.000 | 1.00 000 | .00 |

ρ=0·0 — 0·4    n=12

| r | ρ=0·0 Area | Ordinate | ρ=0·1 Area | Ordinate | ρ=0·2 Area | Ordinate | ρ=0·3 Area | Ordinate | ρ=0·4 Area | Ordinate | r |
|---|---|---|---|---|---|---|---|---|---|---|---|
| −1·00 | ·00 000 | ·00 | ·00 000 | ·00 | ·00 000 | ·00 | ·00 000 | ·00 | ·00 000 | ·00 | −1·00 |
| −·95 | ·00 000 | ·11 | ·00 000 | ·04 | ·00 000 | ·01 | ·00 000 | ·00 | ·00 000 | ·00 | −·95 |
| −·90 | ·00 003 | 1·60 | ·00 001 | ·61 | ·00 000 | ·22 | ·00 000 | ·08 | ·00 000 | ·02 | −·90 |
| −·85 | ·00 023 | 7·30 | ·00 009 | 2·93 | ·00 003 | 1·12 | ·00 001 | ·40 | ·00 000 | ·13 | −·85 |
| −·80 | ·00 089 | 20·67 | ·00 036 | 8·71 | ·00 014 | 3·47 | ·00 005 | 1·28 | ·00 002 | ·43 | −·80 |
| −·75 | ·00 248 | 45·08 | ·00 105 | 19·94 | ·00 042 | 8·29 | ·00 016 | 3·18 | ·00 005 | 1·10 | −·75 |
| −·70 | ·00 563 | 83·24 | ·00 248 | 38·68 | ·00 103 | 16·77 | ·00 040 | 6·68 | ·00 014 | 2·38 | −·70 |
| −·65 | ·01 106 | 136·86 | ·00 508 | 66·80 | ·00 218 | 30·25 | ·00 086 | 12·52 | ·00 031 | 4·62 | −·65 |
| −·60 | ·01 958 | 206·44 | ·00 934 | 105·87 | ·00 416 | 50·10 | ·00 170 | 21·57 | ·00 062 | 8·24 | −·60 |
| −·55 | ·03 196 | 291·24 | ·01 586 | 156·97 | ·00 732 | 77·68 | ·00 309 | 34·81 | ·00 116 | 13·80 | −·55 |
| −·50 | ·04 893 | 389·33 | ·02 525 | 220·58 | ·01 208 | 114·20 | ·00 527 | 53·33 | ·00 204 | 21·95 | −·50 |
| −·45 | ·07 107 | 497·73 | ·03 813 | 296·51 | ·01 891 | 160·71 | ·00 853 | 78·27 | ·00 341 | 33·48 | −·45 |
| −·40 | ·09 881 | 612·62 | ·05 509 | 383·82 | ·02 833 | 217·93 | ·01 322 | 110·79 | ·00 546 | 49·33 | −·40 |
| −·35 | ·13 237 | 729·56 | ·07 667 | 480·84 | ·04 088 | 286·20 | ·01 976 | 152·04 | ·00 844 | 70·53 | −·35 |
| −·30 | ·17 172 | 843·79 | ·10 330 | 585·17 | ·05 713 | 365·35 | ·02 859 | 203·01 | ·01 263 | 98·26 | −·30 |
| −·25 | ·21 662 | 950·51 | ·13 526 | 693·78 | ·07 759 | 454·68 | ·04 023 | 264·55 | ·01 839 | 133·77 | −·25 |
| −·20 | ·26 657 | 1045·09 | ·17 269 | 803·05 | ·10 275 | 552·83 | ·05 523 | 337·18 | ·02 615 | 178·37 | −·20 |
| −·15 | ·32 086 | 1123·41 | ·21 551 | 908·99 | ·13 299 | 657·79 | ·07 414 | 421·03 | ·03 640 | 233·38 | −·15 |
| −·10 | ·37 858 | 1181·98 | ·26 346 | 1007·35 | ·16 859 | 766·83 | ·09 751 | 515·70 | ·04 969 | 299·98 | −·10 |
| −·05 | ·43 868 | 1218·21 | ·31 605 | 1093·82 | ·20 968 | 876·57 | ·12 587 | 620·13 | ·06 661 | 379·19 | −·05 |
| 0·00 | ·50 000 | 1230·47 | ·37 258 | 1164·30 | ·25 619 | 983·02 | ·15 966 | 732·48 | ·08 783 | 471·64 | 0·00 |
| +·05 | ·56 132 | 1218·21 | ·43 215 | 1215·06 | ·30 785 | 1081·67 | ·19 921 | 850·05 | ·11 400 | 577·41 | +·05 |
| +·10 | ·62 142 | 1181·98 | ·49 370 | 1243·05 | ·36 415 | 1167·70 | ·24 469 | 969·15 | ·14 578 | 695·83 | +·10 |
| +·15 | ·67 914 | 1123·41 | ·55 604 | 1246·03 | ·42 433 | 1236·18 | ·29 607 | 1085·08 | ·18 376 | 825·16 | +·15 |
| +·20 | ·73 343 | 1045·09 | ·61 787 | 1222·86 | ·48 740 | 1282·35 | ·35 305 | 1192·20 | ·22 843 | 962·35 | +·20 |
| +·25 | ·78 338 | 950·51 | ·67 789 | 1173·63 | ·55 212 | 1301·99 | ·41 504 | 1284·05 | ·28 005 | 1102·71 | +·25 |
| +·30 | ·82 828 | 843·79 | ·73 482 | 1099·72 | ·61 710 | 1291·80 | ·48 108 | 1353·61 | ·33 864 | 1239·72 | +·30 |
| +·35 | ·86 763 | 729·56 | ·78 749 | 1003·91 | ·68 077 | 1249·80 | ·54 991 | 1393·76 | ·40 382 | 1364·85 | +·35 |
| +·40 | ·90 119 | 612·62 | ·83 491 | 890·31 | ·74 154 | 1175·76 | ·61 986 | 1397·87 | ·47 475 | 1467·64 | +·40 |
| +·45 | ·92 893 | 497·73 | ·87 631 | 764·16 | ·79 784 | 1071·56 | ·68 900 | 1360·66 | ·55 001 | 1536·09 | +·45 |
| +·50 | ·95 107 | 389·33 | ·91 122 | 631·64 | ·84 827 | 941·41 | ·75 518 | 1279·15 | ·62 758 | 1557·56 | +·50 |
| +·55 | ·96 804 | 291·24 | ·93 948 | 499·45 | ·89 166 | 791·98 | ·81 618 | 1153·78 | ·70 479 | 1520·32 | +·55 |
| +·60 | ·98 042 | 206·44 | ·96 129 | 374·33 | ·92 729 | 632·17 | ·86 991 | 989·48 | ·77 848 | 1415·90 | +·60 |
| +·65 | ·98 894 | 136·86 | ·97 714 | 262·47 | ·95 488 | 472·60 | ·91 465 | 796·46 | ·84 521 | 1242·33 | +·65 |
| +·70 | ·99 437 | 83·24 | ·98 784 | 168·90 | ·97 474 | 324·59 | ·94 934 | 590·32 | ·90 168 | 1007·76 | +·70 |
| +·75 | ·99 752 | 45·08 | ·99 439 | 96·80 | ·98 771 | 198·77 | ·97 380 | 391·03 | ·94 532 | 733·76 | +·75 |
| +·80 | ·99 911 | 20·67 | ·99 789 | 46·98 | ·99 513 | 103·20 | ·98 891 | 220·15 | ·97 499 | 456·19 | +·80 |
| +·85 | ·99 977 | 7·30 | ·99 943 | 17·56 | ·99 860 | 41·32 | ·99 659 | 95·84 | ·99 164 | 220·41 | +·85 |
| +·90 | ·99 997 | 1·60 | ·99 991 | 4·09 | ·99 978 | 10·31 | ·99 942 | 26·08 | ·99 842 | 66·93 | +·90 |
| +·95 | 1·00 000 | ·11 | 1·00 000 | ·30 | ·99 999 | ·81 | ·99 998 | 2·25 | ·99 992 | 6·48 | +·95 |
| +1·00 | 1·00 000 | ·00 | 1·00 000 | ·00 | 1·00 000 | ·00 | 1·00 000 | ·00 | 1·00 000 | ·00 | +1·00 |

n = 12

ρ = 0·5 — 0·9

| r | ρ=0·5 Area | ρ=0·5 Ordinate | ρ=0·6 Area | ρ=0·6 Ordinate | ρ=0·7 Area | ρ=0·7 Ordinate | ρ=0·8 Area | ρ=0·8 Ordinate |
|---|---|---|---|---|---|---|---|---|
| −1·00 | ·00 000 | ·00 | ·00 000 | ·00 | ·00 000 | ·00 | ·00 000 | ·00 |
| − ·95 | ·00 000 | ·00 | ·00 000 | ·00 | ·00 000 | ·00 | ·00 000 | ·00 |
| − ·90 | ·00 000 | ·01 | ·00 000 | ·00 | ·00 000 | ·00 | ·00 000 | ·00 |
| − ·85 | ·00 000 | ·04 | ·00 000 | ·01 | ·00 000 | ·00 | ·00 000 | ·00 |
| − ·80 | ·00 001 | ·12 | ·00 000 | ·03 | ·00 000 | ·00 | ·00 000 | ·00 |
| − ·75 | ·00 002 | ·33 | ·00 000 | ·08 | ·00 000 | ·01 | ·00 000 | ·00 |
| − ·70 | ·00 004 | ·73 | ·00 001 | ·18 | ·00 000 | ·03 | ·00 000 | ·00 |
| − ·65 | ·00 010 | 1·46 | ·00 002 | ·37 | ·00 000 | ·07 | ·00 000 | ·01 |
| − ·60 | ·00 020 | 2·69 | ·00 005 | ·70 | ·00 001 | ·13 | ·00 000 | ·01 |
| − ·55 | ·00 038 | 4·65 | ·00 010 | 1·25 | ·00 002 | ·23 | ·00 000 | ·02 |
| − ·50 | ·00 069 | 7·66 | ·00 018 | 2·12 | ·00 003 | ·41 | ·00 000 | ·04 |
| − ·45 | ·00 117 | 12·11 | ·00 032 | 3·46 | ·00 006 | ·69 | ·00 001 | ·07 |
| − ·40 | ·00 192 | 18·52 | ·00 054 | 5·48 | ·00 010 | 1·13 | ·00 001 | ·12 |
| − ·35 | ·00 306 | 27·52 | ·00 088 | 8·45 | ·00 018 | 1·80 | ·00 002 | ·20 |
| − ·30 | ·00 473 | 39·90 | ·00 140 | 12·72 | ·00 029 | 2·80 | ·00 003 | ·32 |
| − ·25 | ·00 712 | 56·63 | ·00 218 | 18·79 | ·00 046 | 4·30 | ·00 005‡ | ·51 |
| − ·20 | ·01 048 | 78·86 | ·00 332 | 27·27 | ·00 073 | 6·49 | ·00 008 | ·80 |
| − ·15 | ·01 512 | 107·94 | ·00 496 | 39·00 | ·00 113 | 9·69 | ·00 013 | 1·24 |
| − ·10 | ·02 143 | 145·45 | ·00 729 | 55·03 | ·00 172 | 14·30 | ·00 021 | 1·91 |
| − ·05 | ·02 985 | 193·12 | ·01 056 | 76·70 | ·00 259 | 20·91 | ·00 033 | 2·93 |
| 0·00 | ·04 093 | 252·88 | ·01 509 | 105·70 | ·00 386 | 30·32 | ·00 051 | 4·46 |
| + ·05 | ·05 536 | 326·69 | ·02 129 | 144·13 | ·00 569 | 43·65 | ·00 079 | 6·79 |
| + ·10 | ·07 387 | 416·49 | ·02 970 | 194·57 | ·00 831 | 62·45 | ·00 121 | 10·30 |
| + ·15 | ·09 730 | 523·93 | ·04 099 | 260·05 | ·01 206 | 88·83 | ·00 185 | 15·62 |
| + ·20 | ·12 657 | 650·07 | ·05 601 | 344·11 | ·01 737 | 125·68 | ·00 282 | 23·70 |
| + ·25 | ·16 262 | 794·98 | ·07 578 | 450·63 | ·02 487 | 176·90 | ·00 429 | 35·99 |
| + ·30 | ·20 636 | 957·09 | ·10 152 | 583·60 | ·03 536 | 247·67 | ·00 653 | 54·78 |
| + ·35 | ·25 856 | 1132·56 | ·13 464 | 746·54 | ·05 007 | 344·77 | ·00 993 | 83·60 |
| + ·40 | ·31 971 | 1314·07 | ·17 671 | 941·64 | ·07 045 | 476·75 | ·01 514 | 127·99 |
| + ·45 | ·38 988 | 1490·50 | ·22 935 | 1168·15 | ·09 850 | 653·92 | ·02 313 | 196·63 |
| + ·50 | ·46 841 | 1645·55 | ·29 395 | 1420·17 | ·13 678 | 887·55 | ·03 543 | 303·03 |
| + ·55 | ·55 372 | 1757·80 | ·37 154 | 1683·36 | ·18 837 | 1187·72 | ·05 441 | 468·00 |
| + ·60 | ·64 306 | 1801·62 | ·46 204 | 1930·97 | ·25 673 | 1558·25 | ·08 373 | 722·68 |
| + ·625 | ·68 799 | 1789·42 | ·51 167 | 2036·21 | ·29 827 | 1767·09 | ·10 389 | 896·86 |
| + ·650 | ·73 229 | 1750·35 | ·56 367 | 2120·06 | ·34 618 | 1986·75 | ·12 890 | 1110·84 |
| + ·675 | ·77 526 | 1681·95 | ·61 742 | 2174·45 | ·39 764 | 2210·30 | ·15 982 | 1371·59 |
| + ·700 | ·81 614 | 1582·81 | ·67 207 | 2190·55 | ·45 564 | 2427·07 | ·19 792 | 1685·48 |
| + ·725 | ·85 415 | 1452·95 | ·72 656 | 2159·36 | ·51 881 | 2621·83 | ·24 458 | 2056·72 |
| + ·750 | ·88 855 | 1294·33 | ·77 958 | 2072·67 | ·58 638 | 2774·00 | ·30 123 | 2484·25 |
| + ·775 | ·91 866 | 1111·34 | ·82 968 | 1924·49 | ·65 694 | 2857·52 | ·36 917 | 2956·77 |
| + ·800 | ·94 397 | 911·23 | ·87 528 | 1713·02 | ·72 843 | 2842·00 | ·44 920 | 3444·82 |
| + ·825 | ·96 417 | 704·19 | ·91 484 | 1443·16 | ·79 796 | 2696·41 | ·54 106 | 3889·71 |
| + ·850 | ·97 923 | 503·13 | ·94 706 | 1129·13 | ·86 196 | 2396·80 | ·64 251 | 4191·11 |
| + ·875 | ·98 949 | 322·52 | ·97 114 | 796·51 | ·91 647 | 1939·75 | ·74 818 | 4200·82 |
| + ·900 | ·99 564 | 176·39 | ·98 704 | 481·91 | ·95 792 | 1361·28 | ·84 864 | 3744·80 |
| + ·925 | ·99 869 | 74·87 | ·99 574 | 227·61 | ·98 429 | 754·99 | ·93 060 | 2719·15 |
| + ·950 | ·99 979 | 19·94 | ·99 921 | 67·87 | ·99 664 | 268·10 | ·98 126 | 1310·98 |
| + ·975 | ·99 999 | 1·69 | ·99 996 | 6·48 | ·99 983 | 30·98 | ·99 872 | 215·03 |
| +1·000 | 1·00 000 | ·00 | 1·00 000 | ·00 | 1·00 000 | ·00 | 1·00 000 | ·00 |

| r | ρ=0·9 Area | ρ=0·9 Ordinate |
|---|---|---|
| −1·00 | ·00 000 | ·00 |
| − ·95 | ·00 000 | ·00 |
| − ·90 | ·00 000 | ·00 |
| − ·85 | ·00 000 | ·00 |
| − ·80 | ·00 000 | ·00 |
| − ·75 | ·00 000 | ·00 |
| − ·70 | ·00 000 | ·00 |
| − ·65 | ·00 000 | ·00 |
| − ·60 | ·00 000 | ·00 |
| − ·55 | ·00 000 | ·00 |
| − ·50 | ·00 000 | ·00 |
| − ·45 | ·00 000 | ·00 |
| − ·40 | ·00 000 | ·00 |
| − ·35 | ·00 000 | ·00 |
| − ·30 | ·00 000 | ·01 |
| − ·25 | ·00 000 | ·01 |
| − ·20 | ·00 000 | ·02 |
| − ·15 | ·00 000 | ·03 |
| − ·10 | ·00 000 | ·05 |
| − ·05 | ·00 001 | ·08 |
| 0·00 | ·00 001 | ·13 |
| + ·05 | ·00 002 | ·21 |
| + ·10 | ·00 004 | ·34 |
| + ·15 | ·00 006 | ·56 |
| + ·20 | ·00 009 | ·91 |
| + ·25 | ·00 015 | 1·49 |
| + ·30 | ·00 025 | 2·49 |
| + ·35 | ·00 041 | 4·20 |
| + ·40 | ·00 069 | 7·20 |
| + ·45 | ·00 117 | 12·58 |
| + ·50 | ·00 202 | 22·49 |
| + ·55 | ·00 357 | 41·25 |
| + ·60 | ·00 644 | 77·95 |
| + ·625 | ·00 875 | 108·51 |
| + ·650 | ·01 198 | 152·41 |
| + ·675 | ·01 654 | 216·13 |
| + ·700 | ·02 303 | 309·54 |
| + ·725 | ·03 239 | 447·86 |
| + ·750 | ·04 599 | 654·55 |
| + ·775 | ·06 598 | 965·69 |
| + ·800 | ·09 559 | 1435·84 |
| + ·810 | ·11 116 | 1685·22 |
| + ·820 | ·12 944 | 1978·64 |
| + ·830 | ·15 091 | 2322·96 |
| + ·840 | ·17 610 | 2725·29 |
| + ·850 | ·20 563 | 3192·59 |
| + ·860 | ·24 018 | 3730·61 |
| + ·870 | ·28 049 | 4342·27 |
| + ·880 | ·32 727 | 5025·11 |
| + ·890 | ·38 119 | 5767·20 |
| + ·900 | ·44 272 | 6541·18 |
| + ·910 | ·51 195 | 7295·98 |
| + ·920 | ·58 830 | 7946·42 |
| + ·930 | ·67 011 | 8362·91 |
| + ·940 | ·75 419 | 8367·60 |
| + ·950 | ·83 538 | 7752·85 |
| + ·955 | ·87 274 | 7155·96 |
| + ·960 | ·90 659 | 6352·77 |
| + ·965 | ·93 594 | 5356·55 |
| + ·970 | ·95 991 | 4210·36 |
| + ·975 | ·97 793 | 2995·94 |
| + ·980 | ·98 996 | 1836·45 |
| + ·985 | ·99 664 | 882·95 |
| + ·990 | ·99 936 | 269·43 |
| + ·995 | ·99 996 | 26·48 |
| +1·000 | 1·00 000 | ·00 |

ρ=0·9

| r | Area | Ordinate |
|---|---|---|
| + ·800 | ·09 559 | 1435·84 |
| + ·825 | ·13 974 | 2141·98 |
| + ·850 | ·20 563 | 3192·59 |
| + ·875 | ·30 302 | 4675·26 |
| + ·900 | ·44 272 | 6541·18 |
| + ·925 | ·62 868 | 8193·95 |
| + ·950 | ·83 538 | 7752·85 |
| + ·975 | ·97 793 | 2995·94 |
| +1·000 | 1·00 000 | ·00 |

ρ=0·0 — 0·4        n=13

| r | ρ=0·0 Area | Ordinate | ρ=0·1 Area | Ordinate | ρ=0·2 Area | Ordinate | ρ=0·3 Area | Ordinate | ρ=0·4 Area | Ordinate | r |
|---|---|---|---|---|---|---|---|---|---|---|---|
| -1·00 | ·00 000 | ·00 | ·00 000 | ·00 | ·00 000 | ·00 | ·00 000 | ·00 | ·00 000 | ·00 | -1·00 |
| - ·95 | ·00 000 | ·04 | ·00 000 | ·01 | ·00 000 | ·00 | ·00 000 | ·00 | ·00 000 | ·00 | - ·95 |
| - ·90 | ·00 001 | ·73 | ·00 000 | ·26 | ·00 000 | ·09 | ·00 000 | ·03 | ·00 000 | ·01 | - ·90 |
| - ·85 | ·00 012 | 4·04 | ·00 004 | 1·49 | ·00 002 | ·52 | ·00 001 | ·17 | ·00 000 | ·05 | - ·85 |
| - ·80 | ·00 051 | 13·03 | ·00 019 | 5·06 | ·00 007 | 1·85 | ·00 002 | ·62 | ·00 001 | ·19 | - ·80 |
| - ·75 | ·00 157 | 31·34 | ·00 062 | 12·84 | ·00 023 | 4·91 | ·00 008 | 1·72 | ·00 002 | ·54 | - ·75 |
| - ·70 | ·00 386 | 62·49 | ·00 158 | 27·00 | ·00 061 | 10·82 | ·00 021 | 3·95 | ·00 007 | 1·28 | - ·70 |
| - ·65 | ·00 809 | 109·33 | ·00 346 | 49·86 | ·00 138 | 20·96 | ·00 050 | 7·99 | ·00 016 | 2·69 | - ·65 |
| - ·60 | ·01 509 | 173·60 | ·00 680 | 83·57 | ·00 280 | 36·87 | ·00 106 | 14·67 | ·00 035 | 5·13 | - ·60 |
| - ·55 | ·02 574 | 255·68 | ·01 203 | 129·97 | ·00 519 | 60·20 | ·00 203 | 25·03 | ·00 070 | 9·10 | - ·55 |
| - ·50 | ·04 093 | 354·43 | ·01 998 | 190·29 | ·00 897 | 92·61 | ·00 364 | 40·28 | ·00 130 | 15·26 | - ·50 |
| - ·45 | ·06 142 | 467·24 | ·03 130 | 265·03 | ·01 463 | 135·62 | ·00 616 | 61·76 | ·00 228 | 24·42 | - ·45 |
| - ·40 | ·08 783 | 590·21 | ·04 672 | 353·79 | ·02 273 | 190·50 | ·00 994 | 90·93 | ·00 381 | 37·56 | - ·40 |
| - ·35 | ·12 053 | 718·39 | ·06 689 | 455·19 | ·03 389 | 258·08 | ·01 541 | 129·26 | ·00 612 | 55·84 | - ·35 |
| - ·30 | ·15 966 | 846·13 | ·09 241 | 566·86 | ·04 875 | 338·66 | ·02 303 | 178·18 | ·00 950 | 80·64 | - ·30 |
| - ·25 | ·20 504 | 967·43 | ·12 369 | 685·47 | ·06 797 | 431·85 | ·03 342 | 238·96 | ·01 432 | 113·45 | - ·25 |
| - ·20 | ·25 619 | 1076·39 | ·16 100 | 806·84 | ·09 213 | 536·45 | ·04 716 | 312·55 | ·02 101 | 155·92 | - ·20 |
| - ·15 | ·31 238 | 1167·55 | ·20 434 | 926·10 | ·12 177 | 650·33 | ·06 490 | 399·46 | ·03 010 | 209·73 | - ·15 |
| - ·10 | ·37 257 | 1236·25 | ·25 349 | 1037·95 | ·15 727 | 770·44 | ·08 732 | 499·56 | ·04 220 | 276·51 | - ·10 |
| - ·05 | ·43 557 | 1278·96 | ·30 792 | 1136·93 | ·19 885 | 892·78 | ·11 506 | 611·89 | ·05 799 | 357·72 | - ·05 |
| 0·00 | ·50 000 | 1293·45 | ·36 688 | 1217·76 | ·24 650 | 1012·46 | ·14 868 | 734·51 | ·07 822 | 454·39 | 0·00 |
| + ·05 | ·56 443 | 1278·96 | ·42 932 | 1275·64 | ·29 996 | 1123·90 | ·18 863 | 864·29 | ·10 369 | 566·93 | + ·05 |
| + ·10 | ·62 743 | 1236·25 | ·49 399 | 1306·66 | ·35 865 | 1221·04 | ·23 516 | 996·83 | ·13 518 | 694·79 | + ·10 |
| + ·15 | ·68 762 | 1167·55 | ·55 949 | 1308·10 | ·42 172 | 1297·68 | ·28 827 | 1126·41 | ·17 340 | 836·11 | + ·15 |
| + ·20 | ·74 381 | 1076·39 | ·62 429 | 1278·72 | ·48 798 | 1347·93 | ·34 764 | 1246·04 | ·21 895 | 987·33 | + ·20 |
| + ·25 | ·79 496 | 967·43 | ·68 685 | 1218·98 | ·55 599 | 1366·66 | ·41 257 | 1347·70 | ·27 220 | 1142·82 | + ·25 |
| + ·30 | ·84 034 | 846·13 | ·74 572 | 1131·13 | ·62 406 | 1350·13 | ·48 196 | 1422·76 | ·33 317 | 1294·58 | + ·30 |
| + ·35 | ·87 947 | 718·39 | ·79 957 | 1019·23 | ·69 038 | 1296·48 | ·55 426 | 1462·65 | ·40 142 | 1432·08 | + ·35 |
| + ·40 | ·91 217 | 590·21 | ·84 733 | 888·97 | ·75 310 | 1206·29 | ·62 751 | 1459·72 | ·47 593 | 1542·51 | + ·40 |
| + ·45 | ·93 858 | 467·24 | ·88 827 | 747·35 | ·81 046 | 1082·96 | ·69 943 | 1408·44 | ·55 498 | 1611·41 | + ·45 |
| + ·50 | ·95 907 | 354·43 | ·92 201 | 602·21 | ·86 095 | 932·90 | ·76 751 | 1306·67 | ·63 613 | 1624·11 | + ·50 |
| + ·55 | ·97 426 | 255·68 | ·94 857 | 461·64 | ·90 345 | 765·35 | ·82 929 | 1157·00 | ·71 624 | 1567·94 | + ·55 |
| + ·60 | ·98 491 | 173·60 | ·96 838 | 333·18 | ·93 738 | 591·84 | ·88 255 | 967·83 | ·79 165 | 1435·55 | + ·60 |
| + ·65 | ·99 191 | 109·33 | ·98 220 | 223·11 | ·96 275 | 425·11 | ·92 566 | 753·80 | ·85 855 | 1228·80 | + ·65 |
| + ·70 | ·99 614 | 62·49 | ·99 107 | 135·64 | ·98 021 | 277·57 | ·95 785 | 534·99 | ·91 353 | 962·72 | + ·70 |
| + ·75 | ·99 843 | 31·34 | ·99 616 | 72·39 | ·99 100 | 159·28 | ·97 946 | 334·58 | ·95 433 | 667·77 | + ·75 |
| + ·80 | ·99 949 | 13·03 | ·99 869 | 32·04 | ·99 673 | 75·91 | ·99 191 | 174·24 | ·98 056 | 387·67 | + ·80 |
| + ·85 | ·99 988 | 4·04 | ·99 968 | 10·58 | ·99 917 | 27·01 | ·99 780 | 67·94 | ·99 414 | 169·42 | + ·85 |
| + ·90 | ·99 999 | ·73 | ·99 996 | 2·05 | ·99 989 | 5·65 | ·99 968 | 15·61 | ·99 907 | 43·90 | + ·90 |
| + ·95 | 1·00 000 | ·04 | 1·00 000 | ·11 | 1·00 000 | ·32 | ·99 999 | ·98 | ·99 997 | 3·14 | + ·95 |
| +1·00 | 1·00 000 | ·00 | 1·00 000 | ·00 | 1·00 000 | ·00 | 1·00 000 | ·00 | 1·00 000 | ·00 | +1·00 |

n = 13      ρ = 0.5 —— 0.9

| r | ρ=0.5 Area | Ordinate | ρ=0.6 Area | Ordinate | ρ=0.7 Area | Ordinate | ρ=0.8 Area | Ordinate | r | ρ=0.9 Area | Ordinate |
|---|---|---|---|---|---|---|---|---|---|---|---|
| −1.00 | .00 000 | .00 | .00 000 | .00 | .00 000 | .00 | .00 000 | .00 | −1.00 | .00 000 | .00 |
| − .95 | .00 000 | .00 | .00 000 | .00 | .00 000 | .00 | .00 000 | .00 | − .95 | .00 000 | .00 |
| − .90 | .00 000 | .00 | .00 000 | .00 | .00 000 | .00 | .00 000 | .00 | − .90 | .00 000 | .00 |
| − .85 | .00 000 | .01 | .00 000 | .00 | .00 000 | .00 | .00 000 | .00 | − .85 | .00 000 | .00 |
| − .80 | .00 000 | .05 | .00 000 | .01 | .00 000 | .00 | .00 000 | .00 | − .80 | .00 000 | .00 |
| − .75 | .00 001 | .14 | .00 000 | .03 | .00 000 | .00 | .00 000 | .00 | − .75 | .00 000 | .00 |
| − .70 | .00 002 | .35 | .00 000 | .08 | .00 000 | .01 | .00 000 | .00 | − .70 | .00 000 | .00 |
| − .65 | .00 004 | .76 | .00 001 | .17 | .00 000 | .03 | .00 000 | .00 | − .65 | .00 000 | .00 |
| − .60 | .00 010 | 1.51 | .00 002 | .35 | .00 000 | .05 | .00 000 | .00 | − .60 | .00 000 | .00 |
| − .55 | .00 020 | 2.78 | .00 005 | .66 | .00 001 | .11 | .00 000 | .01 | − .55 | .00 000 | .00 |
| − .50 | .00 039 | 4.83 | .00 009 | 1.19 | .00 001 | .20 | .00 000 | .02 | − .50 | .00 000 | .00 |
| − .45 | .00 071 | 8.04 | .00 017 | 2.05 | .00 003 | .35 | .00 000 | .03 | − .45 | .00 000 | .00 |
| − .40 | .00 122 | 12.88 | .00 030 | 3.41 | .00 005 | .61 | .00 000 | .05 | − .40 | .00 000 | .00 |
| − .35 | .00 203 | 19.97 | .00 052 | 5.50 | .00 009 | 1.01 | .00 001 | .09 | − .35 | .00 000 | .00 |
| − .30 | .00 327 | 30.14 | .00 087 | 8.65 | .00 016 | 1.66 | .00 001 | .15 | − .30 | .00 000 | .00 |
| − .25 | .00 511 | 44.38 | .00 141 | 13.30 | .00 026 | 2.66 | .00 002 | .26 | − .25 | .00 000 | .00 |
| − .20 | .00 780 | 63.95 | .00 224 | 20.07 | .00 043 | 4.19 | .00 004 | .42 | − .20 | .00 000 | .01 |
| − .15 | .01 162 | 90.38 | .00 347 | 29.75 | .00 070 | 6.51 | .00 007 | .69 | − .15 | .00 000 | .01 |
| − .10 | .01 698 | 125.48 | .00 528 | 43.44 | .00 110 | 9.98 | .00 011 | 1.11 | − .10 | .00 000 | .02 |
| − .05 | .02 385 | 171.31 | .00 790 | 62.54 | .00 172 | 15.15 | .00 018 | 1.77 | − .05 | .00 000 | .04 |
| 0.00 | .03 433 | 230.21 | .01 165 | 88.89 | .00 266 | 22.76 | .00 030 | 2.82 | 0.00 | .00 001 | .06 |
| + .05 | .04 763 | 304.67 | .01 695 | 124.80 | .00 406 | 33.91 | .00 047 | 4.45 | + .05 | .00 001 | .10 |
| + .10 | .06 509 | 397.09 | .02 434 | 173.18 | .00 613 | 50.15 | .00 076 | 7.03 | + .10 | .00 002 | .17 |
| + .15 | .08 768 | 509.76 | .03 454 | 237.58 | .00 919 | 73.66 | .00 120 | 11.07 | + .15 | .00 003 | .29 |
| + .20 | .11 643 | 644.20 | .04 844 | 322.16 | .01 367 | 107.48 | .00 190 | 17.43 | + .20 | .00 005 | .50 |
| + .25 | .15 246 | 800.73 | .06 717 | 431.61 | .02 018 | 155.83 | .00 301 | 27.47 | + .25 | .00 008 | .86 |
| + .30 | .19 685 | 977.68 | .09 209 | 570.83 | .02 959 | 224.46 | .00 475 | 43.35 | + .30 | .00 014 | 1.49 |
| + .35 | .25 050 | 1170.41 | .12 482 | 744.26 | .04 309 | 321.04 | .00 749 | 68.58 | + .35 | .00 024 | 2.63 |
| + .40 | .31 401 | 1370.21 | .16 714 | 954.71 | .06 232 | 455.44 | .01 185 | 108.77 | + .40 | .00 042 | 4.72 |
| + .45 | .38 741 | 1563.16 | .22 090 | 1201.40 | .08 947 | 639.76 | .01 877 | 172.98 | + .45 | .00 074 | 8.65 |
| + .50 | .46 988 | 1729.32 | .28 777 | 1477.09 | .12 736 | 887.39 | .02 979 | 275.75 | + .50 | .00 134 | 16.22 |
| + .55 | .55 947 | 1842.84 | .36 880 | 1763.97 | .17 947 | 1210.31 | .04 736 | 440.01 | + .55 | .00 248 | 31.24 |
| + .60 | .65 279 | 1873.82 | .46 381 | 2029.03 | .24 972 | 1612.76 | .07 539 | 700.88 | + .60 | .00 472 | 62.08 |
| + .625 | .69 939 | 1849.05 | .51 594 | 2137.88 | .29 285 | 1840.09 | .09 510 | 882.66 | + .625 | .00 658 | 88.66 |
| + .650 | .74 498 | 1793.32 | .57 048 | 2220.16 | .34 182 | 2078.63 | .11 990 | 1108.60 | + .650 | .00 926 | 127.80 |
| + .675 | .78 878 | 1704.64 | .62 665 | 2266.55 | .39 681 | 2319.65 | .15 096 | 1386.73 | + .675 | .01 313 | 186.03 |
| + .700 | .82 994 | 1582.52 | .68 343 | 2267.17 | .45 772 | 2549.96 | .18 973 | 1724.34 | + .700 | .01 880 | 273.57 |
| + .725 | .86 764 | 1428.48 | .73 955 | 2212.61 | .52 406 | 2750.98 | .23 772 | 2125.91 | + .725 | .02 718 | 406.44 |
| + .750 | .90 113 | 1246.49 | .79 354 | 2095.16 | .59 482 | 2898.15 | .29 654 | 2589.24 | + .750 | .03 971 | 609.94 |
| + .775 | .92 979 | 1043.43 | .84 375 | 1910.76 | .66 827 | 2961.42 | .36 758 | 3099.30 | + .775 | .05 860 | 923.69 |
| + .800 | .95 320 | 829.19 | .88 853 | 1661.34 | .74 191 | 2907.54 | .45 159 | 3618.62 | + .800 | .08 731 | 1408.68 |
| + .825 | .97 125 | 616.38 | .92 636 | 1357.41 | .81 241 | 2705.85 | .54 800 | 4074.81 | + .810 | .10 267 | 1669.60 |
| + .850 | .98 415 | 419.43 | .95 612 | 1020.26 | .87 582 | 2338.80 | .65 382 | 4348.30 | + .820 | .12 087 | 1978.98 |
| + .875 | .99 247 | 252.58 | .97 736 | 682.31 | .92 806 | 1818.05 | .76 249 | 4272.35 | + .830 | .14 244 | 2344.61 |
| + .900 | .99 714 | 127.20 | .99 056 | 383.80 | .96 592 | 1203.06 | .86 305 | 3673.97 | + .840 | .16 798 | 2774.51 |
|  |  |  |  |  |  |  |  |  | + .850 | .19 817 | 3276.40 |
| + .925 | .99 923 | 48.16 | .99 721 | 163.33 | .98 839 | 610.50 | .94 134 | 2504.28 | + .860 | .23 376 | 3856.36 |
| + .950 | .99 991 | 10.79 | .99 956 | 41.42 | .99 787 | 187.46 | .98 595 | 1074.87 | + .870 | .27 556 | 4516.80 |
| + .975 | 1.00 000 | .67 | .99 999 | 2.92 | .99 993 | 16.26 | .99 926 | 136.86 | + .880 | .32 435 | 5253.25 |
| +1.000 | 1.00 000 | .00 | 1.00 000 | .00 | 1.00 000 | .00 | 1.00 000 | .00 | + .890 | .38 083 | 6049.39 |
|  |  |  |  |  |  |  |  |  | + .900 | .44 543 | 6869.84 |
|  |  |  |  |  |  |  |  |  | + .910 | .51 810 | 7650.78 |
|  |  |  |  |  |  |  |  |  | + .920 | .59 797 | 8288.90 |
|  |  |  |  |  |  |  |  |  | + .930 | .68 290 | 8632.78 |
|  |  |  |  |  |  |  |  |  | + .940 | .76 899 | 8486.04 |
|  |  |  |  |  |  |  |  |  | + .950 | .85 028 | 7642.57 |
|  |  |  |  |  |  |  |  |  | + .955 | .88 676 | 6915.33 |
|  |  |  |  |  |  |  |  |  | + .960 | .91 910 | 5987.17 |
|  |  |  |  |  |  |  |  |  | + .965 | .94 635 | 4890.02 |
|  |  |  |  |  |  |  |  |  | + .970 | .96 783 | 3689.29 |
|  |  |  |  |  |  |  |  |  | + .975 | .98 325 | 2487.61 |
|  |  |  |  |  |  |  |  |  | + .980 | .99 292 | 1417.66 |
|  |  |  |  |  |  |  |  |  | + .985 | .99 786 | 614.46 |
|  |  |  |  |  |  |  |  |  | + .990 | .99 965 | 159.61 |
|  |  |  |  |  |  |  |  |  | + .995 | .99 998 | 11.59 |
|  |  |  |  |  |  |  |  |  | +1.000 | 1.00 000 | .00 |

ρ = 0.9

| r | Area | Ordinate |
|---|---|---|
| + .800 | .08 731 | 1408.68 |
| + .825 | .13 120 | 2152.24 |
| + .850 | .19 817 | 3276.40 |
| + .875 | .29 903 | 4876.19 |
| + .900 | .44 543 | 6869.84 |
| + .925 | .64 000 | 8509.02 |
| + .950 | .85 028 | 7642.57 |
| + .975 | .98 325 | 2487.61 |
| +1.000 | 1.00 000 | .00 |

ρ=0·0 — 0·4      n=14

| r | ρ=0·0 Area | ρ=0·0 Ordinate | ρ=0·1 Area | ρ=0·1 Ordinate | ρ=0·2 Area | ρ=0·2 Ordinate | ρ=0·3 Area | ρ=0·3 Ordinate | ρ=0·4 Area | ρ=0·4 Ordinate | r |
|---|---|---|---|---|---|---|---|---|---|---|---|
| −1·00 | ·00 000 | ·00 | ·00 000 | ·00 | ·00 000 | ·00 | ·00 000 | ·00 | ·00 000 | ·00 | −1·00 |
| − ·95 | ·00 000 | ·01 | ·00 000 | ·00 | ·00 000 | ·00 | ·00 000 | ·00 | ·00 000 | ·00 | − ·95 |
| − ·90 | ·00 001 | ·34 | ·00 000 | ·11 | ·00 000 | ·03 | ·00 000 | ·01 | ·00 000 | ·00 | − ·90 |
| − ·85 | ·00 006 | 2·23 | ·00 002 | ·75 | ·00 001 | ·24 | ·00 000 | ·07 | ·00 000 | ·02 | − ·85 |
| − ·80 | ·00 030 | 8·18 | ·00 011 | 2·93 | ·00 004 | ·98 | ·00 001 | ·30 | ·00 000 | ·08 | − ·80 |
| − ·75 | ·00 100 | 21·69 | ·00 038 | 8·22 | ·00 013 | 2·90 | ·00 004 | ·93 | ·00 001 | ·26 | − ·75 |
| − ·70 | ·00 266 | 46·70 | ·00 101 | 18·77 | ·00 036 | 6·95 | ·00 012 | 2·33 | ·00 003 | ·69 | − ·70 |
| − ·65 | ·00 593 | 86·94 | ·00 237 | 37·04 | ·00 088 | 14·45 | ·00 031 | 5·07 | ·00 009 | 1·55 | − ·65 |
| − ·60 | ·01 165 | 145·33 | ·00 489 | 65·68 | ·00 189 | 27·00 | ·00 066 | 9·93 | ·00 020 | 3·17 | − ·60 |
| − ·55 | ·02 079 | 223·45 | ·00 915 | 107·13 | ·00 369 | 46·44 | ·00 134 | 17·91 | ·00 043 | 5·98 | − ·55 |
| − ·50 | ·03 433 | 321·20 | ·01 585 | 163·42 | ·00 668 | 74·76 | ·00 252 | 30·28 | ·00 083 | 10·57 | − ·50 |
| − ·45 | ·05 317 | 436·63 | ·02 576 | 235·83 | ·01 135 | 113·94 | ·00 446 | 48·51 | ·00 152 | 17·73 | − ·45 |
| − ·40 | ·07 822 | 566·06 | ·03 971 | 324·64 | ·01 828 | 165·76 | ·00 750 | 74·28 | ·00 266 | 28·46 | − ·40 |
| − ·35 | ·10 996 | 704·20 | ·05 849 | 428·97 | ·02 815 | 231·67 | ·01 205 | 109·39 | ·00 445 | 44·02 | − ·35 |
| − ·30 | ·14 868 | 844·64 | ·08 283 | 546·64 | ·04 170 | 312·50 | ·01 862 | 155·68 | ·00 717 | 65·88 | − ·30 |
| − ·25 | ·19 434 | 980·21 | ·11 332 | 674·21 | ·05 965 | 408·32 | ·02 783 | 214·86 | ·01 117 | 95·78 | − ·25 |
| − ·20 | ·24 650 | 1103·62 | ·15 033 | 806·97 | ·08 276 | 518·19 | ·04 035 | 288·41 | ·01 691 | 135·67 | − ·20 |
| − ·15 | ·30 438 | 1207·94 | ·19 400 | 939·26 | ·11 167 | 640·05 | ·05 693 | 377·28 | ·02 494 | 187·62 | − ·15 |
| − ·10 | ·36 688 | 1287·18 | ·24 415 | 1064·65 | ·14 691 | 770·57 | ·07 834 | 481·74 | ·03 591 | 253·73 | − ·10 |
| − ·05 | ·43 261 | 1336·68 | ·30 024 | 1176·41 | ·18 880 | 905·17 | ·10 535 | 601·04 | ·05 058 | 335·93 | − ·05 |
| 0·00 | ·50 000 | 1353·52 | ·36 145 | 1267·92 | ·23 741 | 1038·07 | ·13 866 | 733·22 | ·06 980 | 435·79 | 0·00 |
| + ·05 | ·56 739 | 1336·68 | ·42 660 | 1333·18 | ·29 247 | 1162·50 | ·17 883 | 874·80 | ·09 447 | 554·12 | + ·05 |
| + ·10 | ·63 312 | 1287·18 | ·49 425 | 1367·32 | ·35 339 | 1271·05 | ·22 621 | 1020·68 | ·12 551 | 690·62 | + ·10 |
| + ·15 | ·69 562 | 1207·94 | ·56 275 | 1367·06 | ·41 918 | 1356·11 | ·28 086 | 1164·04 | ·16 380 | 843·38 | + ·15 |
| + ·20 | ·75 350 | 1103·62 | ·63 036 | 1331·09 | ·48 849 | 1410·48 | ·34 243 | 1296·43 | ·21 007 | 1008·40 | + ·20 |
| + ·25 | ·80 566 | 980·21 | ·69 528 | 1260·37 | ·55 961 | 1428·09 | ·41 015 | 1408·13 | ·26 474 | 1179·07 | + ·25 |
| + ·30 | ·85 132 | 844·64 | ·75 587 | 1158·20 | ·63 061 | 1404·74 | ·48 273 | 1488·72 | ·32 790 | 1345·78 | + ·30 |
| + ·35 | ·89 004 | 704·20 | ·81 067 | 1030·12 | ·69 938 | 1338·84 | ·55 834 | 1528·05 | ·39 904 | 1495·87 | + ·35 |
| + ·40 | ·92 178 | 566·06 | ·85 857 | 883·63 | ·76 381 | 1232·03 | ·63 470 | 1517·46 | ·47 695 | 1613·92 | + ·40 |
| + ·45 | ·94 683 | 436·63 | ·89 887 | 727·61 | ·82 198 | 1089·56 | ·70 916 | 1451·35 | ·55 961 | 1682·84 | + ·45 |
| + ·50 | ·96 567 | 321·20 | ·93 133 | 571·57 | ·87 232 | 920·31 | ·77 889 | 1328·79 | ·64 414 | 1685·90 | + ·50 |
| + ·55 | ·97 921 | 223·45 | ·95 619 | 424·77 | ·91 377 | 736·29 | ·84 118 | 1155·02 | ·72 688 | 1609·82 | + ·55 |
| + ·60 | ·98 835 | 145·33 | ·97 410 | 295·23 | ·94 594 | 551·59 | ·89 374 | 942·42 | ·80 370 | 1448·96 | + ·60 |
| + ·65 | ·99 407 | 86·94 | ·98 609 | 188·79 | ·96 916 | 380·68 | ·93 509 | 710·22 | ·87 047 | 1209·98 | + ·65 |
| + ·70 | ·99 734 | 46·70 | ·99 342 | 108·45 | ·98 445 | 236·30 | ·96 484 | 482·68 | ·92 376 | 915·58 | + ·70 |
| + ·75 | ·99 900 | 21·69 | ·99 737 | 53·89 | ·99 338 | 127·07 | ·98 386 | 284·99 | ·96 175 | 605·00 | + ·75 |
| + ·80 | ·99 970 | 8·18 | ·99 918 | 21·76 | ·99 779 | 55·58 | ·99 418 | 137·29 | ·98 483 | 327·97 | + ·80 |
| + ·85 | ·99 994 | 2·23 | ·99 983 | 6·34 | ·99 950 | 17·57 | ·99 857 | 47·94 | ·99 587 | 129·65 | + ·85 |
| + ·90 | ·99 999 | ·34 | ·99 998 | 1·02 | ·99 994 | 3·08 | ·99 982 | 9·30 | ·99 944 | 28·67 | + ·90 |
| + ·95 | 1·00 000 | ·01 | 1·00 000 | ·04 | 1·00 000 | ·13 | ·99 999 | ·43 | ·99 999 | 1·52 | + ·95 |
| +1·00 | 1·00 000 | ·00 | 1·00 000 | ·00 | 1·00 000 | ·00 | 1·00 000 | ·00 | 1·00 000 | ·00 | +1·00 |

$n_2 = 14$

$\rho = 0.5 \longrightarrow 0.9$

| r | $\rho=0.5$ Area | Ordinate | $\rho=0.6$ Area | Ordinate | $\rho=0.7$ Area | Ordinate | $\rho=0.8$ Area | Ordinate | r | $\rho=0.9$ Area | Ordinate |
|---|---|---|---|---|---|---|---|---|---|---|---|
| −1·00 | ·00 000 | ·00 | ·00 000 | ·00 | ·00 000 | ·00 | ·00 000 | ·00 | −1·00 | ·00 000 | ·00 |
| − ·95 | ·00 000 | ·00 | ·00 000 | ·00 | ·00 000 | ·00 | ·00 000 | ·00 | − ·95 | ·00 000 | ·00 |
| − ·90 | ·00 000 | ·00 | ·00 000 | ·00 | ·00 000 | ·00 | ·00 000 | ·00 | − ·90 | ·00 000 | ·00 |
| − ·85 | ·00 000 | ·00 | ·00 000 | ·00 | ·00 000 | ·00 | ·00 000 | ·00 | − ·85 | ·00 000 | ·00 |
| − ·80 | ·00 000 | ·02 | ·00 000 | ·00 | ·00 000 | ·00 | ·00 000 | ·00 | − ·80 | ·00 000 | ·00 |
| − ·75 | ·00 000 | ·06 | ·00 000 | ·01 | ·00 000 | ·00 | ·00 000 | ·00 | − ·75 | ·00 000 | ·00 |
| − ·70 | ·00 001 | ·17 | ·00 000 | ·03 | ·00 000 | ·00 | ·00 000 | ·00 | − ·70 | ·00 000 | ·00 |
| − ·65 | ·00 002 | ·40 | ·00 000 | ·08 | ·00 000 | ·01 | ·00 000 | ·00 | − ·65 | ·00 000 | ·00 |
| − ·60 | ·00 005 | ·84 | ·00 001 | ·17 | ·00 000 | ·02 | ·00 000 | ·00 | − ·60 | ·00 000 | ·00 |
| − ·55 | ·00 012 | 1·65 | ·00 002 | ·35 | ·00 000 | ·05 | ·00 000 | ·00 | − ·55 | ·00 000 | ·00 |
| − ·50 | ·00 023 | 3·04 | ·00 005 | ·66 | ·00 001 | ·09 | ·00 000 | ·01 | − ·50 | ·00 000 | ·00 |
| − ·45 | ·00 043 | 5·31 | ·00 009 | 1·21 | ·00 001 | ·18 | ·00 000 | ·01 | − ·45 | ·00 000 | ·00 |
| − ·40 | ·00 078 | 8·92 | ·00 017 | 2·11 | ·00 003 | ·32 | ·00 000 | ·02 | − ·40 | ·00 000 | ·00 |
| − ·35 | ·00 135 | 14·43 | ·00 031 | 3·57 | ·00 005 | ·57 | ·00 000 | ·04 | − ·35 | ·00 000 | ·00 |
| − ·30 | ·00 227 | 22·66 | ·00 054 | 5·86 | ·00 008 | ·98 | ·00 001 | ·07 | − ·30 | ·00 000 | ·00 |
| − ·25 | ·00 368 | 34·62 | ·00 092 | 9·38 | ·00 015 | 1·64 | ·00 001 | ·13 | − ·25 | ·00 000 | ·00 |
| − ·20 | ·00 581 | 51·63 | ·00 151 | 14·70 | ·00 026 | 2·69 | ·00 002 | ·23 | − ·20 | ·00 000 | ·00 |
| − ·15 | ·00 896 | 75·34 | ·00 243 | 22·59 | ·00 043 | 4·35 | ·00 003 | ·38 | − ·15 | ·00 000 | ·01 |
| − ·10 | ·01 349 | 107·76 | ·00 383 | 34·14 | ·00 071 | 6·94 | ·00 006 | ·64 | − ·10 | ·00 000 | ·01 |
| − ·05 | ·01 992 | 151·28 | ·00 593 | 50·77 | ·00 115 | 10·92 | ·00 010 | 1·07 | − ·05 | ·00 000 | ·02 |
| 0·00 | ·02 885 | 208·62 | ·00 903 | 74·41 | ·00 183 | 17·01 | ·00 017 | 1·77 | 0·00 | ·00 000 | ·03 |
| + ·05 | ·04 106 | 282·80 | ·01 353 | 107·57 | ·00 290 | 26·23 | ·00 029 | 2·91 | + ·05 | ·00 000 | ·05 |
| + ·10 | ·05 746 | 376·89 | ·01 999 | 153·45 | ·00 453 | 40·10 | ·00 047 | 4·77 | + ·10 | ·00 001 | ·09 |
| + ·15 | ·07 913 | 493·74 | ·02 915 | 216·07 | ·00 702 | 60·80 | ·00 078 | 7·81 | + ·15 | ·00 001 | ·15 |
| + ·20 | ·10 725 | 635·51 | ·04 196 | 300·25 | ·01 078 | 91·50 | ·00 128 | 12·76 | + ·20 | ·00 002 | ·27 |
| + ·25 | ·14 311 | 800·90 | ·05 963 | 411·53 | ·01 641 | 136·65 | ·00 211 | 20·87 | + ·25 | ·00 004 | ·47 |
| + ·30 | ·18 794 | 994·23 | ·08 366 | 555·84 | ·02 479 | 202·51 | ·00 345 | 34·16 | + ·30 | ·00 008 | ·89 |
| + ·35 | ·24 284 | 1204·16 | ·11 586 | 738·66 | ·03 715 | 297·60 | ·00 566 | 56·01 | + ·35 | ·00 014 | 1·64 |
| + ·40 | ·30 850 | 1422·35 | ·15 823 | 963·63 | ·05 522 | 433·15 | ·00 929 | 92·02 | + ·40 | ·00 025 | 3·08 |
| + ·45 | ·38 494 | 1632·02 | ·21 291 | 1230·08 | ·08 137 | 623·12 | ·01 525 | 161·50 | + ·45 | ·00 047 | 5·92 |
| + ·50 | ·47 117 | 1809·23 | ·28 180 | 1529·42 | ·11 870 | 883·27 | ·02 508 | 249·81 | + ·50 | ·00 089 | 11·64 |
| + ·55 | ·56 481 | 1923·36 | ·36 606 | 1840·21 | ·17 111 | 1227·85 | ·04 129 | 411·85 | + ·55 | ·00 173 | 23·56 |
| + ·60 | ·66 188 | 1940·22 | ·46 536 | 2122·58 | ·24 299 | 1661·76 | ·06 797 | 676·71 | + ·60 | ·00 347 | 49·22 |
| + ·625 | ·70 998 | 1902·14 | ·51 988 | 2234·62 | ·28 757 | 1907·61 | ·08 715 | 864·84 | + ·625 | ·00 497 | 72·12 |
| + ·650 | ·75 669 | 1829·15 | ·57 682 | 2314·64 | ·33 847 | 2165·10 | ·11 162 | 1101·47 | + ·650 | ·00 717 | 106·69 |
| + ·675 | ·80 113 | 1719·92 | ·63 525 | 2352·03 | ·39 584 | 2423·62 | ·14 280 | 1395·85 | + ·675 | ·01 045 | 159·42 |
| + ·700 | ·84 239 | 1575·19 | ·69 398 | 2336·05 | ·45 953 | 2667·21 | ·18 196 | 1756·31 | + ·700 | ·01 538 | 240·71 |
| + ·725 | ·87 962 | 1398·17 | ·75 153 | 2257·11 | ·52 889 | 2873·71 | ·23 111 | 2187·73 | + ·725 | ·02 286 | 367·23 |
| + ·750 | ·91 208 | 1195·09 | ·80 625 | 2108·50 | ·60 267 | 3014·47 | ·29 192 | 2686·77 | + ·750 | ·03 434 | 565·87 |
| + ·775 | ·93 924 | 975·31 | ·85 636 | 1888·71 | ·67 879 | 3055·52 | ·36 593 | 3234·38 | + ·775 | ·05 211 | 879·64 |
| + ·800 | ·96 081 | 751·18 | ·90 014 | 1604·06 | ·75 431 | 2961·44 | ·45 367 | 3784·47 | + ·800 | ·07 983 | 1375·97 |
| + ·825 | ·97 688 | 537·13 | ·93 616 | 1271·11 | ·82 549 | 2703·33 | ·55 440 | 4249·94 | + ·810 | ·09 491 | 1646·87 |
| + ·850 | ·98 788 | 348·10 | ·96 353 | 917·81 | ·88 803 | 2272·14 | ·66 432 | 4491·55 | + ·820 | ·11 295 | 1970·64 |
| + ·875 | ·99 460 | 196·93 | ·98 219 | 581·89 | ·93 787 | 1696·48 | ·77 559 | 4326·02 | + ·830 | ·13 452 | 2356·09 |
| + ·900 | ·99 811 | 91·32 | ·99 311 | 304·31 | ·97 232 | 1058·56 | ·87 581 | 3588·66 | + ·840 | ·16 030 | 2812·25 |
| | | | | | | | | | + ·850 | ·19 103 | 3347·70 |
| + ·925 | ·99 955 | 30·84 | ·99 816 | 116·69 | ·99 139 | 491·49 | ·95 028 | 2296·29 | + ·860 | ·22 754 | 3968·90 |
| + ·950 | ·99 995 | 5·81 | ·99 976 | 25·17 | ·99 864 | 130·49 | ·98 944 | 877·42 | + ·870 | ·27 070 | 4677·79 |
| + ·975 | 1·00 000 | ·26 | ·99 999 | 1·31 | ·99 997 | 8·50 | ·99 957 | 86·73 | + ·880 | ·32 137 | 5467·74 |
| +1·000 | 1·00 000 | ·00 | 1·00 000 | ·00 | 1·00 000 | ·00 | 1·00 000 | ·00 | + ·890 | ·38 026 | 6317·64 |
| | | | | | | | | | + ·900 | ·44 778 | 7183·43 |
| | | | | | | | | | + ·910 | ·52 373 | 7987·79 |
| | | | | | | | | | + ·920 | ·60 693 | 8608·39 |
| | | | | | | | | | + ·930 | ·69 472 | 8872·48 |
| | | | | | | | | | + ·940 | ·78 249 | 8568·62 |
| | | | | | | | | | + ·950 | ·86 353 | 7501·01 |
| | | | | | | | | | + ·955 | ·89 900 | 6653·65 |
| | | | | | | | | | + ·960 | ·92 975 | 5618·01 |
| | | | | | | | | | + ·965 | ·95 495 | 4444·68 |
| | | | | | | | | | + ·970 | ·97 411 | 3218·63 |
| | | | | | | | | | + ·975 | ·98 724 | 2056·54 |
| | | | | | | | | | + ·980 | ·99 505 | 1089·62 |
| | | | | | | | | | + ·985 | ·99 864 | 425·75 |
| | | | | | | | | | + ·990 | ·99 982 | 94·15 |
| | | | | | | | | | + ·995 | 1·00 000 | 5·05 |
| | | | | | | | | | +1·000 | 1·00 000 | ·00 |

| r | $\rho=0.9$ Area | Ordinate |
|---|---|---|
| + ·800 | ·07 983 | 1375·97 |
| + ·825 | ·12 325 | 2153·07 |
| + ·850 | ·19 103 | 3347·70 |
| + ·875 | ·29 505 | 5063·53 |
| + ·900 | ·44 778 | 7183·43 |
| + ·925 | ·65 049 | 8797·65 |
| + ·950 | ·86 353 | 7501·01 |
| + ·975 | ·98 724 | 2056·54 |
| +1·000 | 1·00 000 | ·00 |

ρ=0·0——0·4                                    n = 15

| r | ρ = 0·0 Area | Ordinate | ρ = 0·1 Area | Ordinate | ρ = 0·2 Area | Ordinate | ρ = 0·3 Area | Ordinate | ρ = 0·4 Area | Ordinate | r |
|---|---|---|---|---|---|---|---|---|---|---|---|
| −1·00 | ·00 000 | ·00 | ·00 000 | ·00 | ·00 000 | ·00 | ·00 000 | ·00 | ·00 000 | ·00 | −1·00 |
| − ·95 | ·00 000 | ·00 | ·00 000 | ·00 | ·00 000 | ·00 | ·00 000 | ·00 | ·00 000 | ·00 | − ·95 |
| − ·90 | ·00 000 | ·15 | ·00 000 | ·04 | ·00 000 | ·01 | ·00 000 | ·00 | ·00 000 | ·00 | − ·90 |
| − ·85 | ·00 003 | 1·22 | ·00 001± | ·38 | ·00 000 | ·11 | ·00 000 | ·03 | ·00 000 | ·01 | − ·85 |
| − ·80 | ·00 017 | 5·12 | ·00 005± | 1·69 | ·00 002 | ·52 | ·00 001 | ·14 | ·00 000 | ·04 | − ·80 |
| − ·75 | ·00 064 | 14·96 | ·00 021 | 5·25 | ·00 007 | 1·70 | ·00 002 | ·50 | ·00 001 | ·13 | − ·75 |
| − ·70 | ·00 183 | 34·77 | ·00 065± | 12·99 | ·00 021 | 4·45 | ·00 006 | 1·37 | ·00 002 | ·37 | − ·70 |
| − ·65 | ·00 436 | 68·88 | ·00 163 | 27·42 | ·00 056 | 9·93 | ·00 017 | 3·21 | ·00 005 | ·90 | − ·65 |
| − ·60 | ·00 903 | 121·21 | ·00 355± | 51·42 | ·00 128 | 19·70 | ·00 041 | 6·69 | ·00 012 | 1·96 | − ·60 |
| − ·55 | ·01 683 | 194·55 | ·00 698 | 87·97 | ·00 263 | 35·70 | ·00 088 | 12·77 | ·00 026 | 3·91 | − ·55 |
| − ·50 | ·02 885± | 289·98 | ·01 260 | 139·82 | ·00 499 | 60·13 | ·00 175± | 22·68 | ·00 053 | 7·29 | − ·50 |
| − ·45 | ·04 618 | 406·50 | ·02 125 | 209·05 | ·00 882 | 95·35 | ·00 324 | 37·96 | ·00 102 | 12·82 | − ·45 |
| − ·40 | ·06 980 | 540·85 | ·03 382 | 296·76 | ·01 474 | 143·69 | ·00 567 | 60·46 | ·00 186 | 21·49 | − ·40 |
| − ·35 | ·10 047 | 687·70 | ·05 123 | 402·73 | ·02 345 | 207·17 | ·00 944 | 92·23 | ·00 324 | 34·56 | − ·35 |
| − ·30 | ·13 866 | 839·97 | ·07 436 | 525·16 | ·03 573 | 287·27 | ·01 508 | 135·50 | ·00 542 | 53·62 | − ·30 |
| − ·25 | ·18 442 | 989·42 | ·10 396 | 660·62 | ·05 246 | 384·61 | ·02 322 | 192·47 | ·00 874 | 80·56 | − ·25 |
| − ·20 | ·23 741 | 1127·28 | ·14 056 | 804·07 | ·07 448 | 498·68 | ·03 459 | 265·12 | ·01 364 | 117·61 | − ·20 |
| − ·15 | ·29 682 | 1245·03 | ·18 440 | 949·02 | ·10 258 | 627·56 | ·05 002 | 354·99 | ·02 071 | 167·21 | − ·15 |
| − ·10 | ·36 145 | 1335·15 | ·23 536 | 1087·93 | ·13 742 | 767·80 | ·07 039 | 462·51 | ·03 062 | 231·94 | − ·10 |
| − ·05 | ·42 977 | 1391·74 | ·29 295± | 1212·67 | ·17 946 | 914·28 | ·09 659 | 588·16 | ·04 420 | 314·29 | − ·05 |
| 0·00 | ·50 000 | 1411·04 | ·35 626 | 1315·18 | ·22 884 | 1060·32 | ·12 947 | 729·17 | ·06 238 | 416·38 | 0·00 |
| + ·05 | ·57 023 | 1391·74 | ·42 398 | 1388·08 | ·28 535± | 1197·90 | ·16 971 | 882·11 | ·08 619 | 539·56 | + ·05 |
| + ·10 | ·63 855± | 1335·15 | ·49 447 | 1425·42 | ·34 834 | 1318·13 | ·21 778 | 1041·17 | ·11 669 | 683·89 | + ·10 |
| + ·15 | ·70 318 | 1245·03 | ·56 586 | 1423·30 | ·41 672 | 1411·83 | ·27 380 | 1198·41 | ·15 490 | 847·52 | + ·15 |
| + ·20 | ·76 259 | 1127·28 | ·63 612 | 1380·40 | ·48 894 | 1470·37 | ·33 743 | 1343·80 | ·20 169 | 1026·06 | + ·20 |
| + ·25 | ·81 558 | 989·42 | ·70 324 | 1298·27 | ·56 306 | 1486·66 | ·40 779 | 1465·74 | ·25 763 | 1211·89 | + ·25 |
| + ·30 | ·86 134 | 839·97 | ·76 537 | 1181·45 | ·63 682 | 1456·06 | ·48 340 | 1551·88 | ·32 281 | 1393·76 | + ·30 |
| + ·35 | ·89 953 | 687·70 | ·82 093 | 1037·21 | ·70 786 | 1377·39 | ·56 218 | 1590·37 | ·39 669 | 1556·64 | + ·35 |
| + ·40 | ·93 020 | 540·85 | ·86 879 | 875·02 | ·77 381 | 1253·60 | ·64 148 | 1571·56 | ·47 786 | 1682·30 | + ·40 |
| + ·45 | ·95 382 | 406·50 | ·90 831 | 705·73 | ·83 259 | 1092·09 | ·71 829 | 1489·95 | ·56 397 | 1750·86 | + ·45 |
| + ·50 | ·97 115± | 289·98 | ·93 943 | 540·45 | ·88 258 | 904·49 | ·78 944 | 1346·21 | ·65 168 | 1743·50 | + ·50 |
| + ·55 | ·98 317 | 194·55 | ·96 260 | 389·38 | ·92 285± | 705·68 | ·85 201 | 1148·73 | ·73 682 | 1646·64 | + ·55 |
| + ·60 | ·99 097 | 121·21 | ·97 874 | 260·61 | ·95 324 | 512·15 | ·90 369 | 914·23 | ·81 479 | 1457·04 | + ·60 |
| + ·65 | ·99 564 | 68·88 | ·98 912 | 159·16 | ·97 441 | 339·61 | ·94 321 | 666·66 | ·88 117 | 1187·01 | + ·65 |
| + ·70 | ·99 817 | 34·77 | ·99 514 | 86·37 | ·98 776 | 200·40 | ·97 060 | 433·86 | ·93 265± | 867·51 | + ·70 |
| + ·75 | ·99 936 | 14·96 | ·99 819 | 39·97 | ·99 512 | 100·99 | ·98 728 | 241·85 | ·96 789 | 546·09 | + ·75 |
| + ·80 | ·99 983 | 5·12 | ·99 949 | 14·72 | ·99 851 | 40·55 | ·99 576 | 107·77 | ·98 813 | 276·43 | + ·80 |
| + ·85 | ·99 997 | 1·22 | ·99 990 | 3·79 | ·99 970 | 11·39 | ·99 907 | 33·71 | ·99 709 | 98·85 | + ·85 |
| + ·90 | 1·00 000 | ·15 | ·99 999 | ·51 | ·99 997 | 1·67 | ·99 990 | 5·52 | ·99 967 | 18·65 | + ·90 |
| + ·95 | 1·00 000 | ·00 | 1·00 000 | ·01 | 1·00 000 | ·05 | 1·00 000 | ·19 | ·99 999 | ·73 | + ·95 |
| +1·00 | 1·00 000 | ·00 | 1·00 000 | ·00 | 1·00 000 | ·00 | 1·00 000 | ·00 | 1·00 000 | ·00 | +1·00 |

n = 15    ρ = 0.5 —— 0.9

| r | ρ = 0.5 Area | Ordinate | ρ = 0.6 Area | Ordinate | ρ = 0.7 Area | Ordinate | ρ = 0.8 Area | Ordinate |
|---|---|---|---|---|---|---|---|---|
| −1·00 | ·00 000 | ·00 | ·00 000 | ·00 | ·00 000 | ·00 | ·00 000 | ·00 |
| − ·95 | ·00 000 | ·00 | ·00 000 | ·00 | ·00 000 | ·00 | ·00 000 | ·00 |
| − ·90 | ·00 000 | ·00 | ·00 000 | ·00 | ·00 000 | ·00 | ·00 000 | ·00 |
| − ·85 | ·00 000 | ·00 | ·00 000 | ·00 | ·00 000 | ·00 | ·00 000 | ·00 |
| − ·80 | ·00 000 | ·01 | ·00 000 | ·00 | ·00 000 | ·00 | ·00 000 | ·00 |
| − ·75 | ·00 000 | ·03 | ·00 000 | ·00 | ·00 000 | ·00 | ·00 000 | ·00 |
| − ·70 | ·00 000 | ·08 | ·00 000 | ·01 | ·00 000 | ·00 | ·00 000 | ·00 |
| − ·65 | ·00 001 | ·21 | ·00 000 | ·04 | ·00 000 | ·00 | ·00 000 | ·00 |
| − ·60 | ·00 003 | ·47 | ·00 000 | ·08 | ·00 000 | ·01 | ·00 000 | ·00 |
| − ·55 | ·00 006 | ·97 | ·00 001 | ·18 | ·00 000 | ·02 | ·00 000 | ·00 |
| − ·50 | ·00 013 | 1·90 | ·00 002 | ·37 | ·00 000 | ·05 | ·00 000 | ·00 |
| − ·45 | ·00 026 | 3·50 | ·00 005 | ·71 | ·00 001 | ·09 | ·00 000 | ·01 |
| − ·40 | ·00 050 | 6·15 | ·00 010 | 1·30 | ·00 001 | ·17 | ·00 000 | ·01 |
| − ·35 | ·00 090 | 10·39 | ·00 019 | 2·30 | ·00 002 | ·32 | ·00 000 | ·02 |
| − ·30 | ·00 158 | 16·97 | ·00 034 | 3·95 | ·00 005 | ·57 | ·00 000 | ·04 |
| − ·25 | ·00 266 | 26·90 | ·00 060 | 6·59 | ·00 008 | 1·00 | ·00 001 | ·07 |
| − ·20 | ·00 434 | 41·52 | ·00 102 | 10·72 | ·00 015 | 1·72 | ·00 001 | ·12 |
| − ·15 | ·00 692 | 62·56 | ·00 171 | 17·09 | ·00 026 | 2·90 | ·00 002 | ·21 |
| − ·10 | ·01 074 | 92·19 | ·00 279 | 26·73 | ·00 045 | 4·80 | ·00 003 | ·37 |
| − ·05 | ·01 632 | 133·08 | ·00 446 | 41·06 | ·00 076 | 7·85 | ·00 006 | ·64 |
| 0·00 | ·02 429 | 188·35 | ·00 700 | 62·06 | ·00 127 | 12·66 | ·00 010 | 1·11 |
| + ·05 | ·03 546 | 261·56 | ·01 082 | 92·37 | ·00 207 | 20·21 | ·00 017 | 1·89 |
| + ·10 | ·05 081 | 356·36 | ·01 645 | 135·46 | ·00 336 | 31·94 | ·00 030 | 3·23 |
| + ·15 | ·07 151 | 476·43 | ·02 465 | 195·77 | ·00 537 | 50·00 | ·00 051 | 5·49 |
| + ·20 | ·09 892 | 624·58 | ·03 641 | 278·78 | ·00 851 | 77·60 | ·00 087 | 9·31 |
| + ·25 | ·13 446 | 800·05 | ·05 301 | 390·92 | ·01 337 | 119·38 | ·00 149 | 15·80 |
| + ·30 | ·17 959 | 1007·27 | ·07 610 | 539·22 | ·02 080 | 182·03 | ·00 253 | 26·81 |
| + ·35 | ·23 555 | 1234·25 | ·10 765 | 730·36 | ·03 207 | 274·85 | ·00 430 | 45·57 |
| + ·40 | ·30 318 | 1470·94 | ·14 993 | 968·99 | ·04 899 | 410·40 | ·00 730 | 77·56 |
| + ·45 | ·38 249 | 1697·55 | ·20 534 | 1254·73 | ·07 409 | 604·66 | ·01 243 | 132·19 |
| + ·50 | ·47 230 | 1885·77 | ·27 604 | 1577·71 | ·11 074 | 875·91 | ·02 116 | 225·46 |
| + ·55 | ·56 983 | 1999·91 | ·36 332 | 1912·58 | ·16 325 | 1241·01 | ·03 605 | 384·06 |
| + ·60 | ·67 041 | 2001·47 | ·46 678 | 2212·18 | ·23 652 | 1705·89 | ·06 136 | 650·97 |
| + ·625 | ·71 987 | 1949·46 | ·52 353 | 2327·06 | ·28 244 | 1970·26 | ·07 995 | 844·25 |
| + ·650 | ·76 756 | 1858·75 | ·58 276 | 2404·17 | ·33 514 | 2246·82 | ·10 402 | 1090·34 |
| + ·675 | ·81 249 | 1728·89 | ·64 332 | 2431·69 | ·39 478 | 2522·87 | ·13 499 | 1399·83 |
| + ·700 | ·85 369 | 1562·05 | ·70 383 | 2398·10 | ·46 112 | 2779·52 | ·17 462 | 1782·27 |
| + ·725 | ·89 032 | 1363·41 | ·76 264 | 2293·96 | ·53 338 | 2990·81 | ·22 477 | 2243·04 |
| + ·750 | ·92 167 | 1141·54 | ·81 790 | 2114·05 | ·61 001 | 3123·86 | ·28 739 | 2777·70 |
| + ·775 | ·94 730 | 908·25 | ·86 772 | 1860·00 | ·68 860 | 3140·97 | ·36 411 | 3362·92 |
| + ·800 | ·96 711 | 677·98 | ·91 037 | 1543·04 | ·76 578 | 3005·21 | ·45 551 | 3943·34 |
| + ·825 | ·98 136 | 466·33 | ·94 453 | 1185·88 | ·83 737 | 2690·86 | ·56 037 | 4416·30 |
| + ·850 | ·99 070 | 287·83 | ·96 961 | 822·59 | ·89 884 | 2199·24 | ·67 411 | 4622·48 |
| + ·875 | ·99 611 | 152·97 | ·98 595 | 494·42 | ·94 623 | 1577·20 | ·78 762 | 4364·28 |
| + ·900 | ·99 875 | 65·32 | ·99 495 | 240·39 | ·97 746 | 927·98 | ·88 715 | 3492·48 |
| + ·925 | ·99 973 | 19·68 | ·99 879 | 83·06 | ·99 360 | 394·23 | ·95 776 | 2097·85 |
| + ·950 | ·99 999 | 3·12 | ·99 986 | 15·23 | ·99 913 | 90·51 | ·99 204 | 713·62 |
| + ·975 | 1·00 000 | ·10 | 1·00 000 | 0·58 | ·99 999 | 4·43 | ·99 976 | 54·76 |
| +1·000 | 1·00 000 | ·00 | 1·00 000 | ·00 | 1·00 000 | ·00 | 1·00 000 | ·00 |

| r | ρ = 0.9 Area | Ordinate |
|---|---|---|
| −1·00 | ·00 000 | ·00 |
| − ·95 | ·00 000 | ·00 |
| − ·90 | ·00 000 | ·00 |
| − ·85 | ·00 000 | ·00 |
| − ·80 | ·00 000 | ·00 |
| − ·75 | ·00 000 | ·00 |
| − ·70 | ·00 000 | ·00 |
| − ·65 | ·00 000 | ·00 |
| − ·60 | ·00 000 | ·00 |
| − ·55 | ·00 000 | ·00 |
| − ·50 | ·00 000 | ·00 |
| − ·45 | ·00 000 | ·00 |
| − ·40 | ·00 000 | ·00 |
| − ·35 | ·00 000 | ·00 |
| − ·30 | ·00 000 | ·00 |
| − ·25 | ·00 000 | ·00 |
| − ·20 | ·00 000 | ·00 |
| − ·15 | ·00 000 | ·00 |
| − ·10 | ·00 000 | ·00 |
| − ·05 | ·00 000 | ·01 |
| 0·00 | ·00 000 | ·01 |
| + ·05 | ·00 000 | ·02 |
| + ·10 | ·00 000 | ·04 |
| + ·15 | ·00 001 | ·08 |
| + ·20 | ·00 001 | ·15 |
| + ·25 | ·00 002 | ·28 |
| + ·30 | ·00 004 | ·53 |
| + ·35 | ·00 008 | 1·02 |
| + ·40 | ·00 015 | 2·01 |
| + ·45 | ·00 030 | 4·04 |
| + ·50 | ·00 059 | 8·33 |
| + ·55 | ·00 121 | 17·70 |
| + ·60 | ·00 255 | 38·88 |
| + ·625 | ·00 375 | 58·45 |
| + ·650 | ·00 557 | 88·74 |
| + ·675 | ·00 833 | 136·12 |
| + ·700 | ·01 260 | 211·01 |
| + ·725 | ·01 925 | 330·58 |
| + ·750 | ·02 973 | 523·05 |
| + ·775 | ·04 639 | 834·62 |
| + ·800 | ·07 306 | 1339·09 |
| + ·810 | ·08 780 | 1618·49 |
| + ·820 | ·10 562 | 1955·15 |
| + ·830 | ·12 713 | 2358·96 |
| + ·840 | ·15 306 | 2840·07 |
| + ·850 | ·18 422 | 3408·02 |
| + ·860 | ·22 153 | 4069·79 |
| + ·870 | ·26 593 | 4826·80 |
| + ·880 | ·31 835 | 5670·18 |
| + ·890 | ·37 954 | 6573·68 |
| + ·900 | ·44 986 | 7483·96 |
| + ·910 | ·52 894 | 8309·19 |
| + ·920 | ·61 529 | 8907·55 |
| + ·930 | ·70 570 | 9085·55 |
| + ·940 | ·79 485 | 8620·44 |
| + ·950 | ·87 535 | 7335·23 |
| + ·955 | ·90 971 | 6378·54 |
| + ·960 | ·93 885 | 5252·40 |
| + ·965 | ·96 207 | 4025·18 |
| + ·970 | ·97 910 | 2797·79 |
| + ·975 | ·99 025 | 1693·97 |
| + ·980 | ·99 645 | 834·43 |
| + ·985 | ·99 913 | 293·92 |
| + ·990 | ·99 990 | 55·33 |
| + ·995 | 1·00 000 | 2·19 |
| +1·000 | 1·00 000 | ·00 |

| r | ρ = 0.9 Area | Ordinate |
|---|---|---|
| + ·800 | ·07 306 | 1339·09 |
| + ·825 | ·11 587 | 2146·01 |
| + ·850 | ·18 422 | 3408·02 |
| + ·875 | ·29 109 | 5238·83 |
| + ·900 | ·44 986 | 7483·96 |
| + ·925 | ·66 027 | 9062·87 |
| + ·950 | ·87 535 | 7335·23 |
| + ·975 | ·99 025 | 1693·97 |
| +1·000 | 1·00 000 | ·00 |

ρ=0.0——0.4      n=16

| r | ρ=0.0 Area | Ordinate | ρ=0.1 Area | Ordinate | ρ=0.2 Area | Ordinate | ρ=0.3 Area | Ordinate | ρ=0.4 Area | Ordinate | r |
|---|---|---|---|---|---|---|---|---|---|---|---|
| −1·00 | ·00 000 | ·00 | ·00 000 | ·00 | ·00 000 | ·00 | ·00 000 | ·00 | ·00 000 | ·00 | −1·00 |
| − ·95 | ·00 000 | ·00 | ·00 000 | ·00 | ·00 000 | ·00 | ·00 000 | ·00 | ·00 000 | ·00 | − ·95 |
| − ·90 | ·00 000 | ·07 | ·00 000 | ·02 | ·00 000 | ·00 | ·00 000 | ·00 | ·00 000 | ·00 | − ·90 |
| − ·85 | ·00 002 | ·67 | ·00 001 | ·19 | ·00 000 | ·05 | ·00 000 | ·01 | ·00 000 | ·00 | − ·85 |
| − ·80 | ·00 010 | 3·19 | ·00 003 | ·97 | ·00 001 | ·27 | ·00 000 | ·07 | ·00 000 | ·02 | − ·80 |
| − ·75 | ·00 041 | 10·28 | ·00 013 | 3·34 | ·00 003 | 1·00 | ·00 001 | ·27 | ·00 000 | ·06 | − ·75 |
| − ·70 | ·00 127 | 25·80 | ·00 042 | 8·97 | ·00 013 | 2·84 | ·00 003 | ·80 | ·00 001 | ·19 | − ·70 |
| − ·65 | ·00 321 | 54·39 | ·00 112 | 20·23 | ·00 036 | 6·80 | ·00 010 | 2·02 | ·00 002 | ·51 | − ·65 |
| − ·60 | ·00 700 | 100·76 | ·00 258 | 40·12 | ·00 087 | 14·33 | ·00 026 | 4·50 | ·00 007 | 1·20 | − ·60 |
| − ·55 | ·01 365 | 168·85 | ·00 533 | 72·01 | ·00 188 | 27·35 | ·00 059 | 9·08 | ·00 015 | 2·55 | − ·55 |
| − ·50 | ·02 429 | 260·97 | ·01 004 | 119·24 | ·00 373 | 48·20 | ·00 122 | 16·93 | ·00 034 | 5·01 | − ·50 |
| − ·45 | ·04 015 | 377·23 | ·01 756 | 184·72 | ·00 687 | 79·55 | ·00 236 | 29·61 | ·00 069 | 9·24 | − ·45 |
| − ·40 | ·06 238 | 515·11 | ·02 885 | 270·42 | ·01 191 | 124·16 | ·00 429 | 49·05 | ·00 131 | 16·17 | − ·40 |
| − ·35 | ·09 193 | 669·43 | ·04 495 | 376·88 | ·01 956 | 184·68 | ·00 741 | 77·51 | ·00 237 | 27·05 | − ·35 |
| − ·30 | ·12 947 | 832·67 | ·06 687 | 502·91 | ·03 068 | 263·24 | ·01 224 | 117·56 | ·00 410 | 43·50 | − ·30 |
| − ·25 | ·17 519 | 995·53 | ·09 551 | 645·25 | ·04 620 | 361·12 | ·01 941 | 171·85 | ·00 684 | 67·54 | − ·25 |
| − ·20 | ·22 884 | 1147·76 | ·13 158 | 798·62 | ·06 711 | 478·36 | ·02 970 | 242·94 | ·01 103 | 101·63 | − ·20 |
| − ·15 | ·28 962 | 1279·16 | ·17 544 | 955·82 | ·09 434 | 613·35 | ·04 402 | 332·95 | ·01 722 | 148·55 | − ·15 |
| − ·10 | ·35 626 | 1380·50 | ·22 708 | 1108·16 | ·12 868 | 762·59 | ·06 333 | 443·20 | ·02 615 | 211·35 | − ·10 |
| − ·05 | ·42 705 | 1444·45 | ·28 601 | 1246·06 | ·17 074 | 920·53 | ·08 868 | 573·72 | ·03 867 | 293·10 | − ·05 |
| 0·00 | ·50 000 | 1466·31 | ·35 128 | 1359·84 | ·22 076 | 1079·59 | ·12 102 | 722·83 | ·05 582 | 396·57 | 0·00 |
| + ·05 | ·57 295 | 1444·45 | ·42 145 | 1440·62 | ·27 856 | 1230·45 | ·16 121 | 886·63 | ·07 872 | 523·71 | + ·05 |
| + ·10 | ·64 374 | 1380·50 | ·49 467 | 1481·25 | ·34 349 | 1362·60 | ·20 982 | 1058·69 | ·10 859 | 675·07 | + ·10 |
| + ·15 | ·71 038 | 1279·16 | ·56 882 | 1477·14 | ·41 433 | 1465·15 | ·26 706 | 1229·86 | ·14 661 | 848·97 | + ·15 |
| + ·20 | ·77 116 | 1147·76 | ·64 162 | 1426·97 | ·48 934 | 1527·93 | ·33 260 | 1388·46 | ·19 379 | 1040·70 | + ·20 |
| + ·25 | ·82 481 | 995·53 | ·71 079 | 1333·04 | ·56 632 | 1542·70 | ·40 547 | 1520·86 | ·25 084 | 1241·67 | + ·25 |
| + ·30 | ·87 053 | 832·67 | ·77 429 | 1201·32 | ·64 273 | 1504·45 | ·48 401 | 1612·58 | ·31 790 | 1438·86 | + ·30 |
| + ·35 | ·90 807 | 669·43 | ·83 044 | 1041·01 | ·71 587 | 1412·54 | ·56 582 | 1649·98 | ·39 437 | 1614·73 | + ·35 |
| + ·40 | ·93 762 | 515·11 | ·87 811 | 863·74 | ·78 316 | 1271·49 | ·64 792 | 1622·42 | ·47 866 | 1748·02 | + ·40 |
| + ·45 | ·95 985 | 377·23 | ·91 675 | 682·33 | ·84 236 | 1091·14 | ·72 689 | 1524·73 | ·56 808 | 1815·85 | + ·45 |
| + ·50 | ·97 571 | 260·97 | ·94 648 | 509·39 | ·89 186 | 886·11 | ·79 926 | 1359·54 | ·65 880 | 1797·36 | + ·50 |
| + ·55 | ·98 635 | 168·85 | ·96 801 | 355·79 | ·93 086 | 674·19 | ·86 191 | 1138·85 | ·74 614 | 1678·96 | + ·55 |
| + ·60 | ·99 300 | 100·76 | ·98 252 | 229·32 | ·95 947 | 474·01 | ·91 257 | 884·08 | ·82 502 | 1460·52 | + ·60 |
| + ·65 | ·99 679 | 54·39 | ·99 146 | 133·74 | ·97 873 | 302·02 | ·95 023 | 623·79 | ·89 083 | 1160·78 | + ·65 |
| + ·70 | ·99 873 | 25·80 | ·99 640 | 68·58 | ·99 034 | 169·43 | ·97 538 | 388·74 | ·94 040 | 819·36 | + ·70 |
| + ·75 | ·99 959 | 10·28 | ·99 876 | 29·55 | ·99 640 | 80·00 | ·98 996 | 204·59 | ·97 319 | 491·36 | + ·75 |
| + ·80 | ·99 990 | 3·19 | ·99 968 | 9·92 | ·99 899 | 29·49 | ·99 691 | 84·33 | ·99 069 | 232·25 | + ·80 |
| + ·85 | ·99 998 | ·67 | ·99 995 | 2·25 | ·99 982 | 7·36 | ·99 939 | 23·62 | ·99 794 | 75·13 | + ·85 |
| + ·90 | 1·00 000 | ·07 | 1·00 000 | ·25 | ·99 998 | ·90 | ·99 995 | 3·27 | ·99 980 | 12·09 | + ·90 |
| + ·95 | 1·00 000 | ·00 | 1·00 000 | ·00 | 1·00 000 | ·02 | 1·00 000 | ·08 | ·99 999 | ·35 | + ·95 |
| +1·00 | 1·00 000 | ·00 | 1·00 000 | ·00 | 1·00 000 | ·00 | 1·00 000 | ·00 | 1·00 000 | ·00 | +1·00 |

$n=16$      $\rho=0.5 \longrightarrow 0.9$

| r | $\rho=0.5$ Area | Ordinate | $\rho=0.6$ Area | Ordinate | $\rho=0.7$ Area | Ordinate | $\rho=0.8$ Area | Ordinate | r | $\rho=0.9$ Area | Ordinate |
|---|---|---|---|---|---|---|---|---|---|---|---|
| −1·00 | ·00 000 | ·00 | ·00 000 | ·00 | ·00 000 | ·00 | ·00 000 | ·00 | −1·00 | ·00 000 | ·00 |
| − ·95 | ·00 000 | ·00 | ·00 000 | ·00 | ·00 000 | ·00 | ·00 000 | ·00 | − ·95 | ·00 000 | ·00 |
| − ·90 | ·00 000 | ·00 | ·00 000 | ·00 | ·00 000 | ·00 | ·00 000 | ·00 | − ·90 | ·00 000 | ·00 |
| − ·85 | ·00 000 | ·00 | ·00 000 | ·00 | ·00 000 | ·00 | ·00 000 | ·00 | − ·85 | ·00 000 | ·00 |
| − ·80 | ·00 000 | ·00 | ·00 000 | ·00 | ·00 000 | ·00 | ·00 000 | ·00 | − ·80 | ·00 000 | ·00 |
| − ·75 | ·00 000 | ·01 | ·00 000 | ·00 | ·00 000 | ·00 | ·00 000 | ·00 | − ·75 | ·00 000 | ·00 |
| − ·70 | ·00 000 | ·04 | ·00 000 | ·01 | ·00 000 | ·00 | ·00 000 | ·00 | − ·70 | ·00 000 | ·00 |
| − ·65 | ·00 000 | ·11 | ·00 000 | ·02 | ·00 000 | ·00 | ·00 000 | ·00 | − ·65 | ·00 000 | ·00 |
| − ·60 | ·00 001 | ·26 | ·00 000 | ·04 | ·00 000 | ·00 | ·00 000 | ·00 | − ·60 | ·00 000 | ·00 |
| − ·55 | ·00 003 | ·57 | ·00 001 | ·09 | ·00 000 | ·01 | ·00 000 | ·00 | − ·55 | ·00 000 | ·00 |
| − ·50 | ·00 008 | 1·18 | ·00 001 | ·20 | ·00 000 | ·02 | ·00 000 | ·00 | − ·50 | ·00 000 | ·00 |
| − ·45 | ·00 016 | 2·29 | ·00 003 | ·41 | ·00 000 | ·05 | ·00 000 | ·00 | − ·45 | ·00 000 | ·00 |
| − ·40 | ·00 032 | 4·23 | ·00 006 | ·80 | ·00 001 | ·09 | ·00 000 | ·00 | − ·40 | ·00 000 | ·00 |
| − ·35 | ·00 060 | 7·46 | ·00 011 | 1·48 | ·00 001 | ·18 | ·00 000 | ·01 | − ·35 | ·00 000 | ·00 |
| − ·30 | ·00 110 | 12·67 | ·00 021 | 2·65 | ·00 002 | ·34 | ·00 000 | ·02 | − ·30 | ·00 000 | ·00 |
| − ·25 | ·00 192 | 20·84 | ·00 039 | 4·61 | ·00 005 | ·61 | ·00 000 | ·03 | − ·25 | ·00 000 | ·00 |
| − ·20 | ·00 325 | 33·28 | ·00 069 | 7·80 | ·00 009 | 1·10 | ·00 000 | ·06 | − ·20 | ·00 000 | ·00 |
| − ·15 | ·00 535 | 51·78 | ·00 120 | 12·89 | ·00 016 | 1·93 | ·00 001 | ·12 | − ·15 | ·00 000 | ·00 |
| − ·10 | ·00 857 | 78·63 | ·00 203 | 20·86 | ·00 029 | 3·32 | ·00 002 | ·21 | − ·10 | ·00 000 | ·00 |
| − ·05 | ·01 340 | 116·70 | ·00 336 | 33·10 | ·00 051 | 5·62 | ·00 003 | ·38 | − ·05 | ·00 000 | ·00 |
| 0·00 | ·02 048 | 169·51 | ·00 544 | 51·59 | ·00 088 | 9·40 | ·00 006 | ·69 | 0·00 | ·00 000 | ·00 |
| + ·05 | ·03 066 | 241·10 | ·00 867 | 79·06 | ·00 149 | 15·52 | ·00 010 | 1·23 | + ·05 | ·00 000 | ·01 |
| + ·10 | ·04 498 | 335·89 | ·01 356 | 119·19 | ·00 249 | 25·36 | ·00 019 | 2·18 | + ·10 | ·00 000 | ·02 |
| + ·15 | ·06 471 | 458·26 | ·02 088 | 176·81 | ·00 412 | 40·99 | ·00 033 | 3·84 | + ·15 | ·00 000 | ·04 |
| + ·20 | ·09 133 | 611·89 | ·03 163 | 258·02 | ·00 674 | 65·60 | ·00 059 | 6·77 | + ·20 | ·00 001 | ·08 |
| + ·25 | ·12 645 | 798·66 | ·04 719 | 370·16 | ·01 091 | 103·97 | ·00 105 | 11·42 | + ·25 | ·00 001 | ·16 |
| + ·30 | ·17 173 | 1017·24 | ·06 930 | 521·43 | ·01 748 | 163·10 | ·00 185 | 20·98 | + ·30 | ·00 002 | ·31 |
| + ·35 | ·22 860 | 1261·08 | ·10 012 | 719·87 | ·02 773 | 253·03 | ·00 326 | 36·96 | + ·35 | ·00 004 | ·63 |
| + ·40 | ·29 799 | 1516·38 | ·14 218 | 971·31 | ·04 353 | 387·62 | ·00 575 | 65·16 | + ·40 | ·00 009 | 1·30 |
| + ·45 | ·38 005 | 1760·11 | ·19 815 | 1275·84 | ·06 764 | 584·87 | ·01 013 | 114·98 | + ·45 | ·00 019 | 2·74 |
| + ·50 | ·47 331 | 1959·32 | ·27 047 | 1622·38 | ·10 311 | 865·86 | ·01 787 | 202·85 | + ·50 | ·00 039 | 5·94 |
| + ·55 | ·57 456 | 2072·94 | ·36 061 | 1981·54 | ·15 585 | 1250·36 | ·03 151 | 357·03 | + ·55 | ·00 085 | 13·25 |
| + ·60 | ·67 845 | 2058·14 | ·46 793 | 2298·30 | ·23 031 | 1745·69 | ·05 545 | 624·23 | + ·60 | ·00 188 | 30·62 |
| + ·625 | ·72 916 | 1991·65 | ·52 694 | 2415·69 | ·27 745 | 2028·58 | ·07 341 | 821·56 | + ·625 | ·00 284 | 47·22 |
| + ·650 | ·77 768 | 1882·87 | ·58 835 | 2489·31 | ·33 185 | 2324·29 | ·09 700 | 1075·93 | + ·650 | ·00 432 | 73·57 |
| + ·675 | ·82 296 | 1732·42 | ·65 093 | 2506·12 | ·39 365 | 2617·94 | ·12 778 | 1399·44 | + ·675 | ·00 665 | 115·85 |
| + ·700 | ·86 399 | 1544·13 | ·71 309 | 2454·04 | ·46 255 | 2887·46 | ·16 764 | 1802·95 | + ·700 | ·01 033 | 184·41 |
| + ·725 | ·89 991 | 1325·33 | ·77 298 | 2324·09 | ·53 758 | 3102·93 | ·21 865 | 2292·56 | + ·725 | ·01 623 | 296·66 |
| + ·750 | ·93 009 | 1086·96 | ·82 861 | 2112·96 | ·61 693 | 3227·08 | ·28 295 | 2862·73 | + ·750 | ·02 577 | 481·97 |
| + ·775 | ·95 421 | 843·14 | ·87 799 | 1825·98 | ·69 782 | 3218·70 | ·36 224 | 3485·65 | + ·775 | ·04 134 | 789·43 |
| + ·800 | ·97 234 | 609·99 | ·91 941 | 1479·67 | ·77 645 | 3040·08 | ·45 714 | 4096·07 | + ·800 | ·06 691 | 1299·15 |
| + ·825 | ·98 494 | 403·58 | ·95 172 | 1102·90 | ·84 822 | 2670·06 | ·56 596 | 4574·86 | + ·810 | ·08 129 | 1585·66 |
| + ·850 | ·99 286 | 237·24 | ·97 463 | 734·93 | ·90 845 | 2121·99 | ·68 331 | 4742·40 | + ·820 | ·09 883 | 1933·75 |
| + ·875 | ·99 720 | 118·45 | ·98 890 | 418·78 | ·95 337 | 1461·73 | ·79 874 | 4389·18 | + ·830 | ·12 021 | 2354·49 |
| + ·900 | ·99 917 | 46·58 | ·99 630 | 189·31 | ·98 161 | 810·97 | ·89 729 | 3388·31 | + ·840 | ·14 620 | 2859·25 |
|  |  |  |  |  |  |  |  |  | + ·850 | ·17 771 | 3458·65 |
| + ·925 | ·99 985 | 12·51 | ·99 921 | 58·94 | ·99 523 | 315·23 | ·96 404 | 1910·61 | + ·860 | ·21 572 | 4160·28 |
| + ·950 | ·99 999 | 1·67 | ·99 992 | 9·19 | ·99 945 | 62·58 | ·99 399 | 578·60 | + ·870 | ·26 126 | 4965·10 |
| + ·975 | 1·00 000 | ·04 | 1·00 000 | ·26 | ·99 998 | 2·30 | ·99 986 | 34·47 | + ·880 | ·31 533 | 5861·87 |
| +1·000 | 1·00 000 | ·00 | 1·00 000 | ·00 | 1·00 000 | ·00 | 1·00 000 | ·00 | + ·890 | ·37 870 | 6818·88 |
|  |  |  |  |  |  |  |  |  | + ·900 | ·45 170 | 7772·88 |
|  |  |  |  |  |  |  |  |  | + ·910 | ·53 380 | 8616·74 |
|  |  |  |  |  |  |  |  |  | + ·920 | ·62 313 | 9188·56 |
|  |  |  |  |  |  |  |  |  | + ·930 | ·71 595 | 9274·92 |
|  |  |  |  |  |  |  |  |  | + ·940 | ·80 624 | 8645·72 |
|  |  |  |  |  |  |  |  |  | + ·950 | ·88 596 | 7150·91 |
|  |  |  |  |  |  |  |  |  | + ·955 | ·91 915 | 6095·88 |
|  |  |  |  |  |  |  |  |  | + ·960 | ·94 667 | 4895·39 |
|  |  |  |  |  |  |  |  |  | + ·965 | ·96 800 | 3633·99 |
|  |  |  |  |  |  |  |  |  | + ·970 | ·98 310 | 2424·44 |
|  |  |  |  |  |  |  |  |  | + ·975 | ·99 254 | 1391·00 |
|  |  |  |  |  |  |  |  |  | + ·980 | ·99 748 | 637·03 |
|  |  |  |  |  |  |  |  |  | + ·985 | ·99 945 | 202·29 |
|  |  |  |  |  |  |  |  |  | + ·990 | ·99 995 | 32·42 |
|  |  |  |  |  |  |  |  |  | + ·995 | 1·00 000 | ·95 |
|  |  |  |  |  |  |  |  |  | +1·000 | 1·00 000 | ·00 |

| r | $\rho=0.9$ Area | Ordinate |
|---|---|---|
| + ·800 | ·06 691 | 1299·15 |
| + ·825 | ·10 899 | 2132·32 |
| + ·850 | ·17 771 | 3458·65 |
| + ·875 | ·28 717 | 5403·38 |
| + ·900 | ·45 170 | 7772·88 |
| + ·925 | ·66 942 | 9307·16 |
| + ·950 | ·88 596 | 7150·91 |
| + ·975 | ·99 254 | 1391·00 |
| +1·000 | 1·00 000 | ·00 |

$\rho = 0.0 — 0.4$      $n = 17$

| r | $\rho = 0.0$ Area | Ordinate | $\rho = 0.1$ Area | Ordinate | $\rho = 0.2$ Area | Ordinate | $\rho = 0.3$ Area | Ordinate | $\rho = 0.4$ Area | Ordinate | r |
|---|---|---|---|---|---|---|---|---|---|---|---|
| −1·00 | ·00 000 | ·00 | ·00 000 | ·00 | ·00 000 | ·00 | ·00 000 | ·00 | ·00 000 | ·00 | −1·00 |
| − ·95 | ·00 000 | ·00 | ·00 000 | ·00 | ·00 000 | ·00 | ·00 000 | ·00 | ·00 000 | ·00 | − ·95 |
| − ·90 | ·00 000 | ·03 | ·00 000 | ·01 | ·00 000 | ·00 | ·00 000 | ·00 | ·00 000 | ·00 | − ·90 |
| − ·85 | ·00 001 | ·37 | ·00 000 | ·10 | ·00 000 | ·02 | ·00 000 | ·01 | ·00 000 | ·00 | − ·85 |
| − ·80 | ·00 006 | 1·98 | ·00 002 | ·56 | ·00 000 | ·14 | ·00 000 | ·03 | ·00 000 | ·01 | − ·80 |
| − ·75 | ·00 026 | 7·05 | ·00 008 | 2·12 | ·00 002 | ·58 | ·00 000 | ·14 | ·00 000 | ·03 | − ·75 |
| − ·70 | ·00 088 | 19·10 | ·00 027 | 6·17 | ·00 007 | 1·81 | ·00 002 | ·47 | ·00 000 | ·10 | − ·70 |
| − ·65 | ·00 237 | 42·84 | ·00 077 | 14·88 | ·00 023 | 4·64 | ·00 006 | 1·27 | ·00 001 | ·29 | − ·65 |
| − ·60 | ·00 545 | 83·54 | ·00 188 | 31·23 | ·00 059 | 10·39 | ·00 016 | 3·02 | ·00 004 | ·74 | − ·60 |
| − ·55 | ·01 109 | 146·13 | ·00 408 | 58·78 | ·00 135 | 20·89 | ·00 040 | 6·43 | ·00 009 | 1·66 | − ·55 |
| − ·50 | ·02 048 | 234·22 | ·00 801 | 101·41 | ·00 279 | 38·54 | ·00 085 | 12·61 | ·00 022 | 3·43 | − ·50 |
| − ·45 | ·03 496 | 349·12 | ·01 453 | 162·78 | ·00 537 | 66·18 | ·00 172 | 23·04 | ·00 046 | 6·64 | − ·45 |
| − ·40 | ·05 582 | 489·26 | ·02 465 | 245·73 | ·00 963 | 106·99 | ·00 326 | 39·68 | ·00 092 | 12·14 | − ·40 |
| − ·35 | ·08 423 | 649·87 | ·03 949 | 351·73 | ·01 634 | 164·17 | ·00 583 | 64·96 | ·00 173 | 21·11 | − ·35 |
| − ·30 | ·12 102 | 823·17 | ·06 020 | 480·28 | ·02 637 | 240·55 | ·00 994 | 101·72 | ·00 312 | 35·19 | − ·30 |
| − ·25 | ·16 658 | 998·93 | ·08 785 | 628·51 | ·04 075 | 338·14 | ·01 625 | 153·03 | ·00 537 | 56·47 | − ·25 |
| − ·20 | ·22 076 | 1165·43 | ·12 329 | 791·03 | ·06 055 | 457·61 | ·02 554 | 222·01 | ·00 893 | 87·57 | − ·20 |
| − ·15 | ·28 277 | 1310·64 | ·16 706 | 960·04 | ·08 685 | 597·82 | ·03 879 | 311·43 | ·01 434 | 131·60 | − ·15 |
| − ·10 | ·35 128 | 1423·48 | ·21 924 | 1125·69 | ·12 062 | 755·35 | ·05 706 | 423·25 | ·02 236 | 192·06 | − ·10 |
| − ·05 | ·42 443 | 1495·05 | ·27 939 | 1276·87 | ·16 258 | 924·29 | ·08 150 | 558·10 | ·03 388 | 272·60 | − ·05 |
| 0·00 | ·50 000 | 1519·58 | ·34 650 | 1402·18 | ·21 310 | 1096·21 | ·11 323 | 714·59 | ·05 001 | 376·67 | 0·00 |
| + ·05 | ·57 557 | 1495·05 | ·41 900 | 1491·07 | ·27 207 | 1260·42 | ·15 325 | 888·75 | ·07 199 | 506·94 | + ·05 |
| + ·10 | ·64 872 | 1423·48 | ·49 485 | 1535·05 | ·33 881 | 1404·71 | ·20 229 | 1073·56 | ·10 116 | 664·54 | + ·10 |
| + ·15 | ·71 723 | 1310·64 | ·57 167 | 1528·81 | ·41 200 | 1516·33 | ·26 062 | 1258·69 | ·13 888 | 848·09 | + ·15 |
| + ·20 | ·77 924 | 1165·43 | ·64 687 | 1471·07 | ·48 970 | 1583·40 | ·32 794 | 1430·69 | ·18 632 | 1052·66 | + ·20 |
| + ·25 | ·83 342 | 998·93 | ·71 796 | 1365·00 | ·56 960 | 1596·49 | ·40 320 | 1573·74 | ·24 434 | 1268·71 | + ·25 |
| + ·30 | ·87 898 | 823·17 | ·78 269 | 1218·19 | ·64 836 | 1550·21 | ·48 455 | 1671·08 | ·31 314 | 1481·38 | + ·30 |
| + ·35 | ·91 577 | 649·87 | ·83 929 | 1041·98 | ·72 347 | 1444·64 | ·56 929 | 1707·15 | ·39 209 | 1670·43 | + ·35 |
| + ·40 | ·94 418 | 489·26 | ·88 663 | 850·26 | ·79 194 | 1286·11 | ·65 405 | 1670·36 | ·47 938 | 1811·36 | + ·40 |
| + ·45 | ·96 504 | 349·12 | ·92 431 | 657·90 | ·85 140 | 1087·22 | ·73 503 | 1556·06 | ·57 198 | 1878·12 | + ·45 |
| + ·50 | ·97 952 | 234·22 | ·95 264 | 478·81 | ·90 028 | 865·73 | ·80 844 | 1369·26 | ·66 557 | 1847·84 | + ·50 |
| + ·55 | ·98 891 | 146·13 | ·97 260 | 324·22 | ·93 794 | 642·35 | ·87 101 | 1125·98 | ·75 492 | 1707·27 | + ·55 |
| + ·60 | ·99 455 | 83·54 | ·98 560 | 201·23 | ·96 483 | 437·52 | ·92 053 | 852·59 | ·83 451 | 1460·03 | + ·60 |
| + ·65 | ·99 763 | 42·84 | ·99 329 | 112·08 | ·98 229 | 267·85 | ·95 631 | 582·09 | ·89 956 | 1132·06 | + ·65 |
| + ·70 | ·99 912 | 19·10 | ·99 733 | 54·30 | ·99 236 | 142·84 | ·97 934 | 347·36 | ·94 717 | 771·78 | + ·70 |
| + ·75 | ·99 974 | 7·05 | ·99 914 | 21·79 | ·99 733 | 63·21 | ·99 206 | 172·59 | ·97 724 | 440·91 | + ·75 |
| + ·80 | ·99 994 | 1·98 | ·99 980 | 6·67 | ·99 931 | 21·38 | ·99 774 | 65·80 | ·99 269 | 194·61 | + ·80 |
| + ·85 | ·99 999 | ·37 | ·99 997 | 1·34 | ·99 989 | 4·74 | ·99 960 | 16·51 | ·99 853 | 56·94 | + ·85 |
| + ·90 | 1·00 000 | ·03 | 1·00 000 | ·12 | ·99 999 | ·49 | ·99 997 | 1·93 | ·99 987 | 7·82 | + ·90 |
| + ·95 | 1·00 000 | ·00 | 1·00 000 | ·00 | 1·00 000 | ·01 | 1·00 000 | ·03 | ·99 999 | ·17 | + ·95 |
| +1·00 | 1·00 000 | ·00 | 1·00 000 | ·00 | 1·00 000 | ·00 | 1·00 000 | ·00 | 1·00 000 | ·00 | +1·00 |

n = 17  ρ = 0.5 — 0.9

| r | ρ=0.5 Area | ρ=0.5 Ordinate | ρ=0.6 Area | ρ=0.6 Ordinate | ρ=0.7 Area | ρ=0.7 Ordinate | ρ=0.8 Area | ρ=0.8 Ordinate |
|---|---|---|---|---|---|---|---|---|
| -1.00 | .00 000 | .00 | .00 000 | .00 | .00 000 | .00 | .00 000 | .00 |
| - .95 | .00 000 | .00 | .00 000 | .00 | .00 000 | .00 | .00 000 | .00 |
| - .90 | .00 000 | .00 | .00 000 | .00 | .00 000 | .00 | .00 000 | .00 |
| - .85 | .00 000 | .00 | .00 000 | .00 | .00 000 | .00 | .00 000 | .00 |
| - .80 | .00 000 | .00 | .00 000 | .00 | .00 000 | .00 | .00 000 | .00 |
| - .75 | .00 000 | .01 | .00 000 | .00 | .00 000 | .00 | .00 000 | .00 |
| - .70 | .00 000 | .02 | .00 000 | .00 | .00 000 | .00 | .00 000 | .00 |
| - .65 | .00 000 | .05 | .00 000 | .01 | .00 000 | .00 | .00 000 | .00 |
| - .60 | .00 001 | .14 | .00 000 | .02 | .00 000 | .00 | .00 000 | .00 |
| - .55 | .00 002 | .34 | .00 000 | .05 | .00 000 | .00 | .00 000 | .00 |
| - .50 | .00 004 | .74 | .00 001 | .11 | .00 000 | .01 | .00 000 | .00 |
| - .45 | .00 010 | 1.50 | .00 001 | .24 | .00 000 | .02 | .00 000 | .00 |
| - .40 | .00 020 | 2.90 | .00 003 | .49 | .00 000 | .05 | .00 000 | .00 |
| - .35 | .00 041 | 5.33 | .00 007 | .95 | .00 001 | .10 | .00 000 | .00 |
| - .30 | .00 077 | 9.43 | .00 013 | 1.78 | .00 001 | .20 | .00 000 | .01 |
| - .25 | .00 139 | 16.10 | .00 026 | 3.22 | .00 003 | .37 | .00 000 | .02 |
| - .20 | .00 244 | 26.61 | .00 047 | 5.66 | .00 005 | .70 | .00 000 | .03 |
| - .15 | .00 415 | 42.74 | .00 085 | 9.69 | .00 010 | 1.28 | .00 000 | .06 |
| - .10 | .00 685 | 66.87 | .00 148 | 16.23 | .00 019 | 2.28 | .00 001 | .12 |
| - .05 | .01 102 | 102.06 | .00 253 | 26.61 | .00 034 | 4.01 | .00 002 | .23 |
| 0.00 | .01 730 | 152.13 | .00 424 | 42.77 | .00 061 | 6.95 | .00 003 | .43 |
| + .05 | .02 655 | 221.67 | .00 695 | 67.49 | .00 107 | 11.89 | .00 006 | .79 |
| + .10 | .03 988 | 315.72 | .01 119 | 104.60 | .00 185 | 20.08 | .00 012 | 1.46 |
| + .15 | .05 862 | 439.58 | .01 770 | 159.26 | .00 316 | 33.51 | .00 022 | 2.68 |
| + .20 | .08 441 | 597.82 | .02 752 | 238.16 | .00 534 | 55.31 | .00 040 | 4.91 |
| + .25 | .11 902 | 793.12 | .04 206 | 349.55 | .00 891 | 90.30 | .00 074 | 8.97 |
| + .30 | .16 432 | 1024.51 | .06 318 | 502.86 | .01 471 | 145.74 | .00 135 | 16.37 |
| + .35 | .22 196 | 1284.98 | .09 321 | 707.60 | .02 400 | 232.31 | .00 248 | 29.89 |
| + .40 | .29 305 | 1558.97 | .13 492 | 970.98 | .03 871 | 365.12 | .00 453 | 54.60 |
| + .45 | .37 765 | 1820.03 | .19 130 | 1293.70 | .06 162 | 564.21 | .00 827 | 99.74 |
| + .50 | .47 421 | 2030.22 | .26 509 | 1663.80 | .09 663 | 853.62 | .01 511 | 182.02 |
| + .55 | .57 905 | 2142.80 | .35 792 | 2047.42 | .14 888 | 1256.38 | .02 758 | 331.00 |
| + .60 | .68 607 | 2110.68 | .46 903 | 2381.31 | .22 433 | 1781.59 | .05 015 | 596.99 |
| + .625 | .73 792 | 2029.25 | .53 014 | 2500.92 | .27 260 | 2082.99 | .06 746 | 797.32 |
| + .650 | .78 716 | 1902.13 | .59 366 | 2570.49 | .32 860 | 2397.95 | .09 053 | 1058.86 |
| + .675 | .83 266 | 1731.26 | .65 813 | 2575.86 | .39 246 | 2709.26 | .12 103 | 1395.26 |
| + .700 | .87 340 | 1522.30 | .72 181 | 2504.51 | .46 382 | 2991.52 | .16 102 | 1818.97 |
| + .725 | .90 853 | 1284.82 | .78 265 | 2348.26 | .54 152 | 3210.57 | .21 277 | 2336.88 |
| + .750 | .93 751 | 1032.18 | .83 850 | 2106.17 | .62 348 | 3324.73 | .27 610 | 2942.45 |
| + .775 | .96 015 | 780.58 | .88 732 | 1787.74 | .70 651 | 3289.48 | .36 036 | 3603.17 |
| + .800 | .97 670 | 547.33 | .92 743 | 1415.08 | .78 639 | 3067.10 | .45 860 | 4243.31 |
| + .825 | .98 781 | 348.34 | .95 791 | 1022.96 | .85 817 | 2642.31 | .57 124 | 4726.42 |
| + .850 | .99 450 | 195.02 | .97 878 | 654.85 | .91 702 | 2042.00 | .69 197 | 4852.41 |
| + .875 | .99 798 | 91.47 | .99 122 | 353.75 | .95 950 | 1351.08 | .80 904 | 4402.40 |
| + .900 | .99 945 | 33.12 | .99 728 | 148.67 | .98 496 | 706.81 | .90 636 | 3278.45 |
| + .925 | .99 991 | 7.94 | .99 947 | 41.71 | .99 644 | 251.38 | .96 932 | 1735.42 |
| + .950 | .99 999 | .89 | .99 996 | 5.53 | .99 965 | 43.15 | .99 545 | 467.87 |
| + .975 | 1.00 000 | .02 | 1.00 000 | .12 | .99 999 | 1.19 | .99 992 | 21.64 |
| +1.000 | 1.00 000 | .00 | 1.00 000 | .00 | 1.00 000 | .00 | 1.00 000 | .00 |

| r | ρ=0.9 Area | ρ=0.9 Ordinate |
|---|---|---|
| -1.00 | .00 000 | .00 |
| - .95 | .00 000 | .00 |
| - .90 | .00 000 | .00 |
| - .85 | .00 000 | .00 |
| - .80 | .00 000 | .00 |
| - .75 | .00 000 | .00 |
| - .70 | .00 000 | .00 |
| - .65 | .00 000 | .00 |
| - .60 | .00 000 | .00 |
| - .55 | .00 000 | .00 |
| - .50 | .00 000 | .00 |
| - .45 | .00 000 | .00 |
| - .40 | .00 000 | .00 |
| - .35 | .00 000 | .00 |
| - .30 | .00 000 | .00 |
| - .25 | .00 000 | .00 |
| - .20 | .00 000 | .00 |
| - .15 | .00 000 | .00 |
| - .10 | .00 000 | .00 |
| - .05 | .00 000 | .00 |
| 0.00 | .00 000 | .00 |
| + .05 | .00 000 | .01 |
| + .10 | .00 000 | .01 |
| + .15 | .00 000 | .02 |
| + .20 | .00 000 | .04 |
| + .25 | .00 001 | .09 |
| + .30 | .00 001 | .18 |
| + .35 | .00 003 | .39 |
| + .40 | .00 006 | .84 |
| + .45 | .00 012 | 1.86 |
| + .50 | .00 026 | 4.22 |
| + .55 | .00 059 | 9.90 |
| + .60 | .00 139 | 24.05 |
| + .625 | .00 215 | 38.05 |
| + .650 | .00 337 | 60.84 |
| + .675 | .00 532 | 98.34 |
| + .700 | .00 849 | 160.72 |
| + .725 | .01 370 | 265.51 |
| + .750 | .02 236 | 442.92 |
| + .775 | .03 687 | 744.69 |
| + .800 | .06 134 | 1257.03 |
| + .810 | .07 532 | 1549.34 |
| + .820 | .09 254 | 1907.48 |
| + .830 | .11 373 | 2343.75 |
| + .840 | .13 972 | 2870.87 |
| + .850 | .17 149 | 3500.67 |
| + .860 | .21 011 | 4241.43 |
| + .870 | .25 669 | 5093.74 |
| + .880 | .31 231 | 6043.88 |
| + .890 | .37 777 | 7054.37 |
| + .900 | .45 330 | 8051.44 |
| + .910 | .53 835 | 8911.86 |
| + .920 | .63 054 | 9453.19 |
| + .930 | .72 558 | 9443.03 |
| + .940 | .81 677 | 8648.01 |
| + .950 | .89 551 | 6952.68 |
| + .955 | .92 748 | 5810.25 |
| + .960 | .95 341 | 4550.52 |
| + .965 | .97 295 | 3272.10 |
| + .970 | .98 630 | 2095.34 |
| + .975 | .99 428 | 1139.19 |
| + .980 | .99 821 | 485.04 |
| + .985 | .99 965 | 138.85 |
| + .990 | .99 997 | 18.94 |
| + .995 | 1.00 000 | .41 |
| +1.000 | 1.00 000 | .00 |

| r | ρ=0.9 Area | ρ=0.9 Ordinate |
|---|---|---|
| + .800 | .06 134 | 1257.03 |
| + .825 | .10 259 | 2113.07 |
| + .850 | .17 149 | 3500.67 |
| + .875 | .28 331 | 5558.23 |
| + .900 | .45 330 | 8051.44 |
| + .925 | .67 807 | 9532.60 |
| + .950 | .89 551 | 6952.68 |
| + .975 | .99 428 | 1139.19 |
| +1.000 | 1.00 000 | .00 |

ρ=0·0—0·4      n₂=18

| r | ρ=0·0 Area | Ordinate | ρ=0·1 Area | Ordinate | ρ=0·2 Area | Ordinate | ρ=0·3 Area | Ordinate | ρ=0·4 Area | Ordinate | r |
|---|---|---|---|---|---|---|---|---|---|---|---|
| −1·00 | ·00 000 | ·00 | ·00 000 | ·00 | ·00 000 | ·00 | ·00 000 | ·00 | ·00 000 | ·00 | −1·00 |
| − ·95 | ·00 000 | ·00 | ·00 000 | ·00 | ·00 000 | ·00 | ·00 000 | ·00 | ·00 000 | ·00 | − ·95 |
| − ·90 | ·00 000 | ·01 | ·00 000 | ·00 | ·00 000 | ·00 | ·00 000 | ·00 | ·00 000 | ·00 | − ·90 |
| − ·85 | ·00 001 | ·20 | ·00 001 | ·05 | ·00 000 | ·01 | ·00 000 | ·00 | ·00 000 | ·00 | − ·85 |
| − ·80 | ·00 003 | 1·23 | ·00 001 | ·32 | ·00 000 | ·08 | ·00 000 | ·02 | ·00 000 | ·00 | − ·80 |
| − ·75 | ·00 019 | 4·82 | ·00 005̄ | 1·34 | ·00 001 | ·34 | ·00 000 | ·08 | ·00 000 | ·01 | − ·75 |
| − ·70 | ·00 061 | 14·10 | ·00 018 | 4·24 | ·00 005̄ | 1·15 | ·00 001 | ·27 | ·00 000 | ·05 | − ·70 |
| − ·65 | ·00 175̄ | 33·66 | ·00 054 | 10·93 | ·00 016 | 3·16 | ·00 004 | ·80 | ·00 001 | ·17 | − ·65 |
| − ·60 | ·00 424 | 69·10 | ·00 138 | 24·24 | ·00 042 | 7·52 | ·00 010 | 2·02 | ·00 002 | ·45 | − ·60 |
| − ·55 | ·00 902 | 126·18 | ·00 313 | 47·87 | ·00 099 | 15·93 | ·00 026 | 4·55 | ·00 006 | 1·08 | − ·55 |
| − ·50 | ·01 730 | 209·71 | ·00 641 | 86·05 | ·00 210 | 30·73 | ·00 059 | 9·36 | ·00 014 | 2·35 | − ·50 |
| − ·45 | ·03 048 | 322·33 | ·01 205̄ | 143·10 | ·00 419 | 54·93 | ·00 126 | 17·88 | ·00 031 | 4·76 | − ·45 |
| − ·40 | ·05 001 | 463·60 | ·02 109 | 222·77 | ·00 781 | 91·98 | ·00 248 | 32·03 | ·00 065̄ | 9·09 | − ·40 |
| − ·35 | ·07 725̄ | 629·39 | ·03 474 | 327·48 | ·01 367 | 145·60 | ·00 459 | 54·32 | ·00 127 | 16·44 | − ·35 |
| − ·30 | ·11 323 | 811·85 | ·05 427 | 457·58 | ·02 270 | 219·30 | ·00 809 | 87·80 | ·00 237 | 28·40 | − ·30 |
| − ·25 | ·15 853 | 999·97 | ·08 089 | 610·75 | ·03 598 | 315·87 | ·01 362 | 135·94 | ·00 422 | 47·10 | − ·25 |
| − ·20 | ·21 310 | 1180·56 | ·11 564 | 781·66 | ·05 469 | 436·73 | ·02 199 | 202·39 | ·00 724 | 75·28 | − ·20 |
| − ·15 | ·27 623 | 1339·70 | ·15 921 | 961·98 | ·08 005̄ | 581·30 | ·03 422 | 290·60 | ·01 196 | 116·32 | − ·15 |
| − ·10 | ·34 650 | 1464·32 | ·21 181 | 1140·77 | ·11 317 | 746·40 | ·05 146 | 403·25 | ·01 915̄ | 174·11 | − ·10 |
| − ·05 | ·42 190 | 1543·76 | ·27 305̄ | 1305·33 | ·15 493 | 925·87 | ·07 498 | 541·62 | ·02 973 | 252·92 | − ·05 |
| 0·00 | ·50 000 | 1571·04 | ·34 189 | 1442·41 | ·20 584 | 1110·44 | ·10 604 | 704·76 | ·04 486 | 356·91 | 0·00 |
| + ·05 | ·57 810 | 1543·76 | ·41 663 | 1539·62 | ·26 586 | 1288·06 | ·14 580 | 888·76 | ·06 589 | 489·54 | + ·05 |
| + ·10 | ·65 350 | 1464·32 | ·49 502 | 1587·04 | ·33 430 | 1444·69 | ·19 514 | 1086·05 | ·09 358 | 652·63 | + ·10 |
| + ·15 | ·72 377 | 1339·70 | ·57 440 | 1578·54 | ·40 973 | 1565·59 | ·25 444 | 1285·14 | ·13 169 | 845·22 | + ·15 |
| + ·20 | ·78 690 | 1180·56 | ·65 192 | 1512·94 | ·49 003 | 1637·00 | ·32 343 | 1470·71 | ·17 925 | 1062·25 | + ·20 |
| + ·25 | ·84 147 | 999·97 | ·72 481 | 1394·42 | ·57 243 | 1648·23 | ·40 098 | 1624·60 | ·23 811 | 1293·27 | + ·25 |
| + ·30 | ·88 677 | 811·85 | ·79 063 | 1232·37 | ·65 375̄ | 1593·57 | ·48 504 | 1727·60 | ·30 854 | 1521·54 | + ·30 |
| + ·35 | ·92 275̄ | 629·39 | ·84 754 | 1040·48 | ·73 070 | 1473·96 | ·57 260 | 1762·12 | ·38 984 | 1723·97 | + ·35 |
| + ·40 | ·94 999 | 463·60 | ·89 445̄ | 835·02 | ·80 020 | 1297·82 | ·65 990 | 1715·64 | ·48 004 | 1872·57 | + ·40 |
| + ·45 | ·96 952 | 322·33 | ·93 110 | 632·84 | ·85 979 | 1080·74 | ·74 274 | 1584·29 | ·57 571 | 1937·95 | + ·45 |
| + ·50 | ·98 270 | 209·71 | ·95 804 | 449·00 | ·90 794 | 843·82 | ·81 703 | 1375·79 | ·67 203 | 1895·25 | + ·50 |
| + ·55 | ·99 098 | 126·18 | ·97 650̄ | 294·75 | ·94 424 | 610·56 | ·87 937 | 1110·63 | ·76 321 | 1731·96 | + ·55 |
| + ·60 | ·99 576 | 69·10 | ·98 812 | 176·17 | ·96 943 | 402·90 | ·92 767 | 820·29 | ·84 333 | 1456·10 | + ·60 |
| + ·65 | ·99 825̄ | 33·66 | ·99 472 | 93·71 | ·98 523 | 236·98 | ·96 160 | 541·90 | ·90 749 | 1101·45 | + ·65 |
| + ·70 | ·99 939 | 14·10 | ·99 802 | 42·89 | ·99 395 | 120·15 | ·98 264 | 309·66 | ·95 311 | 725·26 | + ·70 |
| + ·75 | ·99 981 | 4·82 | ·99 941 | 16·03 | ·99 802 | 49·82 | ·99 371 | 145·26 | ·98 079 | 394·72 | + ·75 |
| + ·80 | ·99 997 | 1·23 | ·99 987 | 4·48 | ·99 953 | 15·47 | ·99 835̄ | 51·23 | ·99 425̄ | 162·68 | + ·80 |
| + ·85 | ·99 999 | ·20 | ·99 998 | ·79 | ·99 994 | 3·05 | ·99 974 | 11·51 | ·99 896 | 43·06 | + ·85 |
| + ·90 | 1·00 000 | ·01 | 1·00 000 | ·05 | 1·00 000 | ·26 | ·99 998 | 1·14 | ·99 992 | 5·05 | + ·90 |
| + ·95 | 1·00 000 | ·00 | 1·00 000 | ·00 | 1·00 000 | ·00 | 1·00 000 | ·01 | 1·00 000 | ·08 | + ·95 |
| +1·00 | 1·00 000 | ·00 | 1·00 000 | ·00 | 1·00 000 | ·00 | 1·00 000 | ·00 | 1·00 000 | ·00 | +1·00 |

$n = 18$ ρ=0.5 — 0.9

| r | ρ=0.5 | | ρ=0.6 | | ρ=0.7 | | ρ=0.8 | |
|---|---|---|---|---|---|---|---|---|
| | Area | Ordinate | Area | Ordinate | Area | Ordinate | Area | Ordinate |
| −1.00 | .00 000 | .00 | .00 000 | .00 | .00 000 | .00 | .00 000 | .00 |
| − .95 | .00 000 | .00 | .00 000 | .00 | .00 000 | .00 | .00 000 | .00 |
| − .90 | .00 000 | .00 | .00 000 | .00 | .00 000 | .00 | .00 000 | .00 |
| − .85 | .00 000 | .00 | .00 000 | .00 | .00 000 | .00 | .00 000 | .00 |
| − .80 | .00 000 | .00 | .00 000 | .00 | .00 000 | .00 | .00 000 | .00 |
| − .75 | .00 000 | .00 | .00 000 | .00 | .00 000 | .00 | .00 000 | .00 |
| − .70 | .00 000 | .01 | .00 000 | .00 | .00 000 | .00 | .00 000 | .00 |
| − .65 | .00 000 | .03 | .00 000 | .00 | .00 000 | .00 | .00 000 | .00 |
| − .60 | .00 000 | .08 | .00 000 | .01 | .00 000 | .00 | .00 000 | .00 |
| − .55 | .00 001 | .20 | .00 000 | .03 | .00 000 | .00 | .00 000 | .00 |
| − .50 | .00 002 | .46 | .00 000 | .06 | .00 000 | .00 | .00 000 | .00 |
| − .45 | .00 006 | .98 | .00 001 | .14 | .00 000 | .01 | .00 000 | .00 |
| − .40 | .00 013 | 1.98 | .00 002 | .30 | .00 000 | .03 | .00 000 | .00 |
| − .35 | .00 027 | 3.80 | .00 004 | .61 | .00 000 | .06 | .00 000 | .00 |
| − .30 | .00 053 | 7.01 | .00 008 | 1.19 | .00 001 | .11 | .00 000 | .00 |
| − .25 | .00 101 | 12.40 | .00 017 | 2.24 | .00 002 | .23 | .00 000 | .01 |
| − .20 | .00 183 | 21.22 | .00 032 | 4.09 | .00 003 | .44 | .00 000 | .02 |
| − .15 | .00 322 | 35.20 | .00 060 | 7.27 | .00 006 | .84 | .00 000 | .03 |
| − .10 | .00 548 | 56.74 | .00 108 | 12.60 | .00 012 | 1.57 | .00 000 | .07 |
| − .05 | .00 907 | 89.04 | .00 191 | 21.34 | .00 023 | 2.86 | .00 001 | .14 |
| 0.00 | .01 463 | 136.21 | .00 331 | 35.38 | .00 042 | 5.13 | .00 002 | .27 |
| + .05 | .02 302 | 203.29 | .00 559 | 57.47 | .00 077 | 9.09 | .00 004 | .51 |
| + .10 | .03 539 | 296.06 | .00 925 | 91.57 | .00 138 | 15.86 | .00 007 | .98 |
| + .15 | .05 316 | 420.66 | .01 503 | 143.10 | .00 243 | 27.33 | .00 014 | 1.87 |
| + .20 | .07 008 | 582.69 | .02 397 | 219.31 | .00 424 | 46.52 | .00 027 | 3.55 |
| + .25 | .11 211 | 785.76 | .03 763 | 329.31 | .00 729 | 78.24 | .00 052 | 6.73 |
| + .30 | .15 733 | 1029.39 | .05 764 | 483.81 | .01 239 | 129.92 | .00 099 | 12.75 |
| + .35 | .21 561 | 1306.25 | .08 683 | 693.90 | .02 080 | 212.79 | .00 189 | 24.12 |
| + .40 | .28 823 | 1598.97 | .12 811 | 968.37 | .03 446 | 343.11 | .00 357 | 45.64 |
| + .45 | .37 528 | 1877.54 | .18 478 | 1308.88 | .05 627 | 543.00 | .00 677 | 86.32 |
| + .50 | .47 503 | 2098.72 | .25 989 | 1702.25 | .09 037 | 839.57 | .01 280 | 162.94 |
| + .55 | .58 332 | 2209.80 | .35 527 | 2110.51 | .14 329 | 1259.46 | .02 416 | 306.15 |
| + .60 | .69 331 | 2159.46 | .47 001 | 2461.52 | .21 859 | 1813.97 | .04 540 | 569.59 |
| + .625 | .74 619 | 2062.69 | .53 317 | 2583.07 | .26 789 | 2133.84 | .06 205 | 772.00 |
| + .650 | .79 604 | 1917.07 | .59 869 | 2648.10 | .32 540 | 2468.15 | .08 455 | 1039.63 |
| + .675 | .84 166 | 1726.03 | .66 498 | 2641.33 | .39 125 | 2797.22 | .11 470 | 1387.86 |
| + .700 | .88 202 | 1497.24 | .73 006 | 2550.03 | .46 498 | 3092.09 | .15 473 | 1830.85 |
| + .725 | .91 631 | 1242.63 | .79 171 | 2367.11 | .54 525 | 3314.19 | .20 710 | 2376.50 |
| + .750 | .94 407 | 977.87 | .84 766 | 2094.49 | .62 970 | 3417.34 | .27 435 | 3017.33 |
| + .775 | .96 527 | 720.96 | .89 580 | 1746.21 | .71 473 | 3353.97 | .35 847 | 3715.97 |
| + .800 | .98 033 | 489.96 | .93 457 | 1350.14 | .79 569 | 3087.13 | .45 993 | 4385.61 |
| + .825 | .99 011 | 299.95 | .96 325 | 946.60 | .86 731 | 2608.74 | .57 624 | 4871.65 |
| + .850 | .99 575 | 159.94 | .98 223 | 582.14 | .92 567 | 1960.44 | .70 014 | 4953.44 |
| + .875 | .99 853 | 70.47 | .99 303 | 298.13 | .96 477 | 1245.89 | .81 868 | 4405.41 |
| + .900 | .99 963 | 23.50 | .99 799 | 116.49 | .98 769 | 614.60 | .91 453 | 3164.79 |
| + .925 | .99 994 | 5.02 | .99 965 | 29.45 | .99 731 | 200.00 | .97 379 | 1572.65 |
| + .950 | .99 999 | .47 | .99 997 | 3.32 | .99 978 | 29.69 | .99 654 | 377.45 |
| + .975 | 1.00 000 | .01 | 1.00 000 | .05 | 1.00 000 | .61 | .99 995 | 13.55 |
| +1.000 | 1.00 000 | .00 | 1.00 000 | .00 | 1.00 000 | .00 | 1.00 000 | .00 |

| r | ρ=0.9 | |
|---|---|---|
| | Area | Ordinate |
| −1.00 | .00 000 | .00 |
| − .95 | .00 000 | .00 |
| − .90 | .00 000 | .00 |
| − .85 | .00 000 | .00 |
| − .80 | .00 000 | .00 |
| − .75 | .00 000 | .00 |
| − .70 | .00 000 | .00 |
| − .65 | .00 000 | .00 |
| − .60 | .00 000 | .00 |
| − .55 | .00 000 | .00 |
| − .50 | .00 000 | .00 |
| − .45 | .00 000 | .00 |
| − .40 | .00 000 | .00 |
| − .35 | .00 000 | .00 |
| − .30 | .00 000 | .00 |
| − .25 | .00 000 | .00 |
| − .20 | .00 000 | .00 |
| − .15 | .00 000 | .00 |
| − .10 | .00 000 | .00 |
| − .05 | .00 000 | .00 |
| 0.00 | .00 000 | .00 |
| + .05 | .00 000 | .00 |
| + .10 | .00 000 | .01 |
| + .15 | .00 000 | .01 |
| + .20 | .00 000 | .02 |
| + .25 | .00 000 | .05 |
| + .30 | .00 001 | .11 |
| + .35 | .00 001 | .24 |
| + .40 | .00 003 | .54 |
| + .45 | .00 007 | 1.26 |
| + .50 | .00 018 | 3.00 |
| + .55 | .00 042 | 7.37 |
| + .60 | .00 103 | 18.84 |
| + .625 | .00 163 | 30.59 |
| + .650 | .00 262 | 50.19 |
| + .675 | .00 425 | 83.28 |
| + .700 | .00 698 | 139.75 |
| + .725 | .01 158 | 237.07 |
| + .750 | .01 942 | 406.10 |
| + .775 | .03 291 | 700.86 |
| + .800 | .05 627 | 1213.45 |
| + .810 | .06 983 | 1510.34 |
| + .820 | .08 671 | 1877.20 |
| + .830 | .10 765 | 2327.65 |
| + .840 | .13 358 | 2875.86 |
| + .850 | .16 554 | 3534.99 |
| + .860 | .20 469 | 4314.15 |
| + .870 | .25 223 | 5213.60 |
| + .880 | .30 931 | 6217.11 |
| + .890 | .37 678 | 7281.10 |
| + .900 | .45 485 | 8320.69 |
| + .910 | .54 265 | 9195.77 |
| + .920 | .63 755 | 9702.95 |
| + .930 | .73 465 | 9591.97 |
| + .940 | .82 657 | 8630.30 |
| + .950 | .90 413 | 6744.32 |
| + .955 | .93 486 | 5525.22 |
| + .960 | .95 924 | 4220.17 |
| + .965 | .97 710 | 2939.44 |
| + .970 | .98 888 | 1806.72 |
| + .975 | .99 560 | 930.81 |
| + .980 | .99 872 | 368.46 |
| + .985 | .99 977 | 95.09 |
| + .990 | .99 998 | 11.04 |
| + .995 | 1.00 000 | .18 |
| +1.000 | 1.00 000 | .00 |

| r | ρ=0.9 | |
|---|---|---|
| | Area | Ordinate |
| + .800 | .05 627 | 1213.45 |
| + .825 | .09 662 | 2089.13 |
| + .850 | .16 554 | 3534.99 |
| + .875 | .27 951 | 5704.28 |
| + .900 | .45 485 | 8320.69 |
| + .925 | .68 623 | 9740.91 |
| + .950 | .90 413 | 6744.32 |
| + .975 | .99 560 | 930.81 |
| +1.000 | 1.00 000 | .00 |

ρ=0·0 —— 0·4                                   n=19

| r | ρ=0·0 Area | ρ=0·0 Ordinate | ρ=0·1 Area | ρ=0·1 Ordinate | ρ=0·2 Area | ρ=0·2 Ordinate | ρ=0·3 Area | ρ=0·3 Ordinate | ρ=0·4 Area | ρ=0·4 Ordinate | r |
|---|---|---|---|---|---|---|---|---|---|---|---|
| −1·00 | ·00 000 | ·00 | ·00 000 | ·00 | ·00 000 | ·00 | ·00 000 | ·00 | ·00 000 | ·00 | −1·00 |
| − ·95 | ·00 000 | ·00 | ·00 000 | ·00 | ·00 000 | ·00 | ·00 000 | ·00 | ·00 000 | ·00 | − ·95 |
| − ·90 | ·00 000 | ·01 | ·00 000 | ·00 | ·00 000 | ·00 | ·00 000 | ·00 | ·00 000 | ·00 | − ·90 |
| − ·85 | ·00 000 | ·11 | ·00 000 | ·02 | ·00 000 | ·00 | ·00 000 | ·00 | ·00 000 | ·00 | − ·85 |
| − ·80 | ·00 002 | ·76 | ·00 001 | ·18 | ·00 000 | ·04 | ·00 000 | ·01 | ·00 000 | ·00 | − ·80 |
| − ·75 | ·00 011 | 3·29 | ·00 003 | ·85 | ·00 001 | ·20 | ·00 000 | ·04 | ·00 000 | ·01 | − ·75 |
| − ·70 | ·00 042 | 10·39 | ·00 012 | 2·90 | ·00 003 | ·73 | ·00 001 | ·16 | ·00 000 | ·03 | − ·70 |
| − ·65 | ·00 130 | 26·39 | ·00 037 | 8·00 | ·00 010 | 2·15 | ·00 002 | ·50 | ·00 000 | ·10 | − ·65 |
| − ·60 | ·00 331 | 57·03 | ·00 101 | 18·78 | ·00 028 | 5·43 | ·00 007 | 1·35 | ·00 001 | ·27 | − ·60 |
| − ·55 | ·00 735 | 108·73 | ·00 240 | 38·90 | ·00 069 | 12·11 | ·00 017 | 3·21 | ·00 004 | ·70 | − ·55 |
| − ·50 | ·01 463 | 187·37 | ·00 513 | 72·86 | ·00 158 | 24·46 | ·00 042 | 6·94 | ·00 009 | 1·60 | − ·50 |
| − ·45 | ·02 660 | 296·98 | ·01 000 | 125·54 | ·00 328 | 45·49 | ·00 092 | 13·84 | ·00 021 | 3·41 | − ·45 |
| − ·40 | ·04 486 | 438·38 | ·01 807 | 201·54 | ·00 633 | 78·91 | ·00 188 | 25·79 | ·00 046 | 6·79 | − ·40 |
| − ·35 | ·07 092 | 608·28 | ·03 060 | 304·27 | ·01 145 | 128·86 | ·00 362 | 45·32 | ·00 093 | 12·77 | − ·35 |
| − ·30 | ·10 604 | 799·03 | ·04 897 | 435·05 | ·01 956 | 199·51 | ·00 660 | 75·63 | ·00 180 | 22·88 | − ·30 |
| − ·25 | ·15 098 | 998·93 | ·07 455 | 592·25 | ·03 181 | 294·45 | ·01 143 | 120·51 | ·00 332 | 39·21 | − ·25 |
| − ·20 | ·20 584 | 1193·40 | ·10 855 | 770·79 | ·04 945 | 415·93 | ·01 896 | 184·13 | ·00 587 | 64·59 | − ·20 |
| − ·15 | ·26 996 | 1366·56 | ·15 184 | 961·93 | ·07 384 | 564·06 | ·03 022 | 270·60 | ·00 999 | 102·59 | − ·15 |
| − ·10 | ·34 189 | 1503·20 | ·20 475 | 1153·66 | ·10 626 | 736·02 | ·04 646 | 383·39 | ·01 641 | 157·51 | − ·10 |
| − ·05 | ·41 946 | 1590·74 | ·26 697 | 1331·66 | ·14 774 | 925·52 | ·06 904 | 524·53 | ·02 611 | 234·18 | − ·05 |
| 0·00 | ·50 000 | 1620·88 | ·33 744 | 1480·70 | ·19 894 | 1122·52 | ·09 938 | 693·62 | ·04 028 | 337·49 | 0·00 |
| + ·05 | ·58 054 | 1590·74 | ·41 432 | 1586·45 | ·25 990 | 1313·57 | ·13 680 | 886·92 | ·06 037 | 471·76 | + ·05 |
| + ·10 | ·65 811 | 1503·20 | ·49 516 | 1637·37 | ·32 993 | 1482·73 | ·18 834 | 1096·42 | ·08 801 | 639·60 | + ·10 |
| + ·15 | ·73 004 | 1366·56 | ·57 703 | 1626·51 | ·40 752 | 1613·08 | ·24 851 | 1309·42 | ·12 489 | 840·60 | + ·15 |
| + ·20 | ·79 416 | 1193·40 | ·65 676 | 1552·77 | ·49 033 | 1688·90 | ·31 906 | 1508·71 | ·17 255 | 1069·70 | + ·20 |
| + ·25 | ·84 902 | 998·93 | ·73 134 | 1421·51 | ·57 530 | 1698·12 | ·39 881 | 1673·64 | ·23 214 | 1315·58 | + ·25 |
| + ·30 | ·89 396 | 799·03 | ·79 814 | 1244·13 | ·65 892 | 1634·75 | ·48 548 | 1782·33 | ·30 408 | 1559·56 | + ·30 |
| + ·35 | ·92 908 | 608·28 | ·85 525 | 1036·82 | ·73 759 | 1500·76 | ·57 577 | 1815·10 | ·38 763 | 1775·54 | + ·35 |
| + ·40 | ·95 514 | 438·38 | ·90 164 | 818·35 | ·80 800 | 1306·92 | 66 550 | 1758·51 | ·48 064 | 1931·83 | + ·40 |
| + ·45 | ·97 340 | 296·98 | ·93 722 | 607·48 | ·86 760 | 1072·08 | ·75 008 | 1609·70 | ·57 927 | 1995·55 | + ·45 |
| + ·50 | ·98 537 | 187·37 | ·96 278 | 420·16 | ·91 493 | 820·76 | ·82 510 | 1379·49 | ·67 819 | 1939·87 | + ·50 |
| + ·55 | ·99 265 | 108·73 | ·97 981 | 267·40 | ·94 984 | 579·15 | ·88 708 | 1093·22 | ·77 106 | 1753·38 | + ·55 |
| + ·60 | ·99 669 | 57·03 | ·99 018 | 153·91 | ·97 339 | 370·23 | ·93 410 | 787·57 | ·85 155 | 1449·19 | + ·60 |
| + ·65 | ·99 870 | 26·39 | ·99 583 | 78·18 | ·98 766 | 209·24 | ·96 620 | 503·43 | ·91 471 | 1069·45 | + ·65 |
| + ·70 | ·99 958 | 10·39 | ·99 852 | 33·81 | ·99 520 | 100·85 | ·98 539 | 275·47 | ·95 834 | 680·14 | + ·70 |
| + ·75 | ·99 989 | 3·29 | ·99 959 | 11·76 | ·99 853 | 39·19 | ·99 500 | 122·01 | ·98 377 | 352·63 | + ·75 |
| + ·80 | ·99 998 | ·76 | ·99 992 | 3·00 | ·99 968 | 11·17 | ·99 879 | 39·80 | ·99 547 | 135·72 | + ·80 |
| + ·85 | 1·00 000 | ·11 | ·99 999 | ·47 | ·99 996 | 1·96 | ·99 983 | 8·01 | ·99 926 | 32·49 | + ·85 |
| + ·90 | 1·00 000 | ·01 | 1·00 000 | ·02 | 1·00 000 | ·14 | ·99 999 | ·67 | ·99 995 | 3·25 | + ·90 |
| + ·95 | 1·00 000 | ·00 | 1·00 000 | ·00 | 1·00 000 | ·00 | 1·00 000 | ·01 | 1·00 000 | ·04 | + ·95 |
| +1·00 | 1·00 000 | ·00 | 1·00 000 | ·00 | 1·00 000 | ·00 | 1·00 000 | ·00 | 1·00 000 | ·00 | +1·00 |

$n = 19$      $\rho = 0.5 — 0.9$

| r | ρ=0.5 Area | ρ=0.5 Ordinate | ρ=0.6 Area | ρ=0.6 Ordinate | ρ=0.7 Area | ρ=0.7 Ordinate | ρ=0.8 Area | ρ=0.8 Ordinate | r | ρ=0.9 Area | ρ=0.9 Ordinate |
|---|---|---|---|---|---|---|---|---|---|---|---|
| −1·00 | ·00 000 | ·00 | ·00 000 | ·00 | ·00 000 | ·00 | ·00 000 | ·00 | −1·00 | ·00 000 | ·00 |
| − ·95 | ·00 000 | ·00 | ·00 000 | ·00 | ·00 000 | ·00 | ·00 000 | ·00 | − ·95 | ·00 000 | ·00 |
| − ·90 | ·00 000 | ·00 | ·00 000 | ·00 | ·00 000 | ·00 | ·00 000 | ·00 | − ·90 | ·00 000 | ·00 |
| − ·85 | ·00 000 | ·00 | ·00 000 | ·00 | ·00 000 | ·00 | ·00 000 | ·00 | − ·85 | ·00 000 | ·00 |
| − ·80 | ·00 000 | ·00 | ·00 000 | ·00 | ·00 000 | ·00 | ·00 000 | ·00 | − ·80 | ·00 000 | ·00 |
| − ·75 | ·00 000 | ·00 | ·00 000 | ·00 | ·00 000 | ·00 | ·00 000 | ·00 | − ·75 | ·00 000 | ·00 |
| − ·70 | ·00 000 | ·00 | ·00 000 | ·00 | ·00 000 | ·00 | ·00 000 | ·00 | − ·70 | ·00 000 | ·00 |
| − ·65 | ·00 000 | ·01 | ·00 000 | ·00 | ·00 000 | ·00 | ·00 000 | ·00 | − ·65 | ·00 000 | ·00 |
| − ·60 | ·00 000 | ·04 | ·00 000 | ·00 | ·00 000 | ·00 | ·00 000 | ·00 | − ·60 | ·00 000 | ·00 |
| − ·55 | ·00 000 | ·12 | ·00 000 | ·01 | ·00 000 | ·00 | ·00 000 | ·00 | − ·55 | ·00 000 | ·00 |
| − ·50 | ·00 001 | ·28 | ·00 000 | ·03 | ·00 000 | ·00 | ·00 000 | ·00 | − ·50 | ·00 000 | ·00 |
| − ·45 | ·00 004 | ·64 | ·00 000 | ·08 | ·00 000 | ·01 | ·00 000 | ·00 | − ·45 | ·00 000 | ·00 |
| − ·40 | ·00 009 | 1·35 | ·00 001 | ·18 | ·00 000 | ·01 | ·00 000 | ·00 | − ·40 | ·00 000 | ·00 |
| − ·35 | ·00 018 | 2·71 | ·00 002 | ·39 | ·00 000 | ·03 | ·00 000 | ·00 | − ·35 | ·00 000 | ·00 |
| − ·30 | ·00 037 | 5·19 | ·00 005 | ·79 | ·00 000 | ·07 | ·00 000 | ·00 | − ·30 | ·00 000 | ·00 |
| − ·25 | ·00 073 | 9·54 | ·00 011 | 1·56 | ·00 001 | ·14 | ·00 000 | ·00 | − ·25 | ·00 000 | ·00 |
| − ·20 | ·00 138 | 16·89 | ·00 022 | 2·96 | ·00 002 | ·28 | ·00 000 | ·01 | − ·20 | ·00 000 | ·00 |
| − ·15 | ·00 250 | 28·93 | ·00 042 | 5·45 | ·00 004 | ·56 | ·00 000 | ·02 | − ·15 | ·00 000 | ·00 |
| − ·10 | ·00 439 | 48·04 | ·00 079 | 9·76 | ·00 008 | 1·07 | ·00 000 | ·04 | − ·10 | ·00 000 | ·00 |
| − ·05 | ·00 747 | 77·52 | ·00 145 | 17·08 | ·00 015 | 2·03 | ·00 001 | ·08 | − ·05 | ·00 000 | ·00 |
| 0·00 | ·01 238 | 121·70 | ·00 258 | 29·20 | ·00 030 | 3·78 | ·00 001 | ·16 | 0·00 | ·00 000 | ·00 |
| + ·05 | ·01 998 | 186·07 | ·00 449 | 48·84 | ·00 056 | 6·93 | ·00 002 | ·33 | + ·05 | ·00 000 | ·00 |
| + ·10 | ·03 143 | 277·05 | ·00 766 | 80·00 | ·00 103 | 12·50 | ·00 005 | ·66 | + ·10 | ·00 000 | ·00 |
| + ·15 | ·04 825 | 401·73 | ·01 278 | 128·32 | ·00 187 | 22·24 | ·00 009 | 1·30 | + ·15 | ·00 000 | ·01 |
| + ·20 | ·07 228 | 566·77 | ·02 090 | 201·53 | ·00 337 | 39·05 | ·00 019 | 2·57 | + ·20 | ·00 000 | ·01 |
| + ·25 | ·10 568 | 776·05 | ·03 351 | 309·60 | ·00 597 | 67·65 | ·00 037 | 5·04 | + ·25 | ·00 000 | ·03 |
| + ·30 | ·15 072 | 1032·16 | ·05 264 | 464·51 | ·01 045 | 115·58 | ·00 073 | 9·90 | + ·30 | ·00 000 | ·06 |
| + ·35 | ·20 953 | 1325·12 | ·08 096 | 679·05 | ·01 804 | 194·50 | ·00 144 | 19·43 | + ·35 | ·00 001 | ·15 |
| + ·40 | ·28 355 | 1636·61 | ·12 172 | 963·78 | ·03 071 | 391·76 | ·00 282 | 38·08 | + ·40 | ·00 002 | ·35 |
| + ·45 | ·37 294 | 1932·88 | ·17 856 | 1321·43 | ·05 143 | 521·0 | ·00 554 | 74·55 | + ·45 | ·00 005 | ·85 |
| + ·50 | ·47 578 | 2165·06 | ·25 486 | 1738·01 | ·08 457 | 824·05 | ·01 085 | 145·56 | + ·50 | ·00 012 | 2·12 |
| + ·55 | ·58 740 | 2274·19 | ·35 266 | 2171·08 | ·13 606 | 1259·96 | ·02 118 | 282·59 | + ·55 | ·00 029 | 5·48 |
| + ·60 | ·70 021 | 2204·82 | ·47 091 | 2539·21 | ·21 305 | 1843·14 | ·04 113 | 542·34 | + ·60 | ·00 076 | 14·73 |
| + ·625 | ·75 403 | 2092·36 | ·53 604 | 2662·44 | ·26 330 | 2181·45 | ·05 711 | 745·94 | + ·625 | ·00 124 | 24·54 |
| + ·650 | ·80 439 | 1928·16 | ·60 350 | 2722·44 | ·32 225 | 2535·20 | ·07 901 | 1018·65 | + ·650 | ·00 205 | 41·32 |
| + ·675 | ·85 005 | 1717·28 | ·67 150 | 2702·90 | ·39 001 | 2882·11 | ·10 876 | 1377·67 | + ·675 | ·00 341 | 70·39 |
| + ·700 | ·88 995 | 1469·57 | ·73 788 | 2591·04 | ·46 603 | 3189·48 | ·14 874 | 1839·04 | + ·700 | ·00 574 | 121·27 |
| + ·725 | ·92 334 | 1199·35 | ·80 023 | 2381·23 | ·54 879 | 3414·15 | ·20 164 | 2411·86 | + ·725 | ·00 979 | 211·25 |
| + ·750 | ·94 988 | 924·51 | ·85 616 | 2078·60 | ·63 562 | 3505·34 | ·27 020 | 3087·80 | + ·750 | ·01 689 | 371·57 |
| + ·775 | ·96 969 | 664·53 | ·90 354 | 1702·14 | ·72 250 | 3412·73 | ·35 656 | 3824·48 | + ·775 | ·02 940 | 658·26 |
| + ·800 | ·98 338 | 437·70 | ·94 093 | 1285·55 | ·80 442 | 3100·93 | ·46 114 | 4523·43 | + ·800 | ·05 165 | 1169·00 |
| + ·825 | ·99 197 | 257·76 | ·96 787 | 874·15 | ·87 568 | 2570·34 | ·58 099 | 5011·10 | + ·810 | ·06 478 | 1469·33 |
| + ·850 | ·99 672 | 130·90 | ·98 509 | 516·43 | ·93 158 | 1878·29 | ·70 792 | 5046·26 | + ·820 | ·08 128 | 1843·64 |
| + ·875 | ·99 893 | 54·18 | ·99 447 | 250·74 | ·96 926 | 1146·55 | ·82 756 | 4399·45 | + ·830 | ·10 195 | 2306·97 |
| + ·900 | ·99 975 | 16·63 | ·99 852 | 91·09 | ·98 990 | 533·33 | ·92 189 | 3048·86 | + ·840 | ·12 777 | 2875·01 |
| + ·925 | ·99 996 | 3·17 | ·99 976 | 20·75 | ·99 796 | 158·79 | ·97 757 | 1422·24 | + ·850 | ·15 985 | 3562·39 |
| + ·950 | ·99 999 | ·25 | ·99 998 | 1·99 | ·99 985 | 20·38 | ·99 737 | 303·89 | + ·860 | ·19 945 | 4379·21 |
| + ·975 | 1·00 000 | ·00 | 1·00 000 | ·02 | 1·00 000 | ·32 | ·99 997 | 8·47 | + ·870 | ·24 787 | 5325·46 |
| +1·000 | 1·00 000 | ·00 | 1·00 000 | ·00 | 1·00 000 | ·00 | 1·00 000 | ·00 | + ·880 | ·30 633 | 6382·33 |
| | | | | | | | | | + ·890 | ·37 572 | 7499·87 |
| | | | | | | | | | + ·900 | ·45 622 | 8581·50 |
| | | | | | | | | | + ·910 | ·54 672 | 9469·48 |
| | | | | | | | | | + ·920 | ·64 422 | 9939·12 |
| | | | | | | | | | + ·930 | ·74 321 | 9723·50 |
| | | | | | | | | | + ·940 | ·83 562 | 8595·18 |
| | | | | | | | | | + ·950 | ·91 194 | 6528·96 |
| | | | | | | | | | + ·955 | ·94 141 | 5243·52 |
| | | | | | | | | | + ·960 | ·96 429 | 3905·87 |
| | | | | | | | | | + ·965 | ·98 058 | 2635·25 |
| | | | | | | | | | + ·970 | ·99 095 | 1554·71 |
| | | | | | | | | | + ·975 | ·99 661 | 759·00 |
| | | | | | | | | | + ·980 | ·99 908 | 279·33 |
| | | | | | | | | | + ·985 | ·99 985 | 64·98 |
| | | | | | | | | | + ·990 | ·99 999 | 6·42 |
| | | | | | | | | | + ·995 | 1·00 000 | ·08 |
| | | | | | | | | | +1·000 | 1·00 000 | ·00 |

| r | ρ=0.9 Area | ρ=0.9 Ordinate |
|---|---|---|
| + ·800 | ·05 165 | 1169·00 |
| + ·825 | ·09 104 | 2061·26 |
| + ·850 | ·15 985 | 3562·39 |
| + ·875 | ·27 578 | 5842·29 |
| + ·900 | ·45 622 | 8581·50 |
| + ·925 | ·69 398 | 9933·60 |
| + ·950 | ·91 194 | 6528·96 |
| + ·975 | ·99 661 | 759·00 |
| +1·000 | 1·00 000 | ·00 |

ρ=0·0—0·4                                        n=20

| r | ρ=0·0 Area | ρ=0·0 Ordinate | ρ=0·1 Area | ρ=0·1 Ordinate | ρ=0·2 Area | ρ=0·2 Ordinate | ρ=0·3 Area | ρ=0·3 Ordinate | ρ=0·4 Area | ρ=0·4 Ordinate | r |
|---|---|---|---|---|---|---|---|---|---|---|---|
| −1·00 | ·00 000 | ·00 | ·00 000 | ·00 | ·00 000 | ·00 | ·00 000 | ·00 | ·00 000 | ·00 | −1·00 |
| − ·95 | ·00 000 | ·00 | ·00 000 | ·00 | ·00 000 | ·00 | ·00 000 | ·00 | ·00 000 | ·00 | − ·95 |
| − ·90 | ·00 000 | ·00 | ·00 000 | ·00 | ·00 000 | ·00 | ·00 000 | ·00 | ·00 000 | ·00 | − ·90 |
| − ·85 | ·00 000 | ·06 | ·00 000 | ·01 | ·00 000 | ·00 | ·00 000 | ·00 | ·00 000 | ·00 | − ·85 |
| − ·80 | ·00 001 | ·47 | ·00 000 | ·10 | ·00 000 | ·02 | ·00 000 | ·00 | ·00 000 | ·00 | − ·80 |
| − ·75 | ·00 007 | 2·24 | ·00 001 | ·53 | ·00 000 | ·11 | ·00 000 | ·02 | ·00 000 | ·00 | − ·75 |
| − ·70 | ·00 029 | 7·64 | ·00 007 | 1·99 | ·00 002 | ·46 | ·00 000 | ·10 | ·00 000 | ·02 | − ·70 |
| − ·65 | ·00 096 | 20·65 | ·00 025 | 5·85 | ·00 006 | 1·46 | ·00 001 | ·31 | ·00 000 | ·05 | − ·65 |
| − ·60 | ·00 258 | 46·98 | ·00 074 | 14·53 | ·00 019 | 3·91 | ·00 004 | ·90 | ·00 001 | ·17 | − ·60 |
| − ·55 | ·00 599 | 93·51 | ·00 185 | 31·56 | ·00 050 | 9·20 | ·00 012 | 2·26 | ·00 002 | ·45 | − ·55 |
| − ·50 | ·01 238 | 167·11 | ·00 411 | 61·57 | ·00 119 | 19·43 | ·00 029 | 5·14 | ·00 006 | 1·09 | − ·50 |
| − ·45 | ·02 325 | 273·13 | ·00 831 | 109·93 | ·00 257 | 37·61 | ·00 067 | 10·70 | ·00 014 | 2·44 | − ·45 |
| − ·40 | ·04 028 | 413·76 | ·01 550 | 181·99 | ·00 514 | 67·57 | ·00 144 | 20·74 | ·00 032 | 5·06 | − ·40 |
| − ·35 | ·06 517 | 586·81 | ·02 698 | 282·18 | ·00 960 | 113·83 | ·00 286 | 37·75 | ·00 069 | 9·91 | − ·35 |
| − ·30 | ·09 938 | 784·96 | ·04 423 | 412·86 | ·01 688 | 181·17 | ·00 538 | 65·03 | ·00 137 | 18·39 | − ·30 |
| − ·25 | ·14 388 | 996·07 | ·06 876 | 573·27 | ·02 814 | 273·98 | ·00 960 | 106·64 | ·00 262 | 32·57 | − ·25 |
| − ·20 | ·19 894 | 1204·17 | ·10 197 | 758·68 | ·04 475 | 395·39 | ·01 636 | 167·20 | ·00 477 | 55·31 | − ·20 |
| − ·15 | ·26 395 | 1391·40 | ·14 490 | 960·11 | ·06 818 | 546·33 | ·02 672 | 251·51 | ·00 835 | 90·32 | − ·15 |
| − ·10 | ·33 744 | 1540·28 | ·19 803 | 1164·55 | ·09 984 | 724·46 | ·04 198 | 363·84 | ·01 409 | 142·24 | − ·10 |
| − ·05 | ·41 709 | 1636·14 | ·26 114 | 1356·03 | ·14 097 | 923·47 | ·06 362 | 507·05 | ·02 295 | 216·43 | − ·05 |
| 0·00 | ·50 000 | 1669·24 | ·33 313 | 1517·23 | ·19 236 | 1132·65 | ·09 321 | 681·41 | ·03 619 | 318·54 | 0·00 |
| + ·05 | ·58 291 | 1636·14 | ·41 208 | 1631·71 | ·25 409 | 1337·14 | ·13 223 | 883·46 | ·05 536 | 453·78 | + ·05 |
| + ·10 | ·66 256 | 1540·28 | ·49 530 | 1686·21 | ·32 570 | 1518·98 | ·18 188 | 1104·85 | ·08 219 | 625·68 | + ·10 |
| + ·15 | ·73 605 | 1391·40 | ·57 957 | 1672·86 | ·40 536 | 1658·98 | ·24 280 | 1331·72 | ·11 854 | 834·48 | + ·15 |
| + ·20 | ·80 106 | 1204·17 | ·66 144 | 1590·74 | ·49 060 | 1739·26 | ·31 482 | 1544·87 | ·16 617 | 1075·23 | + ·20 |
| + ·25 | ·85 612 | 996·07 | ·73 760 | 1446·47 | ·57 806 | 1746·32 | ·39 668 | 1720·99 | ·22 640 | 1335·82 | + ·25 |
| + ·30 | ·90 062 | 784·96 | ·80 527 | 1253·70 | ·66 390 | 1673·93 | ·48 589 | 1835·44 | ·29 974 | 1595·60 | + ·30 |
| + ·35 | ·93 483 | 586·81 | ·86 248 | 1031·28 | ·74 418 | 1525·25 | ·57 882 | 1866·25 | ·38 546 | 1825·32 | + ·35 |
| + ·40 | ·95 972 | 413·76 | ·90 826 | 800·54 | ·81 537 | 1313·68 | ·67 088 | 1799·15 | ·48 118 | 1989·34 | + ·40 |
| + ·45 | ·97 675 | 273·13 | ·94 274 | 582·06 | ·87 487 | 1061·54 | ·75 707 | 1632·51 | ·58 269 | 2051·11 | + ·45 |
| + ·50 | ·98 762 | 167·11 | ·96 695 | 392·47 | ·92 132 | 796·88 | ·83 270 | 1380·68 | ·68 410 | 1981·92 | + ·50 |
| + ·55 | ·99 401 | 93·51 | ·98 264 | 242·15 | ·95 483 | 548·35 | ·89 422 | 1074·11 | ·77 852 | 1771·83 | + ·55 |
| + ·60 | ·99 742 | 46·98 | ·99 188 | 134·22 | ·97 682 | 339·60 | ·93 990 | 754·79 | ·85 923 | 1439·68 | + ·60 |
| + ·65 | ·99 904 | 20·65 | ·99 671 | 65·11 | ·98 969 | 184·41 | ·97 023 | 466·85 | ·92 129 | 1036·50 | + ·65 |
| + ·70 | ·99 970 | 7·64 | ·99 890 | 26·60 | ·99 619 | 84·50 | ·98 769 | 244·62 | ·96 294 | 636·66 | + ·70 |
| + ·75 | ·99 993 | 2·24 | ·99 972 | 8·62 | ·99 891 | 30·77 | ·99 603 | 102·29 | ·98 627 | 314·46 | + ·75 |
| + ·80 | ·99 999 | ·47 | ·99 995 | 2·00 | ·99 978 | 8·05 | ·99 911 | 30·87 | ·99 643 | 113·01 | + ·80 |
| + ·85 | 1·00 000 | ·06 | 1·00 000 | ·28 | ·99 998 | 1·25 | ·99 989 | 5·56 | ·99 947 | 24·47 | + ·85 |
| + ·90 | 1·00 000 | ·00 | 1·00 000 | ·00 | 1·00 000 | ·08 | ·99 999 | ·39 | ·99 996 | 2·09 | + ·90 |
| + ·95 | 1·00 000 | ·00 | 1·00 000 | ·00 | 1·00 000 | ·00 | 1·00 000 | ·00 | 1·00 000 | ·02 | + ·95 |
| +1·00 | 1·00 000 | ·00 | 1·00 000 | ·00 | 1·00 000 | ·00 | 1·00 000 | ·00 | 1·00 000 | ·00 | +1·00 |

$n = 20$    $\rho = 0.5 \text{---} 0.9$

| r | $\rho=0.5$ Area | Ordinate | $\rho=0.6$ Area | Ordinate | $\rho=0.7$ Area | Ordinate | $\rho=0.8$ Area | Ordinate | r | $\rho=0.9$ Area | Ordinate |
|---|---|---|---|---|---|---|---|---|---|---|---|
| −1·00 | ·00 000 | ·00 | ·00 000 | ·00 | ·00 000 | ·00 | ·00 000 | ·00 | −1·00 | ·00 000 | ·00 |
| − ·95 | ·00 000 | ·00 | ·00 000 | ·00 | ·00 000 | ·00 | ·00 000 | ·00 | − ·95 | ·00 000 | ·00 |
| − ·90 | ·00 000 | ·00 | ·00 000 | ·00 | ·00 000 | ·00 | ·00 000 | ·00 | − ·90 | ·00 000 | ·00 |
| − ·85 | ·00 000 | ·00 | ·00 000 | ·00 | ·00 000 | ·00 | ·00 000 | ·00 | − ·85 | ·00 000 | ·00 |
| − ·80 | ·00 000 | ·00 | ·00 000 | ·00 | ·00 000 | ·00 | ·00 000 | ·00 | − ·80 | ·00 000 | ·00 |
| − ·75 | ·00 000 | ·00 | ·00 000 | ·00 | ·00 000 | ·00 | ·00 000 | ·00 | − ·75 | ·00 000 | ·00 |
| − ·70 | ·00 000 | ·00 | ·00 000 | ·00 | ·00 000 | ·00 | ·00 000 | ·00 | − ·70 | ·00 000 | ·00 |
| − ·65 | ·00 000 | ·01 | ·00 000 | ·00 | ·00 000 | ·00 | ·00 000 | ·00 | − ·65 | ·00 000 | ·00 |
| − ·60 | ·00 000 | ·02 | ·00 000 | ·00 | ·00 000 | ·00 | ·00 000 | ·00 | − ·60 | ·00 000 | ·00 |
| − ·55 | ·00 000 | ·07 | ·00 000 | ·01 | ·00 000 | ·00 | ·00 000 | ·00 | − ·55 | ·00 000 | ·00 |
| − ·50 | ·00 001 | ·17 | ·00 000 | ·02 | ·00 000 | ·00 | ·00 000 | ·00 | − ·50 | ·00 000 | ·00 |
| − ·45 | ·00 002 | ·42 | ·00 000 | ·05 | ·00 000 | ·00 | ·00 000 | ·00 | − ·45 | ·00 000 | ·00 |
| − ·40 | ·00 006 | ·92 | ·00 001 | ·11 | ·00 000 | ·01 | ·00 000 | ·00 | − ·40 | ·00 000 | ·00 |
| − ·35 | ·00 012 | 1·93 | ·00 001 | ·25 | ·00 000 | ·02 | ·00 000 | ·00 | − ·35 | ·00 000 | ·00 |
| − ·30 | ·00 026 | 3·84 | ·00 003 | ·53 | ·00 000 | ·04 | ·00 000 | ·00 | − ·30 | ·00 000 | ·00 |
| − ·25 | ·00 053 | 7·32 | ·00 007 | 1·08 | ·00 001 | ·08 | ·00 000 | ·00 | − ·25 | ·00 000 | ·00 |
| − ·20 | ·00 104 | 13·42 | ·00 015 | 2·13 | ·00 001 | ·18 | ·00 000 | ·00 | − ·20 | ·00 000 | ·00 |
| − ·15 | ·00 194 | 23·73 | ·00 030 | 4·07 | ·00 003 | ·37 | ·00 000 | ·01 | − ·15 | ·00 000 | ·00 |
| − ·10 | ·00 352 | 40·60 | ·00 058 | 7·55 | ·00 005 | ·73 | ·00 000 | ·02 | − ·10 | ·00 000 | ·00 |
| − ·05 | ·00 617 | 67·37 | ·00 110 | 13·64 | ·00 010 | 1·44 | ·00 000 | ·05 | − ·05 | ·00 000 | ·00 |
| 0·00 | ·01 050 | 108·54 | ·00 202 | 24·06 | ·00 021 | 2·78 | ·00 001 | ·10 | 0·00 | ·00 000 | ·00 |
| + ·05 | ·01 736 | 169·98 | ·00 362 | 41·43 | ·00 040 | 5·27 | ·00 001 | ·21 | + ·05 | ·00 000 | ·00 |
| + ·10 | ·02 795 | 258·79 | ·00 634 | 69·76 | ·00 077 | 9·84 | ·00 003 | ·44 | + ·10 | ·00 000 | ·00 |
| + ·15 | ·04 383 | 382·94 | ·01 087 | 114·86 | ·00 145 | 18·07 | ·00 006 | ·90 | + ·15 | ·00 000 | ·00 |
| + ·20 | ·06 696 | 550·27 | ·01 824 | 184·86 | ·00 268 | 32·72 | ·00 013 | 1·85 | + ·20 | ·00 000 | ·01 |
| + ·25 | ·09 760 | 766·63 | ·02 995 | 290·54 | ·00 490 | 58·39 | ·00 027 | 3·77 | + ·25 | ·00 000 | ·02 |
| + ·30 | ·14 447 | 1033·05 | ·04 811 | 445·18 | ·00 883 | 102·63 | ·00 054 | 7·68 | + ·30 | ·00 000 | ·04 |
| + ·35 | ·20 369 | 1341·82 | ·07 553 | 663·32 | ·01 567 | 177·47 | ·00 110 | 15·62 | + ·35 | ·00 001 | ·09 |
| + ·40 | ·27 902 | 1672·08 | ·11 571 | 957·45 | ·02 738 | 301·19 | ·00 224 | 31·71 | + ·40 | ·00 001 | ·22 |
| + ·45 | ·37 064 | 1986·22 | ·17 262 | 1331·67 | ·04 703 | 499·96 | ·00 454 | 64·26 | + ·45 | ·00 003 | ·57 |
| + ·50 | ·47 646 | 2229·43 | ·24 999 | 1771·29 | ·07 919 | 807·35 | ·00 920 | 129·80 | + ·50 | ·00 008 | 1·50 |
| + ·55 | ·59 131 | 2336·21 | ·35 008 | 2229·32 | ·13 017 | 1258·17 | ·01 859 | 260·36 | + ·55 | ·00 021 | 4·07 |
| + ·60 | ·70 680 | 2247·04 | ·47 173 | 2614·58 | ·20 771 | 1869·39 | ·03 729 | 515·46 | + ·60 | ·00 056 | 11·50 |
| + ·625 | ·76 149 | 2118·60 | ·53 878 | 2739·27 | ·25 885 | 2226·09 | ·05 260 | 719·46 | + ·625 | ·00 094 | 19·65 |
| + ·650 | ·81 227 | 1935·78 | ·60 811 | 2793·79 | ·31 915 | 2599·35 | ·07 388 | 996·30 | + ·650 | ·00 160 | 33·96 |
| + ·675 | ·85 788 | 1705·47 | ·67 774 | 2760·89 | ·38 875 | 2964·19 | ·10 318 | 1365·09 | + ·675 | ·00 273 | 59·38 |
| + ·700 | ·89 725 | 1439·79 | ·74 532 | 2627·94 | ·46 699 | 3283·99 | ·14 305 | 1843·93 | + ·700 | ·00 473 | 105·05 |
| + ·725 | ·92 971 | 1155·48 | ·80 827 | 2391·09 | ·55 218 | 3510·76 | ·19 638 | 2443·32 | + ·725 | ·00 829 | 187·90 |
| + ·750 | ·95 504 | 872·48 | ·86 408 | 2059·10 | ·64 129 | 3589·11 | ·26 615 | 3154·22 | + ·750 | ·01 469 | 339·37 |
| + ·775 | ·97 352 | 611·41 | ·91 063 | 1656·19 | ·72 995 | 3466·24 | ·35 465 | 3929·07 | + ·775 | ·02 629 | 617·15 |
| + ·800 | ·98 595 | 390·31 | ·94 662 | 1221·83 | ·81 264 | 3109·18 | ·46 224 | 4657·18 | + ·800 | ·04 744 | 1124·15 |
| + ·825 | ·99 347 | 221·09 | ·97 188 | 805·78 | ·88 352 | 2527·94 | ·58 553 | 5145·27 | + ·810 | ·06 014 | 1426·86 |
| + ·850 | ·99 748 | 106·93 | ·98 748 | 457·32 | ·93 776 | 1796·33 | ·71 531 | 5131·57 | + ·820 | ·07 624 | 1807·43 |
| + ·875 | ·99 923 | 41·58 | ·99 560 | 210·50 | ·97 323 | 1053·23 | ·83 593 | 4385·60 | + ·830 | ·09 660 | 2282·36 |
| + ·900 | ·99 984 | 11·76 | ·99 891 | 71·10 | ·99 171 | 461·97 | ·92 854 | 2931·90 | + ·840 | ·12 226 | 2868·99 |
| | | | | | | | | | + ·850 | ·15 441 | 3583·56 |
| + ·925 | ·99 998 | 2·00 | ·99 984 | 14·59 | ·99 851 | 125·85 | ·98 078 | 1283·91 | + ·860 | ·19 439 | 4437·27 |
| + ·950 | 1·00 000 | ·13 | ·99 999 | 1·19 | ·99 990 | 13·97 | ·99 800 | 244·23 | + ·870 | ·24 362 | 5429·95 |
| + ·975 | 1·00 000 | ·00 | 1·00 000 | ·01 | 1·00 000 | ·16 | ·99 998 | 5·28 | + ·880 | ·30 339 | 6540·18 |
| +1·000 | 1·00 000 | ·00 | 1·00 000 | ·00 | 1·00 000 | ·00 | 1·00 000 | ·00 | + ·890 | ·37 463 | 7711·35 |
| | | | | | | | | | + ·900 | ·45 746 | 8834·61 |
| | | | | | | | | | + ·910 | ·55 058 | 9733·86 |
| | | | | | | | | | + ·920 | ·65 058 | 10162·79 |
| | | | | | | | | | + ·930 | ·75 132 | 9839·18 |
| | | | | | | | | | + ·940 | ·84 410 | 8544·86 |
| | | | | | | | | | + ·950 | ·91 903 | 6309·15 |
| | | | | | | | | | + ·955 | ·94 726 | 4967·28 |
| | | | | | | | | | + ·960 | ·96 868 | 3608·52 |
| | | | | | | | | | + ·965 | ·98 352 | 2358·32 |
| | | | | | | | | | + ·970 | ·99 264 | 1335·45 |
| | | | | | | | | | + ·975 | ·99 739 | 617·80 |
| | | | | | | | | | + ·980 | ·99 934 | 211·38 |
| | | | | | | | | | + ·985 | ·99 990 | 44·33 |
| | | | | | | | | | + ·990 | ·99 999 | 3·73 |
| | | | | | | | | | + ·995 | 1·00 000 | ·03 |
| | | | | | | | | | +1·000 | 1·00 000 | ·00 |

$\rho = 0.9$

| r | Area | Ordinate |
|---|---|---|
| + ·800 | ·04 744 | 1124·15 |
| + ·825 | ·08 583 | 2030·11 |
| + ·850 | ·15 441 | 3583·56 |
| + ·875 | ·27 212 | 5972·90 |
| + ·900 | ·45 746 | 8834·61 |
| + ·925 | ·70 135 | 10111·95 |
| + ·950 | ·91 903 | 6309·15 |
| + ·975 | ·99 739 | 617·80 |
| +1·000 | 1·00 000 | ·00 |

$\rho=0\cdot0 - 0\cdot4$      $n=21$

| r | $\rho=0\cdot0$ Area | Ordinate | $\rho=0\cdot1$ Area | Ordinate | $\rho=0\cdot2$ Area | Ordinate | $\rho=0\cdot3$ Area | Ordinate | $\rho=0\cdot4$ Area | Ordinate | r |
|---|---|---|---|---|---|---|---|---|---|---|---|
| −1·00 | ·00 000 | ·00 | ·00 000 | ·00 | ·00 000 | ·00 | ·00 000 | ·00 | ·00 000 | ·00 | −1·00 |
| − ·95 | ·00 000 | ·00 | ·00 000 | ·00 | ·00 000 | ·00 | ·00 000 | ·00 | ·00 000 | ·00 | − ·95 |
| − ·90 | ·00 000 | ·00 | ·00 000 | ·00 | ·00 000 | ·00 | ·00 000 | ·00 | ·00 000 | ·00 | − ·90 |
| − ·85 | ·00 000 | ·03 | ·00 000 | ·01 | ·00 000 | ·00 | ·00 000 | ·00 | ·00 000 | ·00 | − ·85 |
| − ·80 | ·00 001 | ·29 | ·00 000 | ·06 | ·00 000 | ·01 | ·00 000 | ·00 | ·00 000 | ·00 | − ·80 |
| − ·75 | ·00 005 | 1·52 | ·00 001 | ·34 | ·00 000 | ·07 | ·00 000 | ·01 | ·00 000 | ·00 | − ·75 |
| − ·70 | ·00 021 | 5·61 | ·00 005 | 1·36 | ·00 001 | ·29 | ·00 000 | ·05 | ·00 000 | ·01 | − ·70 |
| − ·65 | ·00 071 | 16·13 | ·00 018 | 4·27 | ·00 004 | ·99 | ·00 001 | ·19 | ·00 000 | ·03 | − ·65 |
| − ·60 | ·00 202 | 38·65 | ·00 054 | 11·22 | ·00 013 | 2·82 | ·00 003 | ·60 | ·00 000 | ·10 | − ·60 |
| − ·55 | ·00 490 | 80·30 | ·00 142 | 25·56 | ·00 036 | 6·97 | ·00 008 | 1·59 | ·00 001 | ·29 | − ·55 |
| − ·50 | ·01 049 | 148·80 | ·00 330 | 51·95 | ·00 090 | 15·41 | ·00 020 | 3·79 | ·00 004 | ·74 | − ·50 |
| − ·45 | ·02 034 | 250·78 | ·00 691 | 96·10 | ·00 202 | 31·04 | ·00 049 | 8·26 | ·00 010 | 1·74 | − ·45 |
| − ·40 | ·03 619 | 389·90 | ·01 330 | 164·07 | ·00 418 | 57·77 | ·00 109 | 16·64 | ·00 023 | 3·77 | − ·40 |
| − ·35 | ·05 993 | 565·17 | ·02 381 | 261·27 | ·00 806 | 100·39 | ·00 226 | 31·39 | ·00 050 | 7·67 | − ·35 |
| − ·30 | ·09 321 | 769·89 | ·03 998 | 391·17 | ·01 458 | 164·25 | ·00 439 | 55·82 | ·00 105 | 14·76 | − ·30 |
| − ·25 | ·13 721 | 991·59 | ·06 348 | 553·98 | ·02 493 | 254·51 | ·00 808 | 94·20 | ·00 207 | 27·02 | − ·25 |
| − ·20 | ·19 236 | 1213·06 | ·09 586 | 745·54 | ·04 054 | 375·26 | ·01 413 | 151·59 | ·00 388 | 47·28 | − ·20 |
| − ·15 | ·25 818 | 1414·38 | ·13 836 | 956·73 | ·06 299 | 528·29 | ·02 364 | 233·39 | ·00 699 | 79·38 | − ·15 |
| − ·10 | ·33 313 | 1575·70 | ·19 163 | 1173·63 | ·09 388 | 711·91 | ·03 796 | 344·73 | ·01 210 | 128·23 | − ·10 |
| − ·05 | ·41 479 | 1680·10 | ·25 552 | 1378·59 | ·13 459 | 919·93 | ·05 867 | 489·35 | ·02 019 | 199·70 | − ·05 |
| 0·00 | ·50 000 | 1716·23 | ·32 896 | 1552·13 | ·18 609 | 1141·00 | ·08 747 | 668·33 | ·03 255 | 300·17 | 0·00 |
| + ·05 | ·58 521 | 1680·10 | ·40 990 | 1675·52 | ·24 864 | 1358·90 | ·12 603 | 878·59 | ·05 079 | 435·79 | + ·05 |
| + ·10 | ·66 687 | 1575·70 | ·49 543 | 1733·68 | ·32 160 | 1553·58 | ·17 571 | 1111·53 | ·07 680 | 611·07 | + ·10 |
| + ·15 | ·74 182 | 1414·38 | ·58 203 | 1717·73 | ·40 325 | 1703·39 | ·23 731 | 1352·20 | ·11 258 | 827·05 | + ·15 |
| + ·20 | ·80 764 | 1213·06 | ·66 595 | 1626·97 | ·49 085 | 1788·20 | ·31 081 | 1579·31 | ·16 010 | 1079·03 | + ·20 |
| + ·25 | ·86 279 | 991·59 | ·74 361 | 1469·48 | ·58 073 | 1792·97 | ·39 459 | 1766·81 | ·22 088 | 1354·17 | + ·25 |
| + ·30 | ·90 679 | 769·89 | ·81 205 | 1261·29 | ·66 870 | 1711·25 | ·48 627 | 1887·05 | ·29 554 | 1629·83 | + ·30 |
| + ·35 | ·94 007 | 565·17 | ·86 925 | 1024·10 | ·75 049 | 1547·62 | ·58 176 | 1915·72 | ·38 333 | 1873·44 | + ·35 |
| + ·40 | ·96 381 | 389·90 | ·91 437 | 781·85 | ·82 236 | 1318·32 | ·67 606 | 1837·75 | ·48 168 | 2045·23 | + ·40 |
| + ·45 | ·97 966 | 250·78 | ·94 773 | 556·80 | ·88 165 | 1049·40 | ·76 374 | 1652·96 | ·58 598 | 2104·80 | + ·45 |
| + ·50 | ·98 951 | 148·80 | ·97 063 | 366·00 | ·92 717 | 772·43 | ·83 986 | 1379·62 | ·68 977 | 2021·59 | + ·50 |
| + ·55 | ·99 510 | 80·30 | ·98 506 | 218·92 | ·95 928 | 518·34 | ·90 082 | 1053·63 | ·73 545 | 1787·57 | + ·55 |
| + ·60 | ·99 798 | 38·65 | ·99 327 | 116·85 | ·97 979 | 311·00 | ·94 514 | 722·19 | ·86 641 | 1427·92 | + ·60 |
| + ·65 | ·99 929 | 16·13 | ·99 740 | 54·13 | ·99 137 | 162·26 | ·97 371 | 432·22 | ·92 729 | 1002·93 | + ·65 |
| + ·70 | ·99 979 | 5·61 | ·99 918 | 20·90 | ·99 697 | 70·68 | ·98 962 | 216·87 | ·96 700 | 595·00 | + ·70 |
| + ·75 | ·99 995 | 1·52 | ·99 980 | 6·31 | ·99 919 | 24·12 | ·99 684 | 85·61 | ·98 837 | 279·97 | + ·75 |
| + ·80 | ·99 999 | ·29 | ·99 997 | 1·34 | ·99 985 | 5·79 | ·99 934 | 23·90 | ·99 717 | 93·95 | + ·80 |
| + ·85 | 1·00 000 | ·03 | 1·00 000 | ·16 | ·99 999 | ·80 | ·99 992 | 3·86 | ·99 962 | 18·41 | + ·85 |
| + ·90 | 1·00 000 | ·00 | 1·00 000 | ·00 | 1·00 000 | ·04 | 1·00 000 | ·23 | ·99 997 | 1·34 | + ·90 |
| + ·95 | 1·00 000 | ·00 | 1·00 000 | ·00 | 1·00 000 | ·00 | 1·00 000 | ·00 | 1·00 000 | ·01 | + ·95 |
| +1·00 | 1·00 000 | ·00 | 1·00 000 | ·00 | 1·00 000 | ·00 | 1·00 000 | ·00 | 1·00 000 | ·00 | +1·00 |

n = 21      ρ = 0·5 — 0·9

| r | ρ=0·5 Area | ρ=0·5 Ordinate | ρ=0·6 Area | ρ=0·6 Ordinate | ρ=0·7 Area | ρ=0·7 Ordinate | ρ=0·8 Area | ρ=0·8 Ordinate |
|---|---|---|---|---|---|---|---|---|
| −1·00 | ·00 000 | ·00 | ·00 000 | ·00 | ·00 000 | ·00 | ·00 000 | ·00 |
| − ·95 | ·00 000 | ·00 | ·00 000 | ·00 | ·00 000 | ·00 | ·00 000 | ·00 |
| − ·90 | ·00 000 | ·00 | ·00 000 | ·00 | ·00 000 | ·00 | ·00 000 | ·00 |
| − ·85 | ·00 000 | ·00 | ·00 000 | ·00 | ·00 000 | ·00 | ·00 000 | ·00 |
| − ·80 | ·00 000 | ·00 | ·00 000 | ·00 | ·00 000 | ·00 | ·00 000 | ·00 |
| − ·75 | ·00 000 | ·00 | ·00 000 | ·00 | ·00 000 | ·00 | ·00 000 | ·00 |
| − ·70 | ·00 000 | ·00 | ·00 000 | ·00 | ·00 000 | ·00 | ·00 000 | ·00 |
| − ·65 | ·00 000 | ·00 | ·00 000 | ·00 | ·00 000 | ·00 | ·00 000 | ·00 |
| − ·60 | ·00 000 | ·01 | ·00 000 | ·00 | ·00 000 | ·00 | ·00 000 | ·00 |
| − ·55 | ·00 000 | ·04 | ·00 000 | ·00 | ·00 000 | ·00 | ·00 000 | ·00 |
| − ·50 | ·00 000 | ·11 | ·00 000 | ·01 | ·00 000 | ·00 | ·00 000 | ·00 |
| − ·45 | ·00 001 | ·27 | ·00 000 | ·03 | ·00 000 | ·00 | ·00 000 | ·00 |
| − ·40 | ·00 004 | ·63 | ·00 000 | ·07 | ·00 000 | ·00 | ·00 000 | ·00 |
| − ·35 | ·00 008 | 1·37 | ·00 001 | ·16 | ·00 000 | ·01 | ·00 000 | ·00 |
| − ·30 | ·00 018 | 2·84 | ·00 002 | ·35 | ·00 000 | ·02 | ·00 000 | ·00 |
| − ·25 | ·00 039 | 5·61 | ·00 005 | ·75 | ·00 000 | ·05 | ·00 000 | ·00 |
| − ·20 | ·00 078 | 10·64 | ·00 010 | 1·53 | ·00 001 | ·11 | ·00 000 | ·00 |
| − ·15 | ·00 151 | 19·43 | ·00 021 | 3·04 | ·00 002 | ·24 | ·00 000 | ·01 |
| − ·10 | ·00 283 | 34·26 | ·00 043 | 5·83 | ·00 003 | ·50 | ·00 000 | ·01 |
| − ·05 | ·00 510 | 58·45 | ·00 083 | 10·88 | ·00 007 | 1·02 | ·00 000 | ·03 |
| 0·00 | ·00 890 | 96·65 | ·00 158 | 19·79 | ·00 015 | 2·04 | ·00 000 | ·06 |
| + ·05 | ·01 510 | 155·05 | ·00 292 | 35·09 | ·00 029 | 4·01 | ·00 001 | ·14 |
| + ·10 | ·02 487 | 241·33 | ·00 526 | 60·74 | ·00 057 | 7·73 | ·00 002 | ·29 |
| + ·15 | ·03 984 | 364·44 | ·00 926 | 102·64 | ·00 111 | 14·66 | ·00 004 | ·63 |
| + ·20 | ·06 208 | 533·39 | ·01 593 | 169·29 | ·00 214 | 27·37 | ·00 009 | 1·33 |
| + ·25 | ·09 407 | 755·33 | ·02 679 | 272·20 | ·00 402 | 50·31 | ·00 019 | 2·82 |
| + ·30 | ·13 854 | 1032·26 | ·04 400 | 425·95 | ·00 746 | 90·99 | ·00 040 | 5·95 |
| + ·35 | ·19 809 | 1356·52 | ·07 051 | 646·90 | ·01 362 | 161·66 | ·00 084 | 12·53 |
| + ·40 | ·27 461 | 1705·56 | ·11 006 | 949·63 | ·02 444 | 281·48 | ·00 177 | 26·36 |
| + ·45 | ·36 838 | 2037·72 | ·16 694 | 1359·82 | ·04 305 | 478·52 | ·00 372 | 55·31 |
| + ·50 | ·47 709 | 2291·99 | ·24 526 | 1802·29 | ·07 419 | 789·71 | ·00 781 | 115·56 |
| + ·55 | ·59 507 | 2396·03 | ·34 754 | 2285·43 | ·12 458 | 1254·35 | ·01 632 | 239·50 |
| + ·60 | ·71 311 | 2288·37 | ·47 249 | 2687·85 | ·20 256 | 1892·96 | ·03 383 | 489·12 |
| + ·625 | ·76 859 | 2141·70 | ·54 140 | 2813·76 | ·25 451 | 2267·97 | ·04 847 | 692·81 |
| + ·650 | ·81 972 | 1940·29 | ·61 253 | 2862·39 | ·31 611 | 2660·83 | ·06 911 | 972·87 |
| + ·675 | ·86 520 | 1691·00 | ·68 372 | 2815·57 | ·38 747 | 3043·71 | ·09 792 | 1350·45 |
| + ·700 | ·90 399 | 1408·33 | ·75 241 | 2661·07 | ·46 788 | 3375·86 | ·13 761 | 1845·86 |
| + ·725 | ·93 550 | 1111·42 | ·81 586 | 2397·11 | ·55 541 | 3604·30 | ·19 129 | 2471·22 |
| + ·750 | ·95 963 | 822·04 | ·87 146 | 2036·49 | ·64 673 | 3668·97 | ·26 217 | 3216·90 |
| + ·775 | ·97 684 | 561·62 | ·91 712 | 1608·87 | ·73 702 | 3514·93 | ·35 271 | 4030·04 |
| + ·800 | ·98 810 | 347·48 | ·95 172 | 1159·40 | ·82 039 | 3112·44 | ·46 323 | 4787·20 |
| + ·825 | ·99 468 | 189·34 | ·97 536 | 741·56 | ·89 073 | 2482·24 | ·58 983 | 5274·55 |
| + ·850 | ·99 803 | 87·22 | ·98 947 | 404·32 | ·94 334 | 1715·20 | ·72 237 | 5209·97 |
| + ·875 | ·99 944 | 31·86 | ·99 649 | 176·43 | ·97 663 | 965·95 | ·84 377 | 4364·79 |
| + ·900 | ·99 989 | 8·29 | ·99 919 | 55·40 | ·99 319 | 399·51 | ·93 457 | 2814·91 |
| + ·925 | ·99 999 | 1·26 | ·99 990 | 10·25 | ·99 888 | 99·58 | ·98 353 | 1157·18 |
| + ·950 | 1·00 000 | ·07 | ·99 999 | ·71 | ·99 993 | 9·56 | ·99 848 | 195·96 |
| + ·975 | 1·00 000 | ·00 | 1·00 000 | ·00 | 1·00 000 | ·08 | ·99 998 | 3·29 |
| +1·000 | 1·00 000 | ·00 | 1·00 000 | ·00 | 1·00 000 | ·00 | 1·00 000 | ·00 |

| r | ρ=0·9 Area | ρ=0·9 Ordinate |
|---|---|---|
| −1·00 | ·00 000 | ·00 |
| − ·95 | ·00 000 | ·00 |
| − ·90 | ·00 000 | ·00 |
| − ·85 | ·00 000 | ·00 |
| − ·80 | ·00 000 | ·00 |
| − ·75 | ·00 000 | ·00 |
| − ·70 | ·00 000 | ·00 |
| − ·65 | ·00 000 | ·00 |
| − ·60 | ·00 000 | ·00 |
| − ·55 | ·00 000 | ·00 |
| − ·50 | ·00 000 | ·00 |
| − ·45 | ·00 000 | ·00 |
| − ·40 | ·00 000 | ·00 |
| − ·35 | ·00 000 | ·00 |
| − ·30 | ·00 000 | ·00 |
| − ·25 | ·00 000 | ·00 |
| − ·20 | ·00 000 | ·00 |
| − ·15 | ·00 000 | ·00 |
| − ·10 | ·00 000 | ·00 |
| − ·05 | ·00 000 | ·00 |
| 0·00 | ·00 000 | ·00 |
| + ·05 | ·00 000 | ·00 |
| + ·10 | ·00 000 | ·00 |
| + ·15 | ·00 000 | ·00 |
| + ·20 | ·00 000 | ·00 |
| + ·25 | ·00 000 | ·01 |
| + ·30 | ·00 000 | ·02 |
| + ·35 | ·00 000 | ·06 |
| + ·40 | ·00 001 | ·14 |
| + ·45 | ·00 002 | ·38 |
| + ·50 | ·00 005 | 1·06 |
| + ·55 | ·00 015 | 3·02 |
| + ·60 | ·00 042 | 8·96 |
| + ·625 | ·00 072 | 15·71 |
| + ·650 | ·00 125 | 27·86 |
| + ·675 | ·00 220 | 50·01 |
| + ·700 | ·00 391 | 90·85 |
| + ·725 | ·00 703 | 166·87 |
| + ·750 | ·01 279 | 309·47 |
| + ·775 | ·02 353 | 577·67 |
| + ·800 | ·04 360 | 1079·30 |
| + ·810 | ·05 585 | 1383·41 |
| + ·820 | ·07 154 | 1769·09 |
| + ·830 | ·09 157 | 2254·40 |
| + ·840 | ·11 702 | 2858·41 |
| + ·850 | ·14 919 | 3599·09 |
| + ·860 | ·18 950 | 4488·91 |
| + ·870 | ·23 947 | 5527·65 |
| + ·880 | ·30 048 | 6691·24 |
| + ·890 | ·37 351 | 7916·14 |
| + ·900 | ·45 861 | 9080·68 |
| + ·910 | ·55 428 | 9989·66 |
| + ·920 | ·65 666 | 10374·91 |
| + ·930 | ·75 902 | 9940·36 |
| + ·940 | ·85 199 | 8481·30 |
| + ·950 | ·92 548 | 6087·03 |
| + ·955 | ·95 246 | 4698·09 |
| + ·960 | ·97 249 | 3328·48 |
| + ·965 | ·98 599 | 2107·12 |
| + ·970 | ·99 400 | 1145·29 |
| + ·975 | ·99 798 | 502·07 |
| + ·980 | ·99 952 | 159·71 |
| + ·985 | ·99 993 | 30·20 |
| + ·990 | ·99 999 | 2·16 |
| + ·995 | 1·00 000 | ·01 |
| +1·000 | 1·00 000 | ·00 |

| r | ρ=0·9 Area | ρ=0·9 Ordinate |
|---|---|---|
| + ·800 | ·04 360 | 1079·30 |
| + ·825 | ·08 095 | 1996·23 |
| + ·850 | ·14 919 | 3599·09 |
| + ·875 | ·26 852 | 6096·67 |
| + ·900 | ·45 861 | 9080·68 |
| + ·925 | ·70 838 | 10277·08 |
| + ·950 | ·92 548 | 6087·03 |
| + ·975 | ·99 798 | 502·07 |
| +1·000 | 1·00 000 | ·00 |

ρ=0·0—0·4        n=22

| r | ρ=0·0 Area | Ordinate | ρ=0·1 Area | Ordinate | ρ=0·2 Area | Ordinate | ρ=0·3 Area | Ordinate | ρ=0·4 Area | Ordinate |
|---|---|---|---|---|---|---|---|---|---|---|
| −1·00 | ·00 000 | ·00 | ·00 000 | ·00 | ·00 000 | ·00 | ·00 000 | ·00 | ·00 000 | ·00 |
| − ·95 | ·00 000 | ·00 | ·00 000 | ·00 | ·00 000 | ·00 | ·00 000 | ·00 | ·00 000 | ·00 |
| − ·90 | ·00 000 | ·01 | ·00 000 | ·00 | ·00 000 | ·00 | ·00 000 | ·00 | ·00 000 | ·00 |
| − ·85 | ·00 000 | ·02 | ·00 000 | ·00 | ·00 000 | ·00 | ·00 000 | ·00 | ·00 000 | ·00 |
| − ·80 | ·00 000 | ·18 | ·00 000 | ·03 | ·00 000 | ·01 | ·00 000 | ·00 | ·00 000 | ·00 |
| − ·75 | ·00 003 | 1·03 | ·00 001 | ·21 | ·00 000 | ·04 | ·00 000 | ·01 | ·00 000 | ·00 |
| − ·70 | ·00 014 | 4·11 | ·00 003 | ·92 | ·00 001 | ·18 | ·00 000 | ·03 | ·00 000 | ·00 |
| − ·65 | ·00 053 | 12·59 | ·00 012 | 3·11 | ·00 002 | ·67 | ·00 000 | ·12 | ·00 000 | ·02 |
| − ·60 | ·00 158 | 31·74 | ·00 040 | 8·65 | ·00 009 | 2·02 | ·00 002 | ·40 | ·00 000 | ·06 |
| − ·55 | ·00 400 | 68·85 | ·00 109 | 20·67 | ·00 026 | 5·28 | ·00 005 | 1·12 | ·00 001 | ·19 |
| − ·50 | ·00 890 | 132·30 | ·00 265 | 43·77 | ·00 068 | 12·21 | ·00 014 | 2·80 | ·00 002 | ·50 |
| − ·45 | ·01 780 | 229·92 | ·00 575 | 83·90 | ·00 159 | 25·58 | ·00 036 | 6·36 | ·00 006 | 1·24 |
| − ·40 | ·03 255 | 366·87 | ·01 143 | 147·70 | ·00 341 | 49·31 | ·00 084 | 13·34 | ·00 016 | 2·80 |
| − ·35 | ·05 516 | 543·53 | ·02 103 | 241·56 | ·00 677 | 88·41 | ·00 179 | 26·06 | ·00 037 | 5·93 |
| − ·30 | ·08 747 | 754·00 | ·03 617 | 370·08 | ·01 260 | 148·69 | ·00 359 | 47·84 | ·00 080 | 11·83 |
| − ·25 | ·13 091 | 985·69 | ·05 864 | 534·57 | ·02 209 | 236·09 | ·00 680 | 83·10 | ·00 163 | 22·38 |
| − ·20 | ·18 609 | 1220·22 | ·09 017 | 731·56 | ·03 675 | 355·63 | ·01 222 | 137·23 | ·00 316 | 40·36 |
| − ·15 | ·25 262 | 1435·65 | ·13 219 | 951·98 | ·05 824 | 510·10 | ·02 094 | 216·26 | ·00 586 | 69·67 |
| − ·10 | ·32 896 | 1609·59 | ·18 551 | 1181·06 | ·08 832 | 698·57 | ·03 436 | 326·14 | ·01 040 | 115·44 |
| − ·05 | ·41 256 | 1722·72 | ·25 012 | 1399·48 | ·12 857 | 915·06 | ·05 415 | 471·58 | ·01 778 | 183·99 |
| 0·00 | ·50 000 | 1761·97 | ·32 492 | 1585·51 | ·18 010 | 1147·75 | ·08 214 | 654·54 | ·02 930 | 282·44 |
| + ·05 | ·58 744 | 1722·72 | ·40 777 | 1718·00 | ·24 332 | 1379·01 | ·12 019 | 872·46 | ·04 664 | 417·89 |
| + ·10 | ·67 104 | 1609·59 | ·49 554 | 1779·88 | ·31 761 | 1586·65 | ·16 983 | 1116·63 | ·07 180 | 595·93 |
| + ·15 | ·74 738 | 1435·65 | ·58 441 | 1761·23 | ·40 119 | 1746·45 | ·23 202 | 1370·99 | ·10 698 | 818·50 |
| + ·20 | ·81 391 | 1220·22 | ·67 031 | 1661·60 | ·49 108 | 1835·83 | ·30 671 | 1612·18 | ·15 433 | 1081·27 |
| + ·25 | ·86 909 | 985·69 | ·74 938 | 1490·67 | ·58 332 | 1838·17 | ·39 254 | 1811·21 | ·21 556 | 1370·78 |
| + ·30 | ·91 253 | 754·00 | ·81 849 | 1267·07 | ·67 333 | 1746·86 | ·48 662 | 1937·29 | ·29 144 | 1662·37 |
| + ·35 | ·94 474 | 543·53 | ·87 562 | 1015·49 | ·75 654 | 1568·03 | ·58 460 | 1963·65 | ·38 123 | 1920·03 |
| + ·40 | ·96 745 | 366·87 | ·92 002 | 762·48 | ·82 899 | 1321·06 | ·68 105 | 1874·43 | ·48 215 | 2099·64 |
| + ·45 | ·98 220 | 229·92 | ·95 224 | 531·86 | ·88 800 | 1035·89 | ·77 013 | 1671·23 | ·58 915 | 2156·76 |
| + ·50 | ·99 110 | 132·30 | ·97 387 | 340·82 | ·93 253 | 747·64 | ·84 663 | 1376·56 | ·69 523 | 2059·06 |
| + ·55 | ·99 600 | 68·85 | ·98 713 | 197·63 | ·96 327 | 489·26 | ·90 695 | 1032·04 | ·79 236 | 1800·84 |
| + ·60 | ·99 842 | 31·74 | ·99 442 | 101·58 | ·98 236 | 284·38 | ·94 988 | 690·00 | ·87 315 | 1414·20 |
| + ·65 | ·99 947 | 12·59 | ·99 795 | 44·94 | ·99 277 | 142·57 | ·97 682 | 399·58 | ·93 279 | 969·04 |
| + ·70 | ·99 986 | 4·11 | ·99 939 | 16·39 | ·99 759 | 59·04 | ·99 123 | 191·99 | ·97 059 | 555·25 |
| + ·75 | ·99 997 | 1·03 | ·99 996 | 4·60 | ·99 940 | 18·88 | ·99 749 | 71·56 | ·99 014 | 248·90 |
| + ·80 | 1·00 000 | ·18 | ·99 998 | ·89 | ·99 990 | 4·16 | ·99 952 | 18·48 | ·99 777 | 78·00 |
| + ·85 | 1·00 000 | ·02 | 1·00 000 | ·10 | ·99 999 | ·51 | ·99 995 | 2·67 | ·99 972 | 13·82 |
| + ·90 | 1·00 000 | ·01 | 1·00 000 | ·00 | 1·00 000 | ·02 | 1·00 000 | ·13 | ·99 998 | ·86 |
| + ·95 | 1·00 000 | ·00 | 1 00 000 | ·00 | 1·00 000 | ·00 | 1·00 000 | ·00 | 1·00 000 | ·00 |
| +1·00 | 1·00 000 | ·00 | 1·00 000 | ·00 | 1·00 000 | ·00 | 1·00 000 | ·00 | 1·00 000 | ·00 |

n = 22                                                                                     ρ = 0.5 — 0.9

| r | ρ=0.5 Area | Ordinate | ρ=0.6 Area | Ordinate | ρ=0.7 Area | Ordinate | ρ=0.8 Area | Ordinate |
|---|---|---|---|---|---|---|---|---|
| -1.00 | .00 000 | .00 | .00 000 | .00 | .00 000 | .00 | .00 000 | .00 |
| - .95 | .00 000 | .00 | .00 000 | .00 | .00 000 | .00 | .00 000 | .00 |
| - .90 | .00 000 | .00 | .00 000 | .00 | .00 000 | .00 | .00 000 | .00 |
| - .85 | .00 000 | .00 | .00 000 | .00 | .00 000 | .00 | .00 000 | .00 |
| - .80 | .00 000 | .00 | .00 000 | .00 | .00 000 | .00 | .00 000 | .00 |
| - .75 | .00 000 | .00 | .00 000 | .00 | .00 000 | .00 | .00 000 | .00 |
| - .70 | .00 000 | .00 | .00 000 | .00 | .00 000 | .00 | .00 000 | .00 |
| - .65 | .00 000 | .00 | .00 000 | .00 | .00 000 | .00 | .00 000 | .00 |
| - .60 | .00 000 | .01 | .00 000 | .00 | .00 000 | .00 | .00 000 | .00 |
| - .55 | .00 000 | .02 | .00 000 | .00 | .00 000 | .00 | .00 000 | .00 |
| - .50 | .00 001 | .07 | .00 000 | .01 | .00 000 | .00 | .00 000 | .00 |
| - .45 | .00 001 | .17 | .00 000 | .02 | .00 000 | .00 | .00 000 | .00 |
| - .40 | .00 002 | .43 | .00 000 | .04 | .00 000 | .00 | .00 000 | .00 |
| - .35 | .00 006 | .97 | .00 001 | .10 | .00 000 | .01 | .00 000 | .00 |
| - .30 | .00 013 | 2.09 | .00 001 | .23 | .00 000 | .01 | .00 000 | .00 |
| - .25 | .00 028 | 4.29 | .00 003 | .52 | .00 000 | .03 | .00 000 | .00 |
| - .20 | .00 059 | 8.43 | .00 007 | 1.10 | .00 000 | .07 | .00 000 | .00 |
| - .15 | .00 118 | 15.89 | .00 015 | 2.26 | .00 001 | .16 | .00 000 | .00 |
| - .10 | .00 227 | 28.86 | .00 031 | 4.49 | .00 002 | .34 | .00 000 | .01 |
| - .05 | .00 421 | 50.64 | .00 063 | 8.67 | .00 005 | .72 | .00 000 | .02 |
| 0.00 | .00 756 | 85.93 | .00 124 | 16.25 | .00 010 | 1.50 | .00 000 | .04 |
| + .05 | .01 314 | 141.20 | .00 235 | 29.67 | .00 020 | 3.04 | .00 001 | .09 |
| + .10 | .02 215 | 224.73 | .00 436 | 52.80 | .00 043 | 6.06 | .00 001 | .20 |
| + .15 | .03 624 | 346.33 | .00 789 | 91.59 | .00 086 | 11.87 | .00 003 | .43 |
| + .20 | .05 759 | 516.27 | .01 393 | 154.80 | .00 170 | 22.86 | .00 006 | .96 |
| + .25 | .08 882 | 743.10 | .02 398 | 254.65 | .00 330 | 43.29 | .00 013 | 2.10 |
| + .30 | .13 290 | 1029.97 | .04 027 | 406.96 | .00 631 | 80.55 | .00 029 | 4.60 |
| + .35 | .19 271 | 1369.38 | .06 586 | 629.96 | .01 184 | 147.04 | .00 064 | 10.04 |
| + .40 | .27 033 | 1737.17 | .10 473 | 940.51 | .02 182 | 262.68 | .00 140 | 21.88 |
| + .45 | .36 614 | 2087.52 | .16 150 | 1346.06 | .03 942 | 457.35 | .00 306 | 47.54 |
| + .50 | .47 767 | 2352.90 | .24 068 | 1831.17 | .06 954 | 771.34 | .00 664 | 102.73 |
| + .55 | .59 869 | 2453.82 | .34 504 | 2339.55 | .11 928 | 1248.74 | .01 434 | 219.99 |
| + .60 | .71 916 | 2323.01 | .47 319 | 2759.17 | .19 759 | 1914.05 | .03 071 | 463.46 |
| + .625 | .77 535 | 2161.92 | .54 391 | 2886.10 | .25 029 | 2307.31 | .04 469 | 666.18 |
| + .650 | .82 677 | 1942.00 | .61 679 | 2928.43 | .31 312 | 2719.83 | .06 468 | 948.62 |
| + .675 | .87 205 | 1674.23 | .68 946 | 2867.19 | .38 619 | 3120.85 | .09 298 | 1334.04 |
| + .700 | .91 022 | 1375.57 | .75 918 | 2690.72 | .46 869 | 3465.29 | .13 244 | 1845.14 |
| + .725 | .94 076 | 1067.48 | .82 304 | 2399.69 | .55 850 | 3694.99 | .18 641 | 2495.84 |
| + .750 | .96 372 | 773.40 | .87 835 | 2011.22 | .65 195 | 3745.20 | .25 833 | 3276.10 |
| + .775 | .97 973 | 515.15 | .92 308 | 1560.66 | .74 379 | 3559.18 | .35 083 | 4127.66 |
| + .800 | .98 992 | 308.91 | .95 629 | 1098.57 | .82 770 | 3111.21 | .46 419 | 4913.78 |
| + .825 | .99 567 | 161.91 | .97 839 | 681.47 | .89 741 | 2433.86 | .59 401 | 5399.33 |
| + .850 | .99 849 | 71.03 | .99 114 | 356.95 | .94 837 | 1635.37 | .72 911 | 5281.97 |
| + .875 | .99 959 | 24.38 | .99 720 | 147.66 | .97 958 | 884.63 | .85 114 | 4337.85 |
| + .900 | .99 990 | 5.84 | .99 940 | 43.11 | .99 440 | 345.00 | .94 004 | 2698.71 |
| + .925 | .99 999 | .79 | .99 992 | 7.18 | .99 916 | 78.68 | .98 587 | 1041.46 |
| + .950 | 1.00 000 | .04 | 1.00 000 | .42 | .99 995 | 6.53 | .99 885 | 157.01 |
| + .975 | 1.00 000 | .00 | 1.00 000 | .00 | 1.00 000 | .04 | 1.00 000 | 2.05 |
| +1.000 | 1.00 000 | .00 | 1.00 000 | .00 | 1.00 000 | .00 | 1.00 000 | .00 |

| r | ρ=0.9 Area | Ordinate |
|---|---|---|
| -1.00 | .00 000 | .00 |
| - .95 | .00 000 | .00 |
| - .90 | .00 000 | .00 |
| - .85 | .00 000 | .00 |
| - .80 | .00 000 | .00 |
| - .75 | .00 000 | .00 |
| - .70 | .00 000 | .00 |
| - .65 | .00 000 | .00 |
| - .60 | .00 000 | .00 |
| - .55 | .00 000 | .00 |
| - .50 | .00 000 | .00 |
| - .45 | .00 000 | .00 |
| - .40 | .00 000 | .00 |
| - .35 | .00 000 | .00 |
| - .30 | .00 000 | .00 |
| - .25 | .00 000 | .00 |
| - .20 | .00 000 | .00 |
| - .15 | .00 000 | .00 |
| - .10 | .00 000 | .00 |
| - .05 | .00 000 | .00 |
| 0.00 | .00 000 | .00 |
| + .05 | .00 000 | .00 |
| + .10 | .00 000 | .00 |
| + .15 | .00 000 | .00 |
| + .20 | .00 000 | .00 |
| + .25 | .00 000 | .00 |
| + .30 | .00 000 | .01 |
| + .35 | .00 000 | .03 |
| + .40 | .00 000 | .09 |
| + .45 | .00 001 | .26 |
| + .50 | .00 003 | .76 |
| + .55 | .00 010 | 2.23 |
| + .60 | .00 031 | 6.97 |
| + .625 | .00 055 | 12.54 |
| + .650 | .00 098 | 22.83 |
| + .675 | .00 176 | 42.06 |
| + .700 | .00 322 | 78.45 |
| + .725 | .00 595 | 147.97 |
| + .750 | .01 114 | 281.79 |
| + .775 | .02 105 | 539.94 |
| + .800 | .04 009 | 1034.74 |
| + .810 | .05 190 | 1339.35 |
| + .820 | .06 716 | 1729.09 |
| + .830 | .08 683 | 2223.60 |
| + .840 | .11 205 | 2843.79 |
| + .850 | .14 419 | 3609.52 |
| + .860 | .18 478 | 4534.66 |
| + .870 | .23 542 | 5619.06 |
| + .880 | .29 760 | 6835.99 |
| + .890 | .37 235 | 8114.75 |
| + .900 | .45 966 | 9320.25 |
| + .910 | .55 780 | 10237.52 |
| + .920 | .66 249 | 10576.33 |
| + .930 | .76 635 | 10028.23 |
| + .940 | .85 941 | 8406.19 |
| + .950 | .93 135 | 5864.34 |
| + .955 | .95 711 | 4437.14 |
| + .960 | .97 582 | 3065.80 |
| + .965 | .98 808 | 1880.00 |
| + .970 | .99 510 | 980.80 |
| + .975 | .99 844 | 407.43 |
| + .980 | .99 965 | 120.50 |
| + .985 | .99 995 | 20.54 |
| + .990 | 1.00 000 | 1.25 |
| + .995 | 1.00 000 | .01 |
| +1.000 | 1.00 000 | .00 |

| r | ρ=0.9 Area | Ordinate |
|---|---|---|
| + .800 | .04 009 | 1034.74 |
| + .825 | .07 638 | 1960.11 |
| + .850 | .14 419 | 3609.52 |
| + .875 | .26 499 | 6214.11 |
| + .900 | .45 966 | 9320.25 |
| + .925 | .71 511 | 10429.99 |
| + .950 | .93 135 | 5864.34 |
| + .975 | .99 844 | 407.43 |
| +1.000 | 1.00 000 | .00 |

ρ=0·0 — 0·4          n = 23

| r | ρ=0·0 | | ρ=0·1 | | ρ=0·2 | | ρ=0·3 | | ρ=0·4 | |
|---|---|---|---|---|---|---|---|---|---|---|
| | Area | Ordinate | Area | Ordinate | Area | Ordinate | Area | Ordinate | Area | Ordinate |
| −1·00 | ·00 000 | ·00 | ·00 000 | ·00 | ·00 000 | ·00 | ·00 000 | ·00 | ·00 000 | ·00 |
| − ·95 | ·00 000 | ·00 | ·00 000 | ·00 | ·00 000 | ·00 | ·00 000 | ·00 | ·00 000 | ·00 |
| − ·90 | ·00 000 | ·00 | ·00 000 | ·00 | ·00 000 | ·00 | ·00 000 | ·00 | ·00 000 | ·00 |
| − ·85 | ·00 000 | ·01 | ·00 000 | ·00 | ·00 000 | ·00 | ·00 000 | ·00 | ·00 000 | ·00 |
| − ·80 | ·00 000 | ·11 | ·00 000 | ·02 | ·00 000 | ·00 | ·00 000 | ·00 | ·00 000 | ·00 |
| − ·75 | ·00 002 | ·70 | ·00 000 | ·13 | ·00 000 | ·00 | ·00 000 | ·00 | ·00 000 | ·00 |
| − ·70 | ·00 010 | 3·01 | ·00 002 | ·63 | ·00 000 | ·11 | ·00 000 | ·00 | ·00 000 | ·00 |
| − ·65 | ·00 039 | 9·81 | ·00 008 | 2·26 | ·00 001 | ·45 | ·00 000 | ·08 | ·00 000 | ·01 |
| − ·60 | ·00 124 | 26·04 | ·00 029 | 6·66 | ·00 006 | 1·45 | ·00 001 | ·26 | ·00 000 | ·04 |
| − ·55 | ·00 327 | 58·96 | ·00 084 | 16·69 | ·00 019 | 3·99 | ·00 003 | ·78 | ·00 000 | ·12 |
| − ·50 | ·00 756 | 117·47 | ·00 213 | 36·83 | ·00 051 | 9·65 | ·00 010 | 2·06 | ·00 002 | ·34 |
| − ·45 | ·01 560 | 210·52 | ·00 480 | 73·14 | ·00 124 | 21·06 | ·00 026 | 4·90 | ·00 004 | ·88 |
| − ·40 | ·02 930 | 344·75 | ·00 983 | 132·79 | ·00 277 | 42·04 | ·00 064 | 10·68 | ·00 011 | 2·08 |
| − ·35 | ·05 079 | 522·03 | ·01 859 | 223·04 | ·00 569 | 77·76 | ·00 142 | 21·61 | ·00 027 | 4·58 |
| − ·30 | ·08 214 | 737·47 | ·03 274 | 349·67 | ·01 090 | 134·43 | ·00 295 | 40·95 | ·00 061 | 9·47 |
| − ·25 | ·12 497 | 978·54 | ·05 420 | 515·16 | ·01 960 | 218·71 | ·00 573 | 73·21 | ·00 129 | 18·51 |
| − ·20 | ·18 010 | 1225·82 | ·08 487 | 716·90 | ·03 333 | 336·58 | ·01 057 | 124·07 | ·00 258 | 34·41 |
| − ·15 | ·24 726 | 1455·33 | ·12 635 | 946·00 | ·05 388 | 491·89 | ·01 855 | 200·13 | ·00 491 | 61·07 |
| − ·10 | ·32 492 | 1642·05 | ·17 967 | 1186·97 | ·08 314 | 684·57 | ·03 112 | 308·15 | ·00 895 | 103·79 |
| − ·05 | ·41 038 | 1764·10 | ·24 490 | 1418·82 | ·12 287 | 909·02 | ·05 000 | 453·86 | ·01 567 | 169·29 |
| 0·00 | ·50 000 | 1806·56 | ·32 100 | 1617·48 | ·17 438 | 1153·02 | ·07 718 | 640·19 | ·02 639 | 265·41 |
| + ·05 | ·58 962 | 1764·10 | ·40 570 | 1759·25 | ·23 819 | 1397·58 | ·11 467 | 865·25 | ·04 285 | 400·21 |
| + ·10 | ·67 508 | 1642·05 | ·49 565 | 1824·91 | ·31 374 | 1618·30 | ·16 421 | 1120·27 | ·06 717 | 580·40 |
| + ·15 | ·75 274 | 1455·33 | ·58 673 | 1803·46 | ·39 917 | 1788·25 | ·22 691 | 1388·22 | ·10 170 | 808·98 |
| + ·20 | ·81 990 | 1225·82 | ·67 453 | 1694·75 | ·49 129 | 1882·27 | ·30 283 | 1643·56 | ·14 882 | 1082·09 |
| + ·25 | ·87 503 | 978·54 | ·75 493 | 1510·18 | ·58 582 | 1882·05 | ·39 053 | 1854·28 | ·21 043 | 1385·76 |
| + ·30 | ·91 786 | 737·47 | ·82 464 | 1271·20 | ·67 781 | 1780·86 | ·48 694 | 1986·26 | ·28 746 | 1693·34 |
| + ·35 | ·94 921 | 522·03 | ·88 161 | 1005·63 | ·76 235 | 1586·63 | ·58 735 | 2010·13 | ·37 917 | 1965·21 |
| + ·40 | ·97 070 | 344·75 | ·92 525 | 742·61 | ·83 529 | 1322·06 | ·68 586 | 1909·35 | ·48 258 | 2152·67 |
| + ·45 | ·98 440 | 210·52 | ·95 634 | 507·37 | ·89 394 | 1021·21 | ·77 625 | 1687·50 | ·59 206 | 2207·11 |
| + ·50 | ·99 244 | 117·47 | ·97 674 | 316·95 | ·93 745 | 722·70 | ·85 304 | 1371·71 | ·70 049 | 2094·49 |
| + ·55 | ·99 673 | 58·96 | ·98 890 | 178·18 | ·96 684 | 461·21 | ·91 264 | 1009·57 | ·79 881 | 1811·83 |
| + ·60 | ·99 876 | 26·04 | ·99 537 | 88·20 | ·98 458 | 259·71 | ·95 418 | 658·38 | ·87 947 | 1398·78 |
| + ·65 | ·99 961 | 9·81 | ·99 837 | 37·26 | ·99 393 | 125·10 | ·97 952 | 368·92 | ·93 783 | 935·07 |
| + ·70 | ·99 990 | 3·01 | ·99 954 | 12·84 | ·99 808 | 49·25 | ·99 259 | 169·74 | ·97 376 | 517·49 |
| + ·75 | ·99 998 | ·70 | ·99 990 | 3·36 | ·99 955 | 14·76 | ·99 799 | 59·73 | ·99 163 | 220·99 |
| + ·80 | 1·00 000 | ·11 | ·99 999 | ·59 | ·99 993 | 2·99 | ·99 964 | 14·27 | ·99 823 | 64·66 |
| + ·85 | 1·00 000 | ·01 | 1·00 000 | ·06 | 1·00 000 | ·33 | ·99 997 | 1·85 | ·99 980 | 10·37 |
| + ·90 | 1·00 000 | ·00 | 1·00 000 | ·00 | 1·00 000 | ·01 | 1·00 000 | ·08 | ·99 999 | ·55 |
| + ·95 | 1·00 000 | ·00 | 1·00 000 | ·00 | 1·00 000 | ·00 | 1·00 000 | ·00 | 1·00 000 | ·00 |
| +1·00 | 1·00 000 | ·00 | 1·00 000 | ·00 | 1·00 000 | ·00 | 1·00 000 | ·00 | 1·00 000 | ·00 |

$n = 23$

$\rho = 0.5 - 0.9$

| | $\rho = 0.5$ | | $\rho = 0.6$ | | $\rho = 0.7$ | | $\rho = 0.8$ | | | $\rho = 0.9$ | |
|---|---|---|---|---|---|---|---|---|---|---|---|
| r | Area | Ordinate | Area | Ordinate | Area | Ordinate | Area | Ordinate | r | Area | Ordinate |
| −1·00 | ·00 000 | ·00 | ·00 000 | ·00 | ·00 000 | ·00 | ·00 000 | ·00 | −1·00 | ·00 000 | ·00 |
| − ·95 | ·00 000 | ·00 | ·00 000 | ·00 | ·00 000 | ·00 | ·00 000 | ·00 | − ·95 | ·00 000 | ·00 |
| − ·90 | ·00 000 | ·00 | ·00 000 | ·00 | ·00 000 | ·00 | ·00 000 | ·00 | − ·90 | ·00 000 | ·00 |
| − ·85 | ·00 000 | ·00 | ·00 000 | ·00 | ·00 000 | ·00 | ·00 000 | ·00 | − ·85 | ·00 000 | ·00 |
| − ·80 | ·00 000 | ·00 | ·00 000 | ·00 | ·00 000 | ·00 | ·00 000 | ·00 | − ·80 | ·00 000 | ·00 |
| − ·75 | ·00 000 | ·00 | ·00 000 | ·00 | ·00 000 | ·00 | ·00 000 | ·00 | − ·75 | ·00 000 | ·00 |
| − ·70 | ·00 000 | ·00 | ·00 000 | ·00 | ·00 000 | ·00 | ·00 000 | ·00 | − ·70 | ·00 000 | ·00 |
| − ·65 | ·00 000 | ·00 | ·00 000 | ·00 | ·00 000 | ·00 | ·00 000 | ·00 | − ·65 | ·00 000 | ·00 |
| − ·60 | ·00 000 | ·00 | ·00 000 | ·00 | ·00 000 | ·00 | ·00 000 | ·00 | − ·60 | ·00 000 | ·00 |
| − ·55 | ·00 000 | ·01 | ·00 000 | ·00 | ·00 000 | ·00 | ·00 000 | ·00 | − ·55 | ·00 000 | ·00 |
| − ·50 | ·00 000 | ·04 | ·00 000 | ·00 | ·00 000 | ·00 | ·00 000 | ·00 | − ·50 | ·00 000 | ·00 |
| − ·45 | ·00 000 | ·11 | ·00 000 | ·01 | ·00 000 | ·00 | ·00 000 | ·00 | − ·45 | ·00 000 | ·00 |
| − ·40 | ·00 001 | ·29 | ·00 000 | ·02 | ·00 000 | ·00 | ·00 000 | ·00 | − ·40 | ·00 000 | ·00 |
| − ·35 | ·00 004 | ·69 | ·00 000 | ·06 | ·00 000 | ·00 | ·00 000 | ·00 | − ·35 | ·00 000 | 00 |
| − ·30 | ·00 009 | 1·54 | ·00 001 | ·15 | ·00 000 | ·01 | ·00 000 | ·00 | − ·30 | ·00 000 | ·00 |
| − ·25 | ·00 021 | 3·28 | ·00 002 | ·36 | ·00 000 | ·02 | ·00 000 | ·00 | − ·25 | ·00 000 | ·00 |
| − ·20 | ·00 045 | 6·67 | ·00 005 | ·79 | ·00 000 | ·04 | ·00 000 | ·00 | − ·20 | ·00 000 | ·00 |
| − ·15 | ·00 092 | 12·98 | ·00 011 | 1·68 | ·00 001 | ·10 | ·00 000 | ·00 | − ·15 | ·00 000 | ·00 |
| − ·10 | ·00 183 | 24·29 | ·00 023 | 3·46 | ·00 001 | ·23 | ·00 000 | ·00 | − ·10 | ·00 000 | ·00 |
| − ·05 | ·00 349 | 43·81 | ·00 048 | 6·89 | ·00 003 | ·51 | ·00 000 | ·01 | − ·05 | ·00 000 | ·00 |
| 0·00 | ·00 642 | 76·30 | ·00 097 | 13·33 | ·00 007 | 1·10 | ·00 000 | ·02 | 0·00 | ·00 000 | ·00 |
| + ·05 | ·01 144 | 128·44 | ·00 190 | 25·06 | ·00 015 | 2·30 | ·00 000 | ·06 | + ·05 | ·00 000 | ·00 |
| + ·10 | ·01 974 | 208·99 | ·00 362 | 45·84 | ·00 032 | 4·75 | ·00 001 | ·13 | + ·10 | ·00 000 | ·00 |
| + ·15 | ·03 299 | 328·60 | ·00 673 | 81·62 | ·00 067 | 9·60 | ·00 002 | ·30 | + ·15 | ·00 000 | ·00 |
| + ·20 | ·05 345 | 499·05 | ·01 218 | 141·37 | ·00 136 | 19·07 | ·00 004 | ·69 | + ·20 | ·00 000 | ·00 |
| + ·25 | ·08 391 | 730·12 | ·02 148 | 237·92 | ·00 272 | 37·20 | ·00 009 | 1·56 | + ·25 | ·00 000 | ·00 |
| + ·30 | ·12 756 | 1026·34 | ·03 687 | 388·31 | ·00 534 | 71·21 | ·00 022 | 3·55 | + ·30 | ·00 000 | 01 |
| + ·35 | ·18 752 | 1380·56 | ·06 154 | 612·67 | ·01 030 | 133·58 | ·00 049 | 8·04 | + ·35 | ·00 000 | ·02 |
| + ·40 | ·26 616 | 1767·06 | ·09 970 | 930·25 | ·01 950 | 244·82 | ·00 111 | 18·14 | + ·40 | ·00 000 | ·06 |
| + ·45 | ·36 395 | 2135·75 | ·15 630 | 1350·57 | ·03 612 | 436·53 | ·00 251 | 40·80 | + ·45 | ·00 001 | ·17 |
| + ·50 | ·47 821 | 2412·26 | ·23 623 | 1858·09 | ·06 522 | 752·42 | ·00 564 | 91·21 | + ·50 | ·00 002 | ·52 |
| + ·55 | ·60 210 | 2509·73 | ·34 258 | 2391·84 | ·11 426 | 1241·53 | ·01 265 | 201·81 | + ·55 | ·00 007 | 1·65 |
| + ·60 | ·72 498 | 2357·16 | ·47 383 | 2828·70 | ·19 279 | 1932·86 | ·02 789 | 438·57 | + ·60 | ·00 023 | 5·42 |
| + ·625 | ·78 182 | 2179·48 | ·54 632 | 2956·44 | ·24 618 | 2344·27 | ·04 123 | 639·74 | + ·625 | ·00 042 | 10·00 |
| + ·650 | ·83 345 | 1941·17 | ·62 089 | 2992·09 | ·31 018 | 2776·53 | ·06 058 | 923·77 | + ·650 | ·00 076 | 18·68 |
| + ·675 | ·87 849 | 1655·46 | ·69 499 | 2915·95 | ·38 491 | 3195·79 | ·08 832 | 1316·12 | + ·675 | ·00 142 | 35·33 |
| + ·700 | ·91 599 | 1341·82 | ·76 566 | 2717·16 | ·46 945 | 3552·48 | ·12 750 | 1842·02 | + ·700 | ·00 267 | 67·66 |
| + ·725 | ·94 554 | 1023·95 | ·82 985 | 2399·14 | ·56 148 | 3783·05 | ·15 379 | 2517·46 | + ·725 | ·00 505 | 131·05 |
| + ·750 | ·96 736 | 726·69 | ·88 480 | 1983·68 | ·65 697 | 3818·06 | ·25 454 | 3332·08 | + ·750 | ·00 973 | 256·26 |
| + ·775 | ·98 224 | 471·90 | ·92 856 | 1511·92 | ·75 026 | 3599·30 | ·34 893 | 4222·18 | + ·775 | ·01 887 | 504·03 |
| + ·800 | ·99 144 | 274·26 | ·96 039 | 1039·58 | ·83 463 | 3105·96 | ·46 506 | 5037·18 | + ·800 | ·03 688 | 990·75 |
| + ·825 | ·99 647 | 138·28 | ·98 103 | 625·44 | ·90 364 | 2383·33 | ·59 804 | 5519·91 | + ·810 | ·04 825 | 1295·02 |
| + ·850 | ·99 881 | 57·78 | ·99 253 | 314·71 | ·95 292 | 1557·24 | ·73 557 | 5348·05 | + ·820 | ·06 308 | 1687·81 |
| + ·875 | ·99 970 | 18·63 | ·99 777 | 123·43 | ·98 214 | 809·11 | ·85 807 | 4305·50 | + ·830 | ·08 237 | 2190·38 |
| + ·900 | ·99 995 | 4·11 | ·99 955 | 33·50 | ·99 538 | 297·55 | ·94 501 | 2583·97 | + ·840 | ·10 733 | 2825·60 |
| | | | | | | | | | + ·850 | ·13 940 | 3615·31 |
| + ·925 | 1·00 000 | ·50 | ·99 995 | 5·03 | ·99 936 | 62·09 | ·98 785 | 936·11 | + ·860 | ·18 020 | 4574·96 |
| + ·950 | 1·00 000 | ·02 | 1·00 000 | ·25 | ·99 996 | 4·46 | ·99 911 | 125·64 | + ·870 | ·23 147 | 5704·61 |
| + ·975 | 1·00 000 | ·00 | 1·00 000 | ·00 | 1·00 000 | ·02 | 1·00 000 | 1·27 | + ·880 | ·29 477 | 6974·87 |
| +1·000 | 1·00 000 | ·00 | 1·00 000 | ·00 | 1·00 000 | ·00 | 1·00 000 | ·00 | + ·890 | ·37 118 | 8307·62 |
| | | | | | | | | | + ·900 | ·46 064 | 9553·82 |
| | | | | | | | | | + ·910 | ·56 119 | 10478·01 |
| | | | | | | | | | + ·920 | ·66 809 | 10767·78 |
| | | | | | | | | | + ·930 | ·77 335 | 10103·85 |
| | | | | | | | | | + ·940 | ·86 639 | 8321·01 |
| | | | | | | | | | + ·950 | ·93 672 | 5642·53 |
| | | | | | | | | | + ·955 | ·96 128 | 4185·29 |
| | | | | | | | | | + ·960 | ·97 873 | 2820·21 |
| | | | | | | | | | + ·965 | ·98 985 | 1675·19 |
| | | | | | | | | | + ·970 | ·99 599 | 838·86 |
| | | | | | | | | | + ·975 | ·99 879 | 330·21 |
| | | | | | | | | | + ·980 | ·99 975 | 90·79 |
| | | | | | | | | | + ·985 | ·99 997 | 13·95 |
| | | | | | | | | | + ·990 | 1·00 000 | ·72 |
| | | | | | | | | | + ·995 | 1·00 000 | ·00 |
| | | | | | | | | | +1·000 | 1·00 000 | ·00 |

| | $\rho = 0.9$ | |
|---|---|---|
| r | Area | Ordinate |
| + ·800 | ·03 688 | 990·75 |
| + ·825 | ·07 208 | 1922·15 |
| + ·850 | ·13 940 | 3615·31 |
| + ·875 | ·26 150 | 6325·65 |
| + ·900 | ·46 064 | 9553·82 |
| + ·925 | ·72 204 | 10571·54 |
| + ·950 | ·93 672 | 5642·53 |
| + ·975 | ·99 879 | 330·21 |
| +1·000 | 1·00 000 | ·00 |

$\rho=0\cdot0 - 0\cdot4$      $n=24$

| r | $\rho=0\cdot0$ Area | Ordinate | $\rho=0\cdot1$ Area | Ordinate | $\rho=0\cdot2$ Area | Ordinate | $\rho=0\cdot3$ Area | Ordinate | $\rho=0\cdot4$ Area | Ordinate |
|---|---|---|---|---|---|---|---|---|---|---|
| −1·00 | ·00 000 | ·00 | ·00 000 | ·00 | ·00 000 | ·00 | ·00 000 | ·00 | ·00 000 | ·00 |
| − ·95 | ·00 000 | ·00 | ·00 000 | ·00 | ·00 000 | ·00 | ·00 000 | ·00 | ·00 000 | ·00 |
| − ·90 | ·00 000 | ·00 | ·00 000 | ·00 | ·00 000 | ·00 | ·00 000 | ·00 | ·00 000 | ·00 |
| − ·85 | ·00 000 | ·01 | ·00 000 | ·00 | ·00 000 | ·00 | ·00 000 | ·00 | ·00 000 | ·00 |
| − ·80 | ·00 000 | ·07 | ·00 000 | ·00 | ·00 000 | ·00 | ·00 000 | ·00 | ·00 000 | ·00 |
| − ·75 | ·00 001 | ·48 | ·00 000 | ·08 | ·00 000 | ·01 | ·00 000 | ·00 | ·00 000 | ·00 |
| − ·70 | ·00 007 | 2·20 | ·00 001 | ·43 | ·00 000 | ·07 | ·00 000 | ·00 | ·00 000 | ·00 |
| − ·65 | ·00 029 | 7·63 | ·00 006 | 1·65 | ·00 001 | ·30 | ·00 000 | ·05 | ·00 000 | ·00 |
| − ·60 | ·00 097 | 21·33 | ·00 021± | 5·12 | ·00 004 | 1·04 | ·00 001 | ·17 | ·00 000 | ·02 |
| − ·55 | ·00 268 | 50·42 | ·00 065± | 13·46 | ·00 013 | 3·01 | ·00 002 | ·55 | ·00 000 | ·08 |
| − ·50 | ·00 642 | 104·18 | ·00 171 | 30·95 | ·00 038 | 7·63 | ·00 007 | 1·52 | ·00 001 | ·23 |
| − ·45 | ·01 368 | 192·53 | ·00 400 | 63·69 | ·00 098 | 17·31 | ·00 019 | 3·76 | ·00 003 | ·63 |
| − ·40 | ·02 639 | 323·58 | ·00 846 | 119·24 | ·00 226 | 35·80 | ·00 049 | 8·54 | ·00 008 | 1·54 |
| − ·35 | ·04 681 | 500·79 | ·01 644 | 205·69 | ·00 479 | 68·31 | ·00 112 | 17·90 | ·00 020 | 3·53 |
| − ·30 | ·07 718 | 720·45 | ·02 966 | 329·99 | ·00 943 | 121·39 | ·00 241 | 35·02 | ·00 047 | 7·57 |
| − ·25 | ·11 936 | 970·29 | ·05 013 | 495·86 | ·01 740 | 202·37 | ·00 483 | 64·42 | ·00 102 | 15·29 |
| − ·20 | ·17 438 | 1229·99 | ·07 992 | 701·70 | ·03 025± | 318·18 | ·00 915± | 112·03 | ·00 210 | 29·30 |
| − ·15 | ·24 209 | 1473·52 | ·12 083 | 938·94 | ·04 988 | 473·77 | ·01 646 | 184·97 | ·00 412 | 53·46 |
| − ·10 | ·32 100 | 1673·17 | ·17 407 | 1191·49 | ·07 831 | 670·05 | ·02 820 | 290·81 | ·00 771 | 93·20 |
| − ·05 | ·40 826 | 1804·33 | ·23 986 | 1436·72 | ·11 748 | 901·95 | ·04 620 | 436·29 | ·01 381 | 155·59 |
| 0·00 | ·50 000 | 1850·07 | ·31 718 | 1648·13 | ·16 890 | 1156·93 | ·07 256 | 625·41 | ·02 378 | 249·11 |
| + ·05 | ·59 174 | 1804·33 | ·40 367 | 1799·35 | ·23 323 | 1414·71 | ·10 945± | 857·07 | ·03 940 | 382·82 |
| + ·10 | ·67 900 | 1673·17 | ·49 575‾ | 1868·86 | ·30 997 | 1648·62 | ·15 883 | 1122·60 | ·06 287 | 564·61 |
| + ·15 | ·75 791 | 1473·52 | ·58 898 | 1844·51 | ·39 719 | 1828·88 | ·22 198 | 1403·99 | ·09 673 | 798·61 |
| + ·20 | ·82 562 | 1229·99 | ·67 863 | 1726·50 | ·49 150‾ | 1927·58 | ·29 904 | 1673·57 | ·14 355‾ | 1081·63 |
| + ·25 | ·88 064 | 970·29 | ·76 029 | 1528·14 | ·58 825 | 1924·68 | ·38 856 | 1896·13 | ·20 548 | 1399·26 |
| + ·30 | ·92 282 | 720·45 | ·83 051 | 1273·84 | ·68 215‾ | 1813·38 | ·48 724 | 2034·06 | ·28 359 | 1722·83 |
| + ·35 | ·95 319 | 500·79 | ·88 726 | 994·68 | ·76 794 | 1603·54 | ·59 001 | 2055·28 | ·37 715‾ | 2009·07 |
| + ·40 | ·97 361 | 323·58 | ·93 009 | 722·40 | ·84 129 | 1321·50 | ·69 052 | 1942·61 | ·48 298 | 2204·43 |
| + ·45 | ·98 632 | 192·53 | ·96 006 | 483·43 | ·89 952 | 1005·54 | ·78 212 | 1701·90 | ·59 518 | 2255·95 |
| + ·50 | ·99 358 | 104·18 | ·97 928 | 294·40 | ·94 198 | 697·76 | ·85 910 | 1365·25 | ·70 556 | 2128·00 |
| + ·55 | ·99 732 | 50·42 | ·99 042 | 160·45 | ·97 004 | 434·25 | ·91 794 | 986·41 | ·80 497 | 1820·73 |
| + ·60 | ·99 903 | 21·33 | ·99 616 | 76·48 | ·98 652 | 236·89 | ·95 808 | 627·46 | ·88 542 | 1381·90 |
| + ·65 | ·99 971 | 7·63 | ·99 871 | 30·86 | ·99 491 | 109·64 | ·98 190 | 340·21 | ·94 245± | 901·23 |
| + ·70 | ·99 993 | 2·20 | ·99 966 | 10·05 | ·99 847 | 41·03 | ·99 373 | 149·89 | ·97 658 | 481·72 |
| + ·75 | ·99 999 | ·48 | ·99 993 | 2·45 | ·99 966 | 11·52 | ·99 840 | 49·80 | ·99 289 | 195·98 |
| + ·80 | 1·00 000 | ·07 | ·99 999 | ·39 | ·99 995± | 2·14 | ·99 974 | 11·00 | ·99 859 | 53·55 |
| + ·85 | 1·00 000 | ·01 | 1·00 000 | ·03 | 1·00 000 | ·21 | ·99 998 | 1·28 | ·99 985‾ | 7·76 |
| + ·90 | 1·00 000 | ·00 | 1·00 000 | ·00 | 1·00 000 | ·01 | 1·00 000 | ·04 | ·99 999 | ·35 |
| + ·95 | 1·00 000 | ·00 | 1·00 000 | ·00 | 1·00 000 | ·00 | 1·00 000 | ·00 | 1·00 000 | ·00 |
| +1·00 | 1·00 000 | ·00 | 1·00 000 | ·00 | 1·00 000 | ·00 | 1·00 000 | ·00 | 1·00 000 | ·00 |

n = 24　　　　　　　　　　　　　　　　　　ρ = 0·5 — 0·9

| r | ρ=0·5 Area | Ordinate | ρ=0·6 Area | Ordinate | ρ=0·7 Area | Ordinate | ρ=0·8 Area | Ordinate | r | ρ=0·9 Area | Ordinate |
|---|---|---|---|---|---|---|---|---|---|---|---|
| −1·00 | ·00 000 | ·00 | ·00 000 | ·00 | ·00 000 | ·00 | ·00 000 | ·00 | −1·00 | ·00 000 | ·00 |
| − ·95 | ·00 000 | ·00 | ·00 000 | ·00 | ·00 000 | ·00 | ·00 000 | ·00 | − ·95 | ·00 000 | ·00 |
| − ·90 | ·00 000 | ·00 | ·00 000 | ·00 | ·00 000 | ·00 | ·00 000 | ·00 | − ·90 | ·00 000 | ·00 |
| − ·85 | ·00 000 | ·00 | ·00 000 | ·00 | ·00 000 | ·00 | ·00 000 | ·00 | − ·85 | ·00 000 | ·00 |
| − ·80 | ·00 000 | ·00 | ·00 000 | ·00 | ·00 000 | ·00 | ·00 000 | ·00 | − ·80 | ·00 000 | ·00 |
| − ·75 | ·00 000 | ·00 | ·00 000 | ·00 | ·00 000 | ·00 | ·00 000 | ·00 | − ·75 | ·00 000 | ·00 |
| − ·70 | ·00 000 | ·00 | ·00 000 | ·00 | ·00 000 | ·00 | ·00 000 | ·00 | − ·70 | ·00 000 | ·00 |
| − ·65 | ·00 000 | ·00 | ·00 000 | ·00 | ·00 000 | ·00 | ·00 000 | ·00 | − ·65 | ·00 000 | ·00 |
| − ·60 | ·00 000 | ·00 | ·00 000 | ·00 | ·00 000 | ·00 | ·00 000 | ·00 | − ·60 | ·00 000 | ·00 |
| − ·55 | ·00 000 | ·01 | ·00 000 | ·00 | ·00 000 | ·00 | ·00 000 | ·00 | − ·55 | ·00 000 | ·00 |
| − ·50 | ·00 000 | ·03 | ·00 000 | ·00 | ·00 000 | ·00 | ·00 000 | ·00 | − ·50 | ·00 000 | ·00 |
| − ·45 | ·00 000 | ·08 | ·00 000 | ·01 | ·00 000 | ·00 | ·00 000 | ·00 | − ·45 | ·00 000 | ·00 |
| − ·40 | ·00 001 | ·20 | ·00 000 | ·02 | ·00 000 | ·00 | ·00 000 | ·00 | − ·40 | ·00 000 | ·00 |
| − ·35 | ·00 003 | ·49 | ·00 000 | ·04 | ·00 000 | ·00 | ·00 000 | ·00 | − ·35 | ·00 000 | ·00 |
| − ·30 | ·00 007 | 1·13 | ·00 001 | ·10 | ·00 000 | ·00 | ·00 000 | ·00 | − ·30 | ·00 000 | ·00 |
| − ·25 | ·00 015 | 2·51 | ·00 001 | ·25 | ·00 000 | ·01 | ·00 000 | ·00 | − ·25 | ·00 000 | ·00 |
| − ·20 | ·00 034 | 5·27 | ·00 003 | ·57 | ·00 000 | ·03 | ·00 000 | ·00 | − ·20 | ·00 000 | ·00 |
| − ·15 | ·00 072 | 10·58 | ·00 008 | 1·25 | ·00 000 | ·07 | ·00 000 | ·00 | − ·15 | ·00 000 | ·00 |
| − ·10 | ·00 147 | 20·41 | ·00 017 | 2·66 | ·00 001 | ·16 | ·00 000 | ·00 | − ·10 | ·00 000 | ·00 |
| − ·05 | ·00 279 | 37·86 | ·00 036 | 5·48 | ·00 002 | ·36 | ·00 000 | ·01 | − ·05 | ·00 000 | ·00 |
| 0·00 | ·00 546 | 67·67 | ·00 076 | 10·92 | ·00 005 | ·80 | ·00 000 | ·01 | 0·00 | ·00 000 | ·00 |
| + ·05 | ·00 998 | 116·68 | ·00 154 | 21·14 | ·00 011 | 1·74 | ·00 000 | ·04 | + ·05 | ·00 000 | ·00 |
| + ·10 | ·01 760 | 194·13 | ·00 301 | 39·75 | ·00 024 | 3·71 | ·00 001 | ·09 | + ·10 | ·00 000 | ·00 |
| + ·15 | ·03 005 | 311·57 | ·00 575 | 72·66 | ·00 051 | 7·76 | ·00 001 | ·21 | + ·15 | ·00 000 | ·00 |
| + ·20 | ·04 964 | 481·83 | ·01 067 | 128·95 | ·00 108 | 15·89 | ·00 003 | ·49 | + ·20 | ·00 000 | ·00 |
| + ·25 | ·07 931 | 716·52 | ·01 925 | 222·03 | ·00 223 | 31·93 | ·00 007 | 1·16 | + ·25 | ·00 000 | ·00 |
| + ·30 | ·12 247 | 1021·51 | ·03 378 | 370·08 | ·00 452 | 62·89 | ·00 016 | 2·74 | + ·30 | ·00 000 | ·00 |
| + ·35 | ·18 254 | 1390·19 | ·05 754 | 595·15 | ·00 897 | 121·20 | ·00 039 | 6·42 | + ·35 | ·00 000 | ·01 |
| + ·40 | ·26 211 | 1795·33 | ·09 495 | 919·02 | ·01 744 | 227·90 | ·00 088 | 15·02 | + ·40 | ·00 000 | ·04 |
| + ·45 | ·36 179 | 2182·51 | ·15 130 | 1353·49 | ·03 311 | 416·18 | ·00 206 | 34·98 | + ·45 | ·00 001 | ·12 |
| + ·50 | ·47 872 | 2470·20 | ·23 191 | 1883·18 | ·06 120 | 733·09 | ·00 480 | 80·88 | + ·50 | ·00 002 | ·37 |
| + ·55 | ·60 556 | 2563·87 | ·34 015 | 2442·41 | ·10 948 | 1232·91 | ·01 110 | 184·91 | + ·55 | ·00 005 | 1·22 |
| + ·60 | ·73 059 | 2388·99 | ·47 444 | 2896·56 | ·18 815 | 1949·56 | ·02 535 | 414·53 | + ·60 | ·00 017 | 4·21 |
| + ·625 | ·78 802 | 2194·60 | ·54 864 | 3024·93 | ·24 212 | 2379·03 | ·03 805 | 613·63 | + ·625 | ·00 032 | 7·96 |
| + ·650 | ·83 980 | 1938·06 | ·62 485 | 3053·53 | ·30 727 | 2831·08 | ·05 675 | 898·52 | + ·650 | ·00 060 | 15·27 |
| + ·675 | ·88 453 | 1634·98 | ·70 031 | 2962·05 | ·38 357 | 3267·66 | ·08 393 | 1296·92 | + ·675 | ·00 114 | 29·64 |
| + ·700 | ·92 134 | 1307·36 | ·77 187 | 2740·64 | ·47 009 | 3637·57 | ·12 279 | 1836·75 | + ·700 | ·00 219 | 58·29 |
| + ·725 | ·94 991 | 981·04 | ·83 631 | 2395·77 | ·56 429 | 3868·66 | ·17 712 | 2536·25 | + ·725 | ·00 428 | 115·93 |
| + ·750 | ·97 062 | 682·00 | ·89 085 | 1954·22 | ·66 175 | 3887·76 | ·25 085 | 3385·03 | + ·750 | ·00 848 | 232·77 |
| + ·775 | ·98 442 | 431·77 | ·93 360 | 1462·98 | ·75 641 | 3635·61 | ·34 704 | 4313·80 | + ·775 | ·01 692 | 469·95 |
| + ·800 | ·99 273 | 243·21 | ·96 408 | 982·60 | ·84 113 | 3097·07 | ·46 587 | 5157·63 | + ·800 | ·03 395 | 947·52 |
| + ·825 | ·99 711 | 117·96 | ·98 333 | 573·35 | ·90 934 | 2331·11 | ·60 192 | 5636·57 | + ·810 | ·04 488 | 1250·70 |
| + ·850 | ·99 908 | 46·94 | ·99 370 | 277·15 | ·95 697 | 1481·10 | ·74 177 | 5408·62 | + ·820 | ·05 927 | 1645·60 |
| + ·875 | ·99 973 | 14·22 | ·99 822 | 103·05 | ·98 430 | 739·16 | ·86 461 | 4268·40 | + ·830 | ·07 817 | 2155·14 |
| + ·900 | ·99 995 | 2·89 | ·99 967 | 26·01 | ·99 614 | 256·32 | ·94 953 | 2471·22 | + ·840 | ·10 284 | 2804·24 |
|  |  |  |  |  |  |  |  |  | + ·850 | ·13 480 | 3616·87 |
| + ·925 | 1·00 000 | ·31 | ·99 997 | 3·52 | ·99 949 | 48·94 | ·98 956 | 840·43 | + ·860 | ·17 578 | 4610·23 |
| + ·950 | 1·00 000 | ·01 | 1·00 000 | ·15 | ·99 997 | 3·04 | ·99 932 | 100·42 | + ·870 | ·22 761 | 5784·70 |
| + ·975 | 1·00 000 | ·00 | 1·00 000 | ·00 | 1·00 000 | ·01 | 1·00 000 | ·79 | + ·880 | ·29 198 | 7108·27 |
| +1·000 | 1·00 000 | ·00 | 1·00 000 | ·00 | 1·00 000 | ·00 | 1·00 000 | ·00 | + ·890 | ·37 000 | 8495·15 |
|  |  |  |  |  |  |  |  |  | + ·900 | ·46 156 | 9781·81 |
|  |  |  |  |  |  |  |  |  | + ·910 | ·56 446 | 10711·65 |
|  |  |  |  |  |  |  |  |  | + ·920 | ·67 349 | 10949·90 |
|  |  |  |  |  |  |  |  |  | + ·930 | ·78 004 | 10168·18 |
|  |  |  |  |  |  |  |  |  | + ·940 | ·87 294 | 8227·10 |
|  |  |  |  |  |  |  |  |  | + ·950 | ·94 161 | 5422·78 |
|  |  |  |  |  |  |  |  |  | + ·955 | ·96 500 | 3943·14 |
|  |  |  |  |  |  |  |  |  | + ·960 | ·98 126 | 2591·28 |
|  |  |  |  |  |  |  |  |  | + ·965 | ·99 134 | 1490·96 |
|  |  |  |  |  |  |  |  |  | + ·970 | ·99 671 | 716·62 |
|  |  |  |  |  |  |  |  |  | + ·975 | ·99 905 | 267·31 |
|  |  |  |  |  |  |  |  |  | + ·980 | ·99 981 | 68·33 |
|  |  |  |  |  |  |  |  |  | + ·985 | ·99 997 | 9·47 |
|  |  |  |  |  |  |  |  |  | + ·990 | 1·00 000 | ·42 |
|  |  |  |  |  |  |  |  |  | + ·995 | 1·00 000 | ·00 |
|  |  |  |  |  |  |  |  |  | +1·000 | 1·00 000 | ·00 |

| ρ=0·9 | | |
|---|---|---|
| r | Area | Ordinate |
| + ·800 | ·03 395 | 947·52 |
| + ·825 | ·06 808 | 1882·73 |
| + ·850 | ·13 480 | 3616·87 |
| + ·875 | ·25 814 | 6431·67 |
| + ·900 | ·46 156 | 9781·81 |
| + ·925 | ·72 775 | 10702·52 |
| + ·950 | ·94 161 | 5422·78 |
| + ·975 | ·99 905 | 267·31 |
| +1·000 | 1·00 000 | ·00 |

ρ=0.0 — 0.4                    n = 25

| r | ρ=0.0 Area | Ordinate | ρ=0.1 Area | Ordinate | ρ=0.2 Area | Ordinate | ρ=0.3 Area | Ordinate | ρ=0.4 Area | Ordinate |
|---|---|---|---|---|---|---|---|---|---|---|
| −1·00 | ·00 000 | ·00 | ·00 000 | ·00 | ·00 000 | ·00 | ·00 000 | ·00 | ·00 000 | ·00 |
| − ·95 | ·00 000 | ·00 | ·00 000 | ·00 | ·00 000 | ·00 | ·00 000 | ·00 | ·00 000 | ·00 |
| − ·90 | ·00 000 | ·00 | ·00 000 | ·00 | ·00 000 | ·00 | ·00 000 | ·00 | ·00 000 | ·00 |
| − ·85 | ·00 000 | ·00 | ·00 000 | ·00 | ·00 000 | ·00 | ·00 000 | ·00 | ·00 000 | ·00 |
| − ·80 | ·00 000 | ·04 | ·00 000 | ·00 | ·00 000 | ·00 | ·00 000 | ·00 | ·00 000 | ·00 |
| − ·75 | ·00 001 | ·32 | ·00 000 | ·05 | ·00 000 | ·00 | ·00 000 | ·00 | ·00 000 | ·00 |
| − ·70 | ·00 005 | 1·61 | ·00 001 | ·29 | ·00 000 | ·05 | ·00 000 | ·00 | ·00 000 | ·00 |
| − ·65 | ·00 022 | 5·93 | ·00 005 | 1·20 | ·00 001 | ·21 | ·00 000 | ·03 | ·00 000 | ·00 |
| − ·60 | ·00 076 | 17·46 | ·00 016 | 3·93 | ·00 003 | ·75 | ·00 000 | ·12 | ·00 000 | ·01 |
| − ·55 | ·00 220 | 43·08 | ·00 050 | 10·85 | ·00 010 | 2·27 | ·00 002 | ·38 | ·00 000 | ·05 |
| − ·50 | ·00 546 | 92·30 | ·00 138 | 25·99 | ·00 029 | 6·01 | ·00 005 | 1·11 | ·00 001 | ·16 |
| − ·45 | ·01 200 | 175·88 | ·00 334 | 55·40 | ·00 077 | 14·22 | ·00 014 | 2·89 | ·00 002 | ·44 |
| − ·40 | ·02 378 | 303·38 | ·00 729 | 106·96 | ·00 185 | 30·45 | ·00 037 | 6·82 | ·00 006 | 1·14 |
| − ·35 | ·04 316 | 479·90 | ·01 405 | 189·49 | ·00 404 | 59·94 | ·00 089 | 14·80 | ·00 015 | 2·72 |
| − ·30 | ·07 256 | 703·06 | ·02 689 | 311·08 | ·00 817 | 109·50 | ·00 197 | 29·91 | ·00 036 | 6·05 |
| − ·25 | ·11 405 | 961·07 | ·04 640 | 476·77 | ·01 545 | 187·04 | ·00 408 | 56·62 | ·00 081 | 12·62 |
| − ·20 | ·16 890 | 1232·83 | ·07 530 | 686·08 | ·02 748 | 300·46 | ·00 793 | 101·06 | ·00 171 | 24·92 |
| − ·15 | ·23 710 | 1490·32 | ·11 560 | 930·93 | ·04 620 | 455·82 | ·01 460 | 170·78 | ·00 346 | 46·75 |
| − ·10 | ·31 718 | 1703·04 | ·16 871 | 1194·73 | ·07 379 | 655·13 | ·02 557 | 274·14 | ·00 664 | 83·60 |
| − ·05 | ·40 619 | 1843·49 | ·23 499 | 1453·28 | ·11 238 | 893·96 | ·04 271 | 418·94 | ·01 218 | 142·84 |
| 0·00 | ·50 000 | 1892·58 | ·31 347 | 1677·56 | ·16 364 | 1159·60 | ·06 824 | 610·31 | ·02 145 | 233·56 |
| + ·05 | ·59 381 | 1843·49 | ·40 169 | 1838·37 | ·22 843 | 1430·51 | ·10 452 | 848·05 | ·03 624 | 365·78 |
| + ·10 | ·68 282 | 1703·04 | ·49 584 | 1911·80 | ·30 630 | 1677·69 | ·15 368 | 1123·71 | ·05 887 | 548·65 |
| + ·15 | ·76 290 | 1490·32 | ·59 118 | 1884·46 | ·39 525 | 1868·40 | ·21 721 | 1419·41 | ·09 204 | 787·53 |
| + ·20 | ·83 110 | 1232·83 | ·68 261 | 1756·95 | ·49 168 | 1971·86 | ·29 535 | 1702·29 | ·13 853 | 1080·00 |
| + ·25 | ·88 595 | 961·07 | ·76 545 | 1544·63 | ·59 061 | 1966·16 | ·38 662 | 1936·84 | ·20 070 | 1411·36 |
| + ·30 | ·92 744 | 703·06 | ·83 612 | 1275·10 | ·68 635 | 1844·50 | ·48 752 | 2080·76 | ·27 981 | 1750·96 |
| + ·35 | ·95 684 | 479·90 | ·89 258 | 982·79 | ·77 332 | 1618·89 | ·59 259 | 2099·17 | ·37 515 | 2051·69 |
| + ·40 | ·97 622 | 303·38 | ·93 459 | 701·99 | ·84 700 | 1319·51 | ·69 503 | 1974·33 | ·48 335 | 2255·00 |
| + ·45 | ·98 800 | 175·88 | ·96 344 | 460·12 | ·90 475 | 989·05 | ·78 776 | 1714·58 | ·59 806 | 2303·40 |
| + ·50 | ·99 454 | 92·30 | ·98 152 | 273·17 | ·94 615 | 672·95 | ·86 485 | 1357·37 | ·71 047 | 2159·72 |
| + ·55 | ·99 780 | 43·08 | ·99 173 | 144·33 | ·97 291 | 408·43 | ·92 287 | 962·75 | ·81 086 | 1827·71 |
| + ·60 | ·99 924 | 17·46 | ·99 681 | 66·25 | ·98 820 | 215·85 | ·96 163 | 597·36 | ·89 101 | 1363·75 |
| + ·65 | ·99 978 | 5·93 | ·99 898 | 25·53 | ·99 572 | 95·99 | ·98 398 | 313·40 | ·94 670 | 867·68 |
| + ·70 | ·99 995 | 1·61 | ·99 974 | 7·85 | ·99 877 | 34·15 | ·99 470 | 132·22 | ·97 908 | 447·95 |
| + ·75 | ·99 999 | ·32 | ·99 995 | 1·78 | ·99 974 | 8·99 | ·99 872 | 41·47 | ·99 396 | 173·61 |
| + ·80 | 1·00 000 | ·04 | ·99 999 | ·26 | ·99 996 | 1·53 | ·99 980 | 8·48 | ·99 838 | 44·30 |
| + ·85 | 1·00 000 | ·00 | 1·00 000 | ·02 | ·99 999 | ·13 | ·99 998 | ·88 | ·99 989 | 5·81 |
| + ·90 | 1·00 000 | ·00 | 1·00 000 | ·00 | 1·00 000 | ·00 | 1·00 000 | ·00 | ·99 999 | ·22 |
| + ·95 | 1·00 000 | ·00 | 1·00 000 | ·00 | 1·00 000 | ·00 | 1·00 000 | ·00 | 1·00 000 | ·00 |
| +1·00 | 1·00 000 | ·00 | 1·00 000 | ·00 | 1·00 000 | ·00 | 1·00 000 | ·00 | 1·00 000 | ·00 |

$n=25$ $\qquad$ $\rho=0.5 \text{—} 0.9$

| r | ρ=0.5 Area | ρ=0.5 Ordinate | ρ=0.6 Area | ρ=0.6 Ordinate | ρ=0.7 Area | ρ=0.7 Ordinate | ρ=0.8 Area | ρ=0.8 Ordinate |
|---|---|---|---|---|---|---|---|---|
| −1.00 | ·00 000 | ·00 | ·00 000 | ·00 | ·00 000 | ·00 | ·00 000 | ·00 |
| − ·95 | ·00 000 | ·00 | ·00 000 | ·00 | ·00 000 | ·00 | ·00 000 | ·00 |
| − ·90 | ·00 000 | ·00 | ·00 000 | ·00 | ·00 000 | ·00 | ·00 000 | ·00 |
| − ·85 | ·00 000 | ·00 | ·00 000 | ·00 | ·00 000 | ·00 | ·00 000 | ·00 |
| − ·80 | ·00 000 | ·00 | ·00 000 | ·00 | ·00 000 | ·00 | ·00 000 | ·00 |
| − ·75 | ·00 000 | ·00 | ·00 000 | ·00 | ·00 000 | ·00 | ·00 000 | ·00 |
| − ·70 | ·00 000 | ·00 | ·00 000 | ·00 | ·00 000 | ·00 | ·00 000 | ·00 |
| − ·65 | ·00 000 | ·00 | ·00 000 | ·00 | ·00 000 | ·00 | ·00 000 | ·00 |
| − ·60 | ·00 000 | ·00 | ·00 000 | ·00 | ·00 000 | ·00 | ·00 000 | ·00 |
| − ·55 | ·00 000 | ·00 | ·00 000 | ·00 | ·00 000 | ·00 | ·00 000 | ·00 |
| − ·50 | ·00 000 | ·01 | ·00 000 | ·00 | ·00 000 | ·00 | ·00 000 | ·00 |
| − ·45 | ·00 000 | ·05 | ·00 000 | ·00 | ·00 000 | ·00 | ·00 000 | ·00 |
| − ·40 | ·00 001 | ·13 | ·00 000 | ·01 | ·00 000 | ·00 | ·00 000 | ·00 |
| − ·35 | ·00 002 | ·34 | ·00 000 | ·03 | ·00 000 | ·00 | ·00 000 | ·00 |
| − ·30 | ·00 004 | ·83 | ·00 000 | ·07 | ·00 000 | ·00 | ·00 000 | ·00 |
| − ·25 | ·00 011 | 1·80 | ·00 001 | ·17 | ·00 000 | ·01 | ·00 000 | ·00 |
| − ·20 | ·00 025 | 4·15 | ·00 002 | ·41 | ·00 000 | ·02 | ·00 000 | ·00 |
| − ·15 | ·00 056 | 8·62 | ·00 005 | ·93 | ·00 000 | ·04 | ·00 000 | ·00 |
| − ·10 | ·00 117 | 17·14 | ·00 012 | 2·04 | ·00 001 | ·11 | ·00 000 | ·00 |
| − ·05 | ·00 239 | 32·68 | ·00 028 | 4·35 | ·00 001 | ·25 | ·00 000 | ·00 |
| 0·00 | ·00 465 | 59·95 | ·00 060 | 8·94 | ·00 003 | ·59 | ·00 000 | ·01 |
| + ·05 | ·00 869 | 105·89 | ·00 124 | 17·81 | ·00 008 | 1·32 | ·00 000 | ·02 |
| + ·10 | ·01 570 | 180·12 | ·00 251 | 34·44 | ·00 018 | 2·90 | ·00 000 | ·06 |
| + ·15 | ·02 738 | 295·02 | ·00 471 | 64·59 | ·00 040 | 6·26 | ·00 001 | ·14 |
| + ·20 | ·04 611 | 464·70 | ·00 934 | 117·49 | ·00 087 | 13·22 | ·00 002 | ·35 |
| + ·25 | ·07 406 | 702·41 | ·01 726 | 206·98 | ·00 184 | 27·37 | ·00 005 | ·86 |
| + ·30 | ·11 762 | 1015·61 | ·03 097 | 352·32 | ·00 383 | 55·47 | ·00 012 | 2·11 |
| + ·35 | ·17 772 | 1398·37 | ·05 383 | 577·51 | ·00 782 | 109·85 | ·00 029 | 5·13 |
| + ·40 | ·25 816 | 1822·10 | ·09 046 | 906·95 | ·01 560 | 211·92 | ·00 070 | 12·43 |
| + ·45 | ·35 965 | 2227·89 | ·14 651 | 1354·96 | ·03 037 | 396·34 | ·00 170 | 29·95 |
| + ·50 | ·47 918 | 2526·88 | ·22 771 | 1906·56 | ·05 744 | 713·49 | ·00 409 | 71·65 |
| + ·55 | ·60 883 | 2616·37 | ·33 777 | 2491·37 | ·10 494 | 1223·04 | ·00 977 | 169·24 |
| + ·60 | ·73 598 | 2410·65 | ·47 500 | 2962·87 | ·18 366 | 1964·30 | ·02 305 | 391·39 |
| + ·625 | ·79 395 | 2207·44 | ·55 088 | 3091·69 | ·23 826 | 2411·71 | ·03 513 | 587·96 |
| + ·650 | ·84 618 | 1932·87 | ·62 869 | 3112·90 | ·30 445 | 2883·60 | ·05 319 | 873·02 |
| + ·675 | ·89 021 | 1613·01 | ·70 545 | 3005·66 | ·38 231 | 3338·98 | ·07 977 | 1276·64 |
| + ·700 | ·92 630 | 1272·41 | ·77 782 | 2761·35 | ·47 078 | 3720·72 | ·11 827 | 1829·54 |
| + ·725 | ·95 389 | 938·92 | ·84 245 | 2389·84 | ·56 709 | 3951·97 | ·17 268 | 2552·48 |
| + ·750 | ·97 353 | 639·37 | ·89 652 | 1923·13 | ·66 647 | 3954·50 | ·24 724 | 3435·16 |
| + ·775 | ·98 632 | 394·64 | ·93 825 | 1414·11 | ·76 240 | 3668·36 | ·34 518 | 4402·71 |
| + ·800 | ·99 381 | 215·45 | ·96 741 | 927·75 | ·84 739 | 3084·92 | ·46 662 | 5275·33 |
| + ·825 | ·99 763 | 100·51 | ·98 535 | 525·03 | ·91 475 | 2277·61 | ·60 564 | 5749·57 |
| + ·850 | ·99 927 | 38·09 | ·99 469 | 243·81 | ·96 073 | 1407·18 | ·74 773 | 5464·05 |
| + ·875 | ·99 983 | 10·84 | ·99 857 | 85·94 | ·98 628 | 674·55 | ·87 077 | 4227·11 |
| + ·900 | ·99 997 | 2·03 | ·99 975 | 20·16 | ·99 684 | 220·57 | ·95 365 | 2360·87 |
| + ·925 | 1·00 000 | ·19 | ·99 998 | 2·46 | ·99 964 | 38·53 | ·99 101 | 753·73 |
| + ·950 | 1·00 000 | ·01 | 1·00 000 | ·09 | ·99 997 | 2·07 | ·99 948 | 80·18 |
| + ·975 | 1·00 000 | ·00 | 1·00 000 | ·00 | 1·00 000 | ·01 | 1·00 000 | ·49 |
| +1·000 | 1·00 000 | ·00 | 1·00 000 | ·00 | 1·00 000 | ·00 | 1·00 000 | ·00 |

| r | ρ=0.9 Area | ρ=0.9 Ordinate |
|---|---|---|
| −1·00 | ·00 000 | ·00 |
| − ·95 | ·00 000 | ·00 |
| − ·90 | ·00 000 | ·00 |
| − ·85 | ·00 000 | ·00 |
| − ·80 | ·00 000 | ·00 |
| − ·75 | ·00 000 | ·00 |
| − ·70 | ·00 000 | ·00 |
| − ·65 | ·00 000 | ·00 |
| − ·60 | ·00 000 | ·00 |
| − ·55 | ·00 000 | ·00 |
| − ·50 | ·00 000 | ·00 |
| − ·45 | ·00 000 | ·00 |
| − ·40 | ·00 000 | ·00 |
| − ·35 | ·00 000 | ·00 |
| − ·30 | ·00 000 | ·00 |
| − ·25 | ·00 000 | ·00 |
| − ·20 | ·00 000 | ·00 |
| − ·15 | ·00 000 | ·00 |
| − ·10 | ·00 000 | ·00 |
| − ·05 | ·00 000 | ·00 |
| 0·00 | ·00 000 | ·00 |
| + ·05 | ·00 000 | ·00 |
| + ·10 | ·00 000 | ·00 |
| + ·15 | ·00 000 | ·00 |
| + ·20 | ·00 000 | ·00 |
| + ·25 | ·00 000 | ·00 |
| + ·30 | ·00 000 | ·00 |
| + ·35 | ·00 000 | ·01 |
| + ·40 | ·00 000 | ·02 |
| + ·45 | ·00 000 | ·08 |
| + ·50 | ·00 001 | ·26 |
| + ·55 | ·00 004 | ·90 |
| + ·60 | ·00 013 | 3·26 |
| + ·625 | ·00 026 | 6·33 |
| + ·650 | ·00 050 | 12·46 |
| + ·675 | ·00 094 | 24·84 |
| + ·700 | ·00 184 | 50·16 |
| + ·725 | ·00 364 | 102·44 |
| + ·750 | ·00 743 | 211·20 |
| + ·775 | ·01 520 | 437·71 |
| + ·800 | ·03 129 | 905·21 |
| + ·810 | ·04 178 | 1206·61 |
| + ·820 | ·05 574 | 1602·73 |
| + ·830 | ·07 423 | 2118·22 |
| + ·840 | ·09 859 | 2780·08 |
| + ·850 | ·13 041 | 3614·59 |
| + ·860 | ·17 152 | 4640·84 |
| + ·870 | ·22 387 | 5859·69 |
| + ·880 | ·28 925 | 7236·52 |
| + ·890 | ·36 883 | 8677·69 |
| + ·900 | ·46 244 | 10004·61 |
| + ·910 | ·56 762 | 10938·87 |
| + ·920 | ·67 872 | 11123·29 |
| + ·930 | ·78 645 | 10222·05 |
| + ·940 | ·87 911 | 8125·62 |
| + ·950 | ·94 612 | 5206·06 |
| + ·955 | ·96 838 | 3711·06 |
| + ·960 | ·98 350 | 2378·40 |
| + ·965 | ·99 263 | 1325·58 |
| + ·970 | ·99 733 | 611·55 |
| + ·975 | ·99 928 | 216·17 |
| + ·980 | ·99 988 | 51·37 |
| + ·985 | ·99 999 | 6·42 |
| + ·990 | 1·00 000 | ·24 |
| + ·995 | 1·00 000 | ·00 |
| +1·000 | 1·00 000 | ·00 |

| r | ρ=0.9 Area | ρ=0.9 Ordinate |
|---|---|---|
| + ·800 | ·03 129 | 905·21 |
| + ·825 | ·06 433 | 1842·16 |
| + ·850 | ·13 041 | 3614·59 |
| + ·875 | ·25 483 | 6532·52 |
| + ·900 | ·46 244 | 10004·61 |
| + ·925 | ·73 372 | 10823·62 |
| + ·950 | ·94 612 | 5206·06 |
| + ·975 | ·99 928 | 216·17 |
| +1·000 | 1·00 000 | ·00 |

ρ=0.0 — 0.4      n=50

| r | ρ=0.0 Area | Ordinate | ρ=0.1 Area | Ordinate | ρ=0.2 Area | Ordinate | ρ=0.3 Area | Ordinate | ρ=0.4 Area | Ordinate |
|---|---|---|---|---|---|---|---|---|---|---|
| − .75 | .00 000 | .00 | .00 000 | .00 | .00 000 | .00 | .00 000 | .00 | .00 000 | .00 |
| − .70 | .00 000 | .00 | .00 000 | .00 | .00 000 | .00 | .00 000 | .00 | .00 000 | .00 |
| − .65 | .00 000 | .01 | .00 000 | .00 | .00 000 | .00 | .00 000 | .00 | .00 000 | .00 |
| − .60 | .00 000 | .10 | .00 000 | .00 | .00 000 | .00 | .00 000 | .00 | .00 000 | .00 |
| − .55 | .00 002 | .69 | .00 000 | .04 | .00 000 | .00 | .00 000 | .00 | .00 000 | .00 |
| − .50 | .00 011 | 3.68 | .00 001 | .27 | .00 000 | .01 | .00 000 | .00 | .00 000 | .00 |
| − .45 | .00 052 | 15.10 | .00 004 | 1.40 | .00 000 | .08 | .00 000 | .00 | .00 000 | .00 |
| − .40 | .00 200 | 49.85 | .00 020 | 5.82 | .00 001 | .44 | .00 000 | .02 | .00 000 | .00 |
| − .35 | .00 636 | 136.13 | .00 078 | 20.06 | .00 006 | 1.88 | .00 000 | .11 | .00 000 | .00 |
| − .30 | .01 714 | 314.21 | .00 260 | 58.57 | .00 026 | 6.85 | .00 001 | .48 | .00 000 | .02 |
| − .25 | .03 998 | 623.17 | .00 747 | 147.07 | .00 091 | 21.50 | .00 007 | 1.85 | .00 000 | .09 |
| − .20 | .08 188 | 1075.24 | .01 875 | 321.70 | .00 278 | 59.02 | .00 025 | 6.32 | .00 001 | .36 |
| − .15 | .14 923 | 1629.13 | .04 169 | 618.58 | .00 757 | 142.88 | .00 083 | 19.11 | .00 005 | 1.35 |
| − .10 | .24 480 | 2182.12 | .08 292 | 1052.79 | .01 841 | 307.18 | .00 248 | 51.61 | .00 018 | 4.55 |
| − .05 | .36 511 | 2595.77 | .14 877 | 1593.19 | .04 026 | 589.27 | .00 667 | 125.03 | .00 060 | 13.87 |
| 0.00 | .50 000 | 2749.60 | .24 253 | 2149.47 | .07 969 | 1011.38 | .01 622 | 272.76 | .00 182 | 38.38 |
| + .05 | .63 489 | 2595.77 | .36 174 | 2587.70 | .14 346 | 1554.59 | .03 588 | 535.97 | .00 499 | 96.53 |
| + .10 | .75 520 | 2182.12 | .49 712 | 2777.44 | .23 586 | 2138.36 | .07 232 | 948.41 | .01 256 | 220.60 |
| + .15 | .85 077 | 1629.13 | .63 416 | 2650.80 | .35 568 | 2625.45 | .13 321 | 1507.84 | .02 893 | 457.26 |
| + .20 | .91 812 | 1075.24 | .75 741 | 2239.38 | .49 424 | 2864.46 | .22 445 | 2144.80 | .06 101 | 856.50 |
| + .25 | .96 002 | 623.17 | .85 534 | 1663.32 | .63 632 | 2758.76 | .34 655 | 2712.27 | .11 772 | 1441.46 |
| + .30 | .98 286 | 314.21 | .92 357 | 1076.19 | .76 457 | 2323.98 | .49 135 | 3022.24 | .20 748 | 2161.66 |
| + .35 | .99 394 | 136.13 | .96 482 | 599.07 | .86 541 | 1691.43 | .64 201 | 2932.02 | .33 351 | 2856.13 |
| + .40 | .99 800 | 49.85 | .98 618 | 282.26 | .93 353 | 1046.45 | .77 772 | 2437.43 | .48 847 | 3274.68 |
| + .45 | .99 948 | 15.10 | .99 548 | 110.16 | .97 245 | 538.66 | .88 169 | 1700.04 | .65 250 | 3192.62 |
| + .50 | .99 989 | 3.68 | .99 882 | 34.61 | .99 064 | 224.22 | .94 786 | 967.26 | .79 865 | 2575.61 |
| + .55 | .99 998 | .69 | .99 977 | 8.42 | .99 753 | 72.63 | .98 177 | 432.21 | .90 500 | 1656.86 |
| + .60 | 1.00 000 | .10 | .99 997 | 1.51 | .99 953 | 17.37 | .99 520 | 143.93 | .96 565 | 807.35 |
| + .65 | 1.00 000 | .01 | 1.00 000 | .18 | .99 995 | 2.84 | .99 911 | 33.16 | .99 119 | 276.92 |
| + .70 | 1.00 000 | .00 | 1.00 000 | .01 | 1.00 000 | .29 | .99 992 | 4.73 | .99 847 | 59.97 |
| + .75 | 1.00 000 | .00 | 1.00 000 | .00 | 1.00 000 | .01 | 1.00 000 | .35 | .99 982 | 6.91 |
| + .80 | 1.00 000 | .00 | 1.00 000 | .00 | 1.00 000 | .00 | 1.00 000 | .01 | 1.00 000 | .32 |
| + .85 | 1.00 000 | .00 | 1.00 000 | .00 | 1.00 000 | .00 | 1.00 000 | .00 | 1.00 000 | .00 |
| + .90 | 1.00 000 | .00 | 1.00 000 | .00 | 1.00 000 | .00 | 1.00 000 | .00 | 1.00 000 | .00 |
| + .95 | 1.00 000 | .00 | 1.00 000 | .00 | 1.00 000 | .00 | 1.00 000 | .00 | 1.00 000 | .00 |
| +1.00 | 1.00 000 | .00 | 1.00 000 | .00 | 1.00 000 | .00 | 1.00 000 | .00 | 1.00 000 | .00 |

$n=50$                                        $\rho=0.5 \longrightarrow 0.9$

| r | $\rho=0.5$ Area | Ordinate | $\rho=0.6$ Area | Ordinate | $\rho=0.7$ Area | Ordinate | r | $\rho=0.8$ Area | Ordinate | r | $\rho=0.9$ Area | Ordinate |
|---|---|---|---|---|---|---|---|---|---|---|---|---|
| − ·25 | ·00 000 | ·00 | ·00 000 | ·00 | ·00 000 | ·00 | + ·22 | ·00 000 | ·00 | + ·61 | ·00 000 | ·0 |
| − ·20 | ·00 000 | ·00 | ·00 000 | ·00 | ·00 000 | ·00 | + ·24 | ·00 000 | ·00 | + ·62 | ·00 000 | ·0 |
| − ·15 | ·00 000 | ·04 | ·00 000 | ·00 | ·00 000 | ·00 | + ·26 | ·00 000 | ·00 | + ·63 | ·00 000 | ·0 |
| − ·10 | ·00 001 | ·18 | ·00 000 | ·00 | ·00 000 | ·00 | + ·28 | ·00 000 | ·00 | + ·64 | ·00 000 | ·0 |
| − ·05 | ·00 002 | ·68 | ·00 000 | ·01 | ·00 000 | ·00 | + ·30 | ·00 000 | ·00 | + ·65 | ·00 000 | ·0 |
| ·00 | ·00 009 | 2·39 | ·00 000 | ·05 | ·00 000 | ·00 | + ·32 | ·00 000 | ·00 | + ·66 | ·00 000 | ·0 |
| + ·05 | ·00 032 | 7·70 | ·00 001 | ·20 | ·00 000 | ·00 | + ·34 | ·00 000 | ·01 | + ·67 | ·00 000 | ·2 |
| + ·10 | ·00 102 | 22·82 | ·00 003 | ·78 | ·00 000 | ·01 | + ·36 | ·00 000 | ·02 | + ·68 | ·00 000 | ·3 |
| + ·15 | ·00 300 | 62·11 | ·00 011 | 2·82 | ·00 000 | ·02 | + ·38 | ·00 000 | ·04 | + ·69 | ·00 001 | ·6 |
| + ·20 | ·00 811 | 154·86 | ·00 038 | 9·46 | ·00 000 | ·11 | + ·40 | ·00 000 | ·09 | + ·70 | ·00 001 | 1·0 |
| + ·25 | ·02 021 | 351·85 | ·00 127 | 29·47 | ·00 001 | ·48 | + ·42 | ·00 000 | ·18 | + ·71 | ·00 003 | 1·7 |
| + ·30 | ·04 621 | 723·72 | ·00 391 | 84·89 | ·00 007 | 1·99 | + ·44 | ·00 001 | ·36 | + ·72 | ·00 005 | 2·9 |
| + ·35 | ·09 659 | 1333·93 | ·01 113 | 224·22 | ·00 028 | 7·75 | + ·46 | ·00 002 | ·72 | + ·73 | ·00 009 | 5·1 |
| + ·40 | ·18 352 | 2172·97 | ·02 919 | 536·92 | ·00 108 | 28·37 | + ·48 | ·00 004 | 1·44 | + ·74 | ·00 016 | 8·8 |
| + ·45 | ·31 499 | 3070·40 | ·06 979 | 1147·17 | ·00 388 | 96·36 | + ·50 | ·00 008 | 2·85 | + ·75 | ·00 028 | 15·3 |
| + ·50 | ·48 559 | 3668·50 | ·15 040 | 2138·79 | ·01 289 | 298·71 | + ·52 | ·00 016 | 5·61 | + ·76 | ·00 048 | 26·7 |
| + ·55 | ·67 013 | 3578·51 | ·28 800 | 3372·39 | ·03 906 | 824·60 | + ·54 | ·00 033 | 10·96 | + ·77 | ·00 084 | 46·4 |
| + ·60 | ·83 013 | 2713·26 | ·48 277 | 4300·25 | ·10 547 | 1954·75 | + ·56 | ·00 064 | 21·20 | + ·78 | ·00 145 | 80·4 |
| + ·65 | ·93 545 | 1489·70 | ·69 964 | 4152·55 | ·24 645 | 3764·57 | + ·58 | ·00 124 | 40·60 | + ·79 | ·00 253 | 138·9 |
| + ·70 | ·98 410 | 532·70 | ·87 585 | 2747·94 | ·48 020 | 5398·21 | + ·60 | ·00 237 | 76·77 | + ·80 | ·00 437 | 238·4 |
| + ·75 | ·99 784 | 104·96 | ·96 976 | 1061·91 | ·75 163 | 4990·96 | + ·62 | ·00 450 | 142·90 | + ·81 | ·00 752 | 406·0 |
| + ·80 | ·99 943 | 8·58 | ·99 799 | 181·98 | ·93 593 | 2306·06 | + ·64 | ·00 843 | 261·37 | + ·82 | ·01 285 | 683·7 |
| + ·85 | 1·00 000 | ·17 | ·99 979 | 8·16 | ·99 510 | 322·77 | + ·66 | ·01 554 | 467·29 | + ·83 | ·02 177 | 1134·8 |
| + ·90 | 1·00 000 | ·00 | 1·00 000 | ·03 | ·99 989 | 4·26 | + ·68 | ·02 805 | 812·94 | + ·84 | ·03 643 | 1848·2 |
| + ·95 | 1·00 000 | ·00 | 1·00 000 | ·00 | 1·00 000 | ·00 | + ·70 | ·04 944 | 1367·63 | + ·85 | ·05 998 | 2936·7 |
| +1·00 | 1·00 000 | ·00 | 1·00 000 | ·00 | 1·00 000 | ·00 | + ·72 | ·08 466 | 2207·36 | + ·86 | ·09 681 | 4519·5 |
|  |  |  |  |  |  |  | + ·74 | ·14 000 | 3393·36 | + ·87 | ·15 229 | 6673·8 |
|  |  |  |  |  |  |  | + ·76 | ·22 204 | 4860·05 | + ·88 | ·23 202 | 9340·1 |
|  |  |  |  |  |  |  | + ·78 | ·33 507 | 6429·16 | + ·89 | ·33 974 | 12187·5 |
|  |  |  |  |  |  |  | + ·80 | ·47 693 | 7650·44 | + ·90 | ·47 403 | 14502·0 |
|  |  |  |  |  |  |  | + ·82 | ·63 478 | 7929·47 | + ·91 | ·62 459 | 15261·3 |
|  |  |  |  |  |  |  | + ·84 | ·78 484 | 6841·10 | + ·92 | ·77 108 | 13599·9 |
|  |  |  |  |  |  |  | + ·86 | ·90 054 | 4599·39 | + ·93 | ·88 871 | 9630·6 |
|  |  |  |  |  |  |  | + ·88 | ·96 761 | 2180·85 | + ·94 | ·96 114 | 4918·7 |
|  |  |  |  |  |  |  | + ·90 | ·99 373 | 621·81 | + ·95 | ·99 174 | 1550·4 |
|  |  |  |  |  |  |  | + ·92 | ·99 946 | 81·04 | + ·96 | ·99 920 | 230·3 |
|  |  |  |  |  |  |  | + ·94 | ·99 999 | 2·85 | + ·97 | ·99 998 | 9·6 |
|  |  |  |  |  |  |  | + ·96 | 1·00 000 | ·00 | + ·98 | 1·00 000 | ·3 |
|  |  |  |  |  |  |  | + ·98 | 1·00 000 | ·00 | + ·99 | 1·00 000 | ·0 |
|  |  |  |  |  |  |  | +1·00 | 1·00 000 | ·00 | +1·00 | 1·00 000 | ·0 |

ρ=0·0 — 0·4    n =100

| r | ρ=0·0 | | ρ=0·1 | | ρ=0·2 | | ρ=0·3 | | ρ=0·4 | |
|---|---|---|---|---|---|---|---|---|---|---|
|  | Area | Ordinate | Area | Ordinate | Area | Ordinate | Area | Ordinate | Area | Ordinate |
| − ·50 | ·00 000 | ·00 | ·00 000 | ·00 | ·00 000 | ·00 | ·00 000 | ·00 | ·00 000 | ·00 |
| − ·45 | ·00 000 | ·08 | ·00 000 | ·00 | ·00 000 | ·00 | ·00 000 | ·00 | ·00 000 | ·00 |
| − ·40 | ·00 002 | ·91 | ·00 000 | ·01 | ·00 000 | ·00 | ·00 000 | ·00 | ·00 000 | ·00 |
| − ·35 | ·00 018 | 7·43 | ·00 000 | ·15 | ·00 000 | ·00 | ·00 000 | ·00 | ·00 000 | ·00 |
| − ·30 | ·00 121 | 42·60 | ·00 003 | 1·41 | ·00 000 | ·02 | ·00 000 | ·00 | ·00 000 | ·00 |
| − ·25 | ·00 605 | 177·84 | ·00 025 | 9·50 | ·00 000 | ·18 | ·00 000 | ·00 | ·00 000 | ·00 |
| − ·20 | ·02 294 | 555·18 | ·00 146 | 48·00 | ·00 004 | 1·55 | ·00 000 | ·02 | ·00 000 | ·00 |
| − ·15 | ·06 813 | 1321·36 | ·00 665 | 185·37 | ·00 026 | 9·53 | ·00 001 | ·16 | ·00 000 | ·00 |
| − ·10 | ·16 111 | 2431·67 | ·02 384 | 554·86 | ·00 144 | 45·84 | ·00 004 | 1·24 | ·00 000 | ·01 |
| − ·05 | ·31 057 | 3493·29 | ·06 850 | 1299·62 | ·00 633 | 173·78 | ·00 022 | 7·57 | ·00 003 | ·09 |
| 0·00 | ·50 000 | 3939·27 | ·15 993 | 2395·29 | ·02 246 | 522·21 | ·00 116 | 36·98 | ·00 014 | ·70 |
| + ·05 | ·68 943 | 3493·29 | ·30 800 | 3480·22 | ·06 489 | 1246·23 | ·00 519 | 145·32 | ·00 074 | 4·56 |
| + ·10 | ·83 889 | 2431·67 | ·49 785 | 3979·11 | ·15 382 | 2358·16 | ·01 906 | 458·62 | ·00 353 | 24·21 |
| + ·15 | ·93 187 | 1321·36 | ·69 011 | 3560·42 | ·30 176 | 3519·22 | ·05 751 | 1156·81 | ·01 414 | 104·66 |
| + ·20 | ·97 706 | 555·18 | ·84 276 | 2469·56 | ·49 585 | 4103·61 | ·14 264 | 2311·65 | ·04 666 | 365·76 |
| + ·25 | ·99 395 | 177·84 | ·93 642 | 1309·30 | ·69 479 | 3687·36 | ·29 146 | 3611·32 | ·12 588 | 1021·01 |
| + ·30 | ·99 879 | 42·60 | ·97 989 | 520·41 | ·85 175 | 2504·77 | ·49 389 | 4329·36 | ·27 609 | 2237·43 |
| + ·35 | ·99 982 | 7·43 | ·99 500 | 151·11 | ·94 483 | 1253·64 | ·70 409 | 3884·32 | ·49 198 | 3759·08 |
| + ·40 | ·99 998 | ·91 | ·99 918 | 30·98 | ·98 471 | 446·88 | ·86 672 | 2522·86 | ·71 946 | 4690·55 |
| + ·45 | 1·00 000 | ·08 | ·99 999 | 4·29 | ·99 663 | 108·49 | ·95 706 | 1135·08 | ·88 841 | 4166·27 |
| + ·50 | 1·00 000 | ·00 | 1·00 000 | ·38 | ·99 954 | 16·90 | ·99 056 | 333·60 | ·97 228 | 2488·44 |
| + ·55 | 1·00 000 | ·00 | 1·00 000 | ·02 | 1·00 000 | 1·56 | ·99 794 | 59·16 | ·99 620 | 925·05 |
| + ·60 | 1·00 000 | ·00 | 1·00 000 | ·00 | 1·00 000 | ·08 | ·99 973 | 5·68 | ·99 797 | 192·25 |
| + ·65 | 1·00 000 | ·00 | 1·00 000 | ·00 | 1·00 000 | ·00 | 1·00 000 | ·25 | ·99 984 | 19·17 |
| + ·70 | 1·00 000 | ·00 | 1·00 000 | ·00 | 1·00 000 | ·00 | 1·00 000 | ·00 | 1·00 000 | ·73 |
| + ·75 | 1·00 000 | ·00 | 1·00 000 | ·00 | 1·00 000 | ·00 | 1·00 000 | ·00 | 1·00 000 | ·01 |
| + ·80 | 1·00 000 | ·00 | 1·00 000 | ·00 | 1·00 000 | ·00 | 1·00 000 | ·00 | 1·00 000 | ·00 |
| + ·85 | 1·00 000 | ·00 | 1·00 000 | ·00 | 1·00 000 | ·00 | 1·00 000 | ·00 | 1·00 000 | ·00 |
| + ·90 | 1·00 000 | ·00 | 1·00 000 | ·00 | 1·00 000 | ·00 | 1·00 000 | ·00 | 1·00 000 | ·00 |
| + ·95 | 1·00 000 | ·00 | 1·00 000 | ·00 | 1·00 000 | ·00 | 1·00 000 | ·00 | 1·00 000 | ·00 |
| +1·00 | 1·00 000 | ·00 | 1·00 000 | ·00 | 1·00 000 | ·00 | 1·00 000 | ·00 | 1·00 000 | ·00 |

n = 100                               ρ = 0.5 — 0.9

| r | ρ=0.5 Area | Ordinate | ρ=0.6 Area | Ordinate | ρ=0.7 Area | Ordinate | r | ρ=0.8 Area | Ordinate | r | ρ=0.9 Area | Ordinate |
|---|---|---|---|---|---|---|---|---|---|---|---|---|
| 0.00 | ·00 000 | ·00 | ·00 000 | ·00 | ·00 000 | ·00 | + ·50 | ·00 000 | ·00 | + ·740 | ·00 000 | ·0 |
| + ·02 | ·00 000 | ·01 | ·00 000 | ·00 | ·00 000 | ·00 | ·51 | ·00 000 | ·01 | + ·745 | ·00 000 | ·0 |
| + ·04 | ·00 000 | ·02 | ·00 000 | ·00 | ·00 000 | ·00 | + ·52 | ·00 000 | ·01 | + ·750 | ·00 000 | ·1 |
| + ·06 | ·00 000 | ·04 | ·00 000 | ·00 | ·00 000 | ·00 | + ·53 | ·00 000 | ·02 | + ·755 | ·00 000 | ·1 |
| + ·08 | ·00 000 | ·11 | ·00 000 | ·00 | ·00 000 | ·00 | + ·54 | ·00 000 | ·04 | + ·760 | ·00 000 | ·2 |
| + ·10 | ·00 000 | ·25 | ·00 000 | ·00 | ·00 000 | ·00 | + ·55 | ·00 000 | ·08 | + ·765 | ·00 000 | ·3 |
| + ·12 | ·00 001 | ·57 | ·00 000 | ·00 | ·00 000 | ·00 | + ·56 | ·00 000 | ·16 | + ·770 | ·00 000 | ·5 |
| + ·14 | ·00 003 | 1·26 | ·00 000 | ·00 | ·00 000 | ·00 | + ·57 | ·00 000 | ·31 | + ·775 | ·00 001 | ·8 |
| + ·16 | ·00 007 | 2·74 | ·00 000 | ·01 | ·00 000 | ·00 | + ·58 | ·00 001 | ·58 | + ·780 | ·00 001 | 1·4 |
| + ·18 | ·00 015 | 5·74 | ·00 000 | ·02 | ·00 000 | ·00 | + ·59 | ·00 002 | 1·0 | + ·785 | ·00 002 | 2·3 |
| + ·20 | ·00 032 | 11·67 | ·00 000 | ·04 | ·00 000 | ·00 | + ·60 | ·00 003 | 2·01 | + ·790 | ·00 004 | 4·0 |
| + ·22 | ·00 065 | 23·01 | ·00 000 | ·11 | ·00 000 | ·00 | + ·61 | ·00 006 | 3·70 | + ·795 | ·00 006 | 6·7 |
| + ·24 | ·00 130 | 43·93 | ·00 000 | ·26 | ·00 000 | ·00 | + ·62 | ·00 011 | 6·75 | + ·800 | ·00 011 | 11·3 |
| + ·26 | ·00 252 | 81·15 | ·00 001 | ·63 | ·00 000 | ·00 | + ·63 | ·00 020 | 12·20 | + ·805 | ·00 018 | 18·8 |
| + ·28 | ·00 472 | 144·87 | ·00 003 | 1·47 | ·00 000 | ·00 | + ·64 | ·00 037 | 21·79 | + ·810 | ·00 030 | 31·3 |
| + ·30 | ·00 859 | 249·60 | ·00 008 | 3·35 | ·00 000 | ·00 | + ·65 | ·00 066 | 38·45 | + ·815 | ·00 051 | 51·7 |
| + ·32 | ·01 511 | 414·48 | ·00 018 | 7·40 | ·00 000 | ·01 | + ·66 | ·00 118 | 66·92 | + ·820 | ·00 084 | 84·7 |
| + ·34 | ·02 572 | 662·20 | ·00 040 | 15·87 | ·00 000 | ·02 | + ·67 | ·00 206 | 114·74 | + ·825 | ·00 139 | 137·7 |
| + ·36 | ·04 230 | 1015·99 | ·00 087 | 32·95 | ·00 000 | ·04 | + ·68 | ·00 357 | 193·51 | + ·830 | ·00 227 | 221·8 |
| + ·38 | ·06 719 | 1493·74 | ·00 183 | 66·09 | ·00 000 | ·13 | + ·69 | ·00 609 | 320·43 | + ·835 | ·00 368 | 353·4 |
| + ·40 | ·10 291 | 2099·42 | ·00 371 | 127·86 | ·00 001 | ·35 | + ·70 | ·01 022 | 519·97 | + ·840 | ·00 592 | 556·2 |
| + ·42 | ·15 189 | 2813·06 | ·00 726 | 237·98 | ·00 002 | ·93 | + ·71 | ·01 684 | 825·03 | + ·845 | ·00 942 | 863·6 |
| + ·44 | ·21 581 | 3582·41 | ·01 374 | 425·04 | ·00 005 | 2·42 | + ·72 | ·02 721 | 1276·80 | + ·850 | ·01 481 | 1320·0 |
| + ·46 | ·29 499 | 4320·92 | ·02 502 | 726·22 | ·00 013 | 6·14 | + ·73 | ·04 302 | 1921·73 | + ·855 | ·02 296 | 1982·6 |
| + ·48 | ·38 040 | 4916·59 | ·04 382 | 1182·84 | ·00 033 | 15·05 | + ·74 | ·06 644 | 2803·98 | + ·860 | ·03 509 | 2919·0 |
| + ·50 | ·48 727 | 5254·02 | ·07 361 | 1809·25 | ·00 081 | 35·59 | + ·75 | ·09 999 | 3951·47 | + ·865 | ·05 273 | 4201·9 |
| + ·52 | ·59 554 | 5246·12 | ·11 832 | 2673·66 | ·00 192 | 80·89 | + ·76 | ·14 633 | 5355·44 | ·870 | ·07 779 | 5895·9 |
| + ·54 | ·69 727 | 4865·93 | ·18 161 | 3673·87 | ·00 437 | 175·94 | + ·77 | ·20 773 | 6946·35 | ·875 | ·11 251 | 8036·0 |
| + ·56 | ·78 802 | 4164·48 | ·26 558 | 4716·97 | ·00 957 | 364·39 | + ·78 | ·28 539 | 8573·94 | ·880 | ·15 886 | 10597·3 |
| + ·58 | ·86 251 | 3263·37 | ·36 936 | 5618·68 | ·02 002 | 714·34 | + ·79 | ·37 856 | 10004·61 | ·885 | ·21 892 | 13458·8 |
| + ·60 | ·91 828 | 2320·57 | ·48 790 | 6157·92 | ·03 982 | 1316·29 | + ·80 | ·48 387 | 10951·32 | ·890 | ·29 355 | 16373·8 |
| + ·62 | ·95 604 | 1481·89 | ·61 094 | 6149·47 | ·07 497 | 2260·96 | + ·81 | ·59 506 | 11143·96 | ·895 | ·38 212 | 18963·2 |
| + ·64 | ·97 888 | 839·50 | ·72 978 | 5531·72 | ·13 282 | 3584·36 | + ·82 | ·70 370 | 10429·23 | ·900 | ·48 185 | 20754·4 |
| + ·66 | ·99 109 | 415·85 | ·82 995 | 4421·61 | ·22 022 | 5182·22 | + ·83 | ·80 079 | 8862·13 | ·905 | ·58 754 | 21280·7 |
| + ·68 | ·99 675 | 177·05 | ·90 516 | 3089·62 | ·33 985 | 6734·64 | + ·84 | ·87 905 | 6732·90 | ·910 | ·69 201 | 20233·5 |
| + ·70 | ·99 900 | 63·47 | ·95 417 | 1850·41 | ·48 588 | 7728·87 | + ·85 | ·93 506 | 4488·65 | ·915 | ·78 725 | 17621·0 |
| + ·72 | ·99 975 | 18·68 | ·98 131 | 927·26 | ·64 184 | 7663·43 | + ·86 | ·96 992 | 2566·31 | ·920 | ·86 627 | 13848·4 |
| + ·74 | ·99 996 | 4·38 | ·99 377 | 377·39 | ·78 420 | 6388·17 | + ·87 | ·98 834 | 1222·94 | ·925 | ·92 501 | 9645·8 |
| + ·76 | ·99 999 | ·79 | ·99 837 | 120·18 | ·89 206 | 4324·78 | + ·88 | ·99 635 | 468·62 | ·930 | ·96 338 | 5823·1 |
| + ·78 | 1·00 000 | ·10 | ·99 969 | 28·56 | ·95 740 | 2274·51 | + ·89 | ·99 912 | 137·93 | ·935 | ·98 487 | 2963·7 |
| + ·80 | 1·00 000 | ·01 | ·99 997 | 4·76 | ·98 753 | 876·76 | + ·90 | ·99 985 | 29·37 | ·940 | ·99 490 | 1227·8 |
| + ·82 | 1·00 000 | ·00 | 1·00 000 | ·51 | ·99 743 | 229·15 | + ·91 | ·99 999 | 4·18 | + ·945 | ·99 866 | 395·9 |
| + ·84 | 1·00 000 | ·00 | 1·00 000 | ·03 | ·99 961 | 36·48 | + ·92 | 1·00 000 | ·36 | + ·950 | ·99 975 | 93·7 |
| + ·86 | 1·00 000 | ·00 | 1·00 000 | ·00 | 1·00 000 | 3·04 | + ·93 | 1·00 000 | ·02 | + ·955 | ·99 997 | 15·0 |
| + ·88 | 1·00 000 | ·00 | 1·00 000 | ·00 | 1·00 000 | ·11 | + ·94 | 1·00 000 | ·00 | + ·960 | 1·00 000 | 1·5 |
| + ·90 | 1·00 000 | ·00 | 1·00 000 | ·00 | 1·00 000 | ·00 | + ·95 | 1·00 000 | ·00 | + ·965 | 1·00 000 | ·1 |
| + ·92 | 1·00 000 | ·00 | 1·00 000 | ·00 | 1·00 000 | ·00 | + ·96 | 1·00 000 | ·00 | + ·970 | 1·00 000 | ·0 |
| + ·94 | 1·00 000 | ·00 | 1·00 000 | ·00 | 1·00 000 | ·00 | + ·97 | 1·00 000 | ·00 | + ·975 | 1·00 000 | ·0 |
| + ·96 | 1·00 000 | ·00 | 1·00 000 | ·00 | 1·00 000 | ·00 | + ·98 | 1·00 000 | ·00 | + ·980 | 1·00 000 | ·0 |
| + ·98 | 1·00 000 | ·00 | 1·00 000 | ·00 | 1·00 000 | ·00 | + ·99 | 1·00 000 | ·00 | + ·985 | 1·00 000 | ·0 |
| +1·00 | 1·00 000 | ·00 | 1·00 000 | ·00 | 1·00 000 | ·00 | +1·00 | 1·00 000 | ·00 | + ·990 | 1·00 000 | ·0 |

ρ=0.0 — 0.4     n = 200

| r | ρ = 0·0 Area | ρ = 0·0 Ordinate | ρ = 0·1 Area | ρ = 0·1 Ordinate | ρ = 0·2 Area | ρ = 0·2 Ordinate | ρ = 0·3 Area | ρ = 0·3 Ordinate | ρ = 0·4 Area | ρ = 0·4 Ordinate |
|---|---|---|---|---|---|---|---|---|---|---|
| − ·40 | ·00 000 | ·00 | ·00 000 | ·00 | ·00 000 | ·00 | ·00 000 | ·00 | ·00 000 | ·00 |
| − ·38 | ·00 000 | ·00 | ·00 000 | ·00 | ·00 000 | ·00 | ·00 000 | ·00 | ·00 000 | ·00 |
| − ·36 | ·00 000 | ·01 | ·00 000 | ·00 | ·00 000 | ·00 | ·00 000 | ·00 | ·00 000 | ·00 |
| − ·34 | ·00 000 | ·03 | ·00 000 | ·00 | ·00 000 | ·00 | ·00 000 | ·00 | ·00 000 | ·00 |
| − ·32 | ·00 000 | ·14 | ·00 000 | ·00 | ·00 000 | ·00 | ·00 000 | ·00 | ·00 000 | ·00 |
| − ·30 | ·00 001 | ·54 | ·00 000 | ·00 | ·00 000 | ·00 | ·00 000 | ·00 | ·00 000 | ·00 |
| − ·28 | ·00 003 | 1·88 | ·00 000 | ·00 | ·00 000 | ·00 | ·00 000 | ·00 | ·00 000 | ·00 |
| − ·26 | ·00 010 | 5·88 | ·00 000 | ·01 | ·00 000 | ·00 | ·00 000 | ·00 | ·00 000 | ·00 |
| − ·24 | ·00 031 | 16·74 | ·00 000 | ·06 | ·00 000 | ·00 | ·00 000 | ·00 | ·00 000 | ·00 |
| − ·22 | ·00 087 | 43·37 | ·00 000 | ·21 | ·00 000 | ·00 | ·00 000 | ·00 | ·00 000 | ·00 |
| − ·20 | ·00 226 | 102·63 | ·00 001 | ·74 | ·00 000 | ·00 | ·00 000 | ·00 | ·00 000 | ·00 |
| − ·18 | ·00 538 | 222·27 | ·00 004 | 2·37 | ·00 000 | ·00 | ·00 000 | ·00 | ·00 000 | ·00 |
| − ·16 | ·01 181 | 441·52 | ·00 013 | 6·95 | ·00 000 | ·01 | ·00 000 | ·00 | ·00 000 | ·00 |
| − ·14 | ·02 401 | 805·78 | ·00 037 | 18·77 | ·00 000 | ·06 | ·00 000 | ·00 | ·00 000 | ·00 |
| − ·12 | ·04 527 | 1353·20 | ·00 098 | 46·64 | ·00 000 | ·21 | ·00 000 | ·00 | ·00 000 | ·00 |
| − ·10 | ·07 943 | 2093·84 | ·00 244 | 106·87 | ·00 001 | ·71 | ·00 000 | ·00 | ·00 000 | ·00 |
| − ·08 | ·13 006 | 2988·34 | ·00 565 | 226·05 | ·00 004 | 2·20 | ·00 000 | ·00 | ·00 000 | ·00 |
| − ·06 | ·19 934 | 3937·30 | ·01 213 | 441·77 | ·00 012 | 6·35 | ·00 000 | ·01 | ·00 000 | ·00 |
| − ·04 | ·28 693 | 4792·28 | ·02 426 | 798·18 | ·00 033 | 16·97 | ·00 000 | ·04 | ·00 000 | ·00 |
| − ·02 | ·38 932 | 5390·97 | ·04 526 | 1333·91 | ·00 089 | 42·03 | ·00 000 | ·14 | ·00 000 | ·00 |
| 0·00 | ·50 000 | 5606·53 | ·07 891 | 2062·51 | ·00 221 | 96·53 | ·00 001 | ·47 | ·00 000 | ·00 |
| + ·02 | ·61 068 | 5390·97 | ·12 884 | 2950·91 | ·00 511 | 205·67 | ·00 003 | 1·50 | ·00 000 | ·00 |
| + ·04 | ·71 307 | 4792·28 | ·19 740 | 3906·31 | ·01 104 | 406·41 | ·00 008 | 4·42 | ·00 000 | ·00 |
| + ·06 | ·80 066 | 3937·30 | ·28 457 | 4783·05 | ·02 229 | 744·62 | ·00 023 | 12·18 | ·00 000 | ·01 |
| + ·08 | ·86 994 | 2988·34 | ·38 708 | 5414·60 | ·04 204 | 1264·43 | ·00 064 | 31·21 | ·00 000 | ·06 |
| + ·10 | ·92 056 | 2093·84 | ·49 858 | 5663·20 | ·07 424 | 1988·65 | ·00 164 | 74·37 | ·00 000 | ·20 |
| + ·12 | ·95 473 | 1353·20 | ·61 063 | 5467·82 | ·12 282 | 2894·46 | ·00 393 | 164·70 | ·00 001 | ·69 |
| + ·14 | ·97 599 | 805·78 | ·71 458 | 4868·06 | ·19 066 | 3894·58 | ·00 878 | 338·68 | ·00 004 | 2·19 |
| + ·16 | ·98 819 | 441·52 | ·80 350 | 3991·41 | ·27 823 | 4838·22 | ·01 836 | 645·87 | ·00 012 | 6·49 |
| + ·18 | ·99 462 | 222·27 | ·87 353 | 3009·30 | ·38 258 | 5541·09 | ·03 587 | 1140·60 | ·00 034 | 17·96 |
| + ·20 | ·99 774 | 102·63 | ·92 425 | 2082·57 | ·49 716 | 5840·28 | ·06 551 | 1862·26 | ·00 094 | 46·26 |
| + ·22 | ·99 913 | 43·37 | ·95 795 | 1320·30 | ·61 292 | 5653·75 | ·11 186 | 2805·50 | ·00 243 | 110·68 |
| + ·24 | ·99 969 | 16·74 | ·97 845 | 765·02 | ·72 028 | 5015·54 | ·17 870 | 3891·49 | ·00 584 | 245·55 |
| + ·26 | ·99 990 | 5·88 | ·98 985 | 404·08 | ·81 146 | 4066·90 | ·26 740 | 4957·56 | ·01 307 | 503·96 |
| + ·28 | ·99 997 | 1·88 | ·99 563 | 194·00 | ·88 221 | 3005·51 | ·37 540 | 5784·24 | ·02 728 | 954·34 |
| + ·30 | ·99 999 | ·54 | ·99 828 | 84·39 | ·93 219 | 2017·83 | ·49 575 | 6161·37 | ·05 297 | 1662·53 |
| + ·32 | 1·00 000 | ·14 | ·99 939 | 33·13 | ·96 424 | 1226·20 | ·61 804 | 5970·69 | ·09 569 | 2655·31 |
| + ·34 | 1·00 000 | ·03 | ·99 980 | 11·70 | ·98 283 | 671·77 | ·73 096 | 5242·80 | ·16 072 | 3873·63 |
| + ·36 | 1·00 000 | ·01 | ·99 994 | 3·69 | ·99 254 | 330·28 | ·82 532 | 4153·07 | ·25 098 | 5139·62 |
| + ·38 | 1·00 000 | ·00 | ·99 998 | 1·04 | ·99 709 | 145·00 | ·89 724 | 2953·16 | ·36 475 | 6172·83 |
| + ·40 | 1·00 000 | ·00 | 1·00 000 | ·26 | ·99 899 | 56·52 | ·94 428 | 1874·61 | ·49 433 | 6675·12 |
| + ·42 | 1·00 000 | ·00 | 1·00 000 | ·06 | ·99 969 | 19·44 | ·97 309 | 1055·67 | ·62 693 | 6460·26 |
| + ·44 | 1·00 000 | ·00 | 1·00 000 | ·01 | ·99 991 | 5·86 | ·98 843 | 523·76 | ·74 811 | 5558·16 |
| + ·46 | 1·00 000 | ·00 | 1·00 000 | ·00 | ·99 998 | 1·53 | ·99 562 | 227·16 | ·84 632 | 4218·96 |
| + ·48 | 1·00 000 | ·00 | 1·00 000 | ·00 | 1·00 000 | ·35 | ·99 856 | 85·37 | ·91 636 | 2801·28 |
| + ·50 | 1·00 000 | ·00 | 1·00 000 | ·00 | 1·00 000 | ·07 | ·99 959 | 27·52 | ·95 992 | 1611·32 |
| + ·52 | 1·00 000 | ·00 | 1·00 000 | ·00 | 1·00 000 | ·01 | ·99 990 | 7·53 | ·98 332 | 794·15 |
| + ·54 | 1·00 000 | ·00 | 1·00 000 | ·00 | 1·00 000 | ·00 | ·99 998 | 1·72 | ·99 406 | 331·21 |
| + ·56 | 1·00 000 | ·00 | 1·00 000 | ·00 | 1·00 000 | ·00 | 1·00 000 | ·33 | ·99 823 | 115·23 |
| + ·58 | 1·00 000 | ·00 | 1·00 000 | ·00 | 1·00 000 | ·00 | 1·00 000 | ·05 | ·99 957 | 32·90 |
| + ·60 | 1·00 000 | ·00 | 1·00 000 | ·00 | 1·00 000 | ·00 | 1·00 000 | ·01 | ·99 992 | 7·56 |
| + ·62 | 1·00 000 | ·00 | 1·00 000 | ·00 | 1·00 000 | ·00 | 1·00 000 | ·00 | ·99 998 | 1·37 |
| + ·64 | 1·00 000 | ·00 | 1·00 000 | ·00 | 1·00 000 | ·00 | 1·00 000 | ·00 | 1·00 000 | ·19 |
| + ·66 | 1·00 000 | ·00 | 1·00 000 | ·00 | 1·00 000 | ·00 | 1·00 000 | ·00 | 1·00 000 | ·02 |
| + ·68 | 1·00 000 | ·00 | 1·00 000 | ·00 | 1·00 000 | ·00 | 1·00 000 | ·00 | 1·00 000 | ·00 |
| + ·70 | 1·00 000 | ·00 | 1·00 000 | ·00 | 1·00 000 | ·00 | 1·00 000 | ·00 | 1·00 000 | ·00 |

n = 200  ρ = 0·5 — 0·9

| r | ρ=0·5 Area | ρ=0·5 Ordinate | r | ρ=0·6 Area | ρ=0·6 Ordinate | ρ=0·7 Area | ρ=0·7 Ordinate | ρ=0·8 Area | ρ=0·8 Ordinate | r | ρ=0·9 Area | ρ=0·9 Ordinate |
|---|---|---|---|---|---|---|---|---|---|---|---|---|
| + ·15 | ·00 000 | ·00 | + ·30 | ·00 000 | ·00 | ·00 000 | ·00 | ·00 000 | ·0 | + ·8025 | ·00 000 | ·0 |
| + ·16 | ·00 000 | ·00 | + ·31 | ·00 000 | ·01 | ·00 000 | ·00 | ·00 000 | ·0 | + ·8050 | ·00 000 | ·0 |
| + ·17 | ·00 000 | ·01 | + ·32 | ·00 000 | ·02 | ·00 000 | ·00 | ·00 000 | ·0 | + ·8075 | ·00 000 | ·1 |
| + ·18 | ·00 000 | ·01 | + ·33 | ·00 000 | ·04 | ·00 000 | ·00 | ·00 000 | ·0 | + ·8100 | ·00 000 | ·1 |
| + ·19 | ·00 000 | ·02 | + ·34 | ·00 000 | ·08 | ·00 000 | ·00 | ·00 000 | ·0 | + ·8125 | ·00 000 | ·2 |
| + ·20 | ·00 000 | ·05 | + ·35 | ·00 000 | ·17 | ·00 000 | ·00 | ·00 000 | ·0 | + ·8150 | ·00 000 | ·3 |
| + ·21 | ·00 000 | ·09 | + ·36 | ·00 000 | ·34 | ·00 000 | ·00 | ·00 000 | ·0 | + ·8175 | ·00 000 | ·6 |
| + ·22 | ·00 000 | ·18 | + ·37 | ·00 001 | ·69 | ·00 000 | ·00 | ·00 000 | ·0 | + ·8200 | ·00 000 | ·9 |
| + ·23 | ·00 000 | ·34 | + ·38 | ·00 002 | 1·36 | ·00 000 | ·00 | ·00 000 | ·0 | + ·8225 | ·00 001 | 1·5 |
| + ·24 | ·00 001 | ·65 | + ·39 | ·00 004 | 2·64 | ·00 000 | ·00 | ·00 000 | ·0 | + ·8250 | ·00 001 | 2·3 |
| + ·25 | ·00 002 | 1·21 | + ·40 | ·00 008 | 5·03 | ·00 000 | ·00 | ·00 000 | ·0 | + ·8275 | ·00 002 | 3·7 |
| + ·26 | ·00 004 | 2·21 | + ·41 | ·00 015 | 9·38 | ·00 000 | ·00 | ·00 000 | ·0 | + ·8300 | ·00 003 | 5·9 |
| + ·27 | ·00 007 | 3·96 | + ·42 | ·00 027 | 17·16 | ·00 000 | ·00 | ·00 000 | ·0 | + ·8325 | ·00 005 | 9·3 |
| + ·28 | ·00 012 | 6·99 | + ·43 | ·00 051 | 30·70 | ·00 000 | ·00 | ·00 000 | ·0 | + ·8350 | ·00 008 | 14·5 |
| + ·29 | ·00 021 | 12·11 | + ·44 | ·00 092 | 53·75 | ·00 000 | ·00 | ·00 000 | ·0 | + ·8375 | ·00 012 | 22·6 |
| + ·30 | ·00 037 | 20·59 | + ·45 | ·00 163 | 91·98 | ·00 000 | ·00 | ·00 000 | ·0 | + ·8400 | ·00 019 | 35·0 |
| + ·31 | ·00 064 | 34·36 | + ·46 | ·00 284 | 153·72 | ·00 000 | ·01 | ·00 000 | ·0 | + ·8425 | ·00 030 | 53·7 |
| + ·32 | ·00 109 | 56·24 | + ·47 | ·00 482 | 250·67 | ·00 000 | ·03 | ·00 000 | ·0 | + ·8450 | ·00 047 | 81·8 |
| + ·33 | ·00 181 | 90·23 | + ·48 | ·00 802 | 398·45 | ·00 000 | ·06 | ·00 000 | ·0 | + ·8475 | ·00 072 | 123·6 |
| + ·34 | ·00 295 | 141·88 | + ·49 | ·01 303 | 616·71 | ·00 000 | ·15 | ·00 000 | ·0 | + ·8500 | ·00 111 | 185·2 |
| + ·35 | ·00 473 | 218·49 | + ·50 | ·02 067 | 928·39 | ·00 000 | ·35 | ·00 000 | ·0 | + ·8525 | ·00 167 | 274·9 |
| + ·36 | ·00 743 | 329·41 | + ·51 | ·03 199 | 1357·65 | ·00 001 | ·80 | ·00 000 | ·0 | + ·8550 | ·00 251 | 404·1 |
| + ·37 | ·01 147 | 485·90 | + ·52 | ·04 828 | 1926·17 | ·00 002 | 1·77 | ·00 000 | ·0 | + ·8575 | ·00 374 | 587·8 |
| + ·38 | ·01 735 | 700·81 | + ·53 | ·07 102 | 2647·53 | ·00 005 | 3·85 | ·00 000 | ·0 | + ·8600 | ·00 551 | 845·6 |
| + ·39 | ·02 572 | 987·72 | + ·54 | ·10 174 | 3520·22 | ·00 011 | 8·16 | ·00 000 | ·0 | + ·8625 | ·00 805 | 1202·3 |
| + ·40 | ·03 758 | 1357·10 | + ·55 | ·14 186 | 4520·42 | ·00 022 | 16·89 | ·00 000 | ·0 | + ·8650 | ·01 163 | 1688·3 |
| + ·41 | ·05 323 | 1825·75 | + ·56 | ·19 241 | 5596·32 | ·00 047 | 34·02 | ·00 000 | ·0 | + ·8675 | ·01 663 | 2339·2 |
| + ·42 | ·07 423 | 2390·95 | + ·57 | ·25 377 | 6666·99 | ·00 096 | 66·61 | ·00 000 | ·0 | + ·8700 | ·02 350 | 3195·7 |
| + ·43 | ·10 136 | 3050·63 | + ·58 | ·32 538 | 7627·36 | ·00 189 | 126·55 | ·00 000 | ·0 | + ·8725 | ·03 281 | 4300·3 |
| + ·44 | ·13 551 | 3788·98 | + ·59 | ·40 556 | 8361·42 | ·00 364 | 232·87 | ·00 000 | ·0 | + ·8750 | ·04 524 | 5694·4 |
| + ·45 | ·17 731 | 4576·95 | + ·60 | ·49 149 | 8762·19 | ·00 680 | 414·22 | ·00 000 | ·0 | + ·8775 | ·06 156 | 7412·6 |
| + ·46 | ·22 707 | 5371·71 | + ·61 | ·57 944 | 8754·91 | ·01 231 | 710·59 | ·00 000 | ·0 | + ·8800 | ·08 259 | 9474·4 |
| + ·47 | ·28 459 | 6119·18 | + ·62 | ·66 515 | 8317·22 | ·02 156 | 1172·86 | ·00 000 | ·0 | + ·8825 | ·10 921 | 11876·0 |
| + ·48 | ·34 909 | 6758·24 | + ·63 | ·74 448 | 7489·76 | ·03 651 | 1857·62 | ·00 000 | ·0 | + ·8850 | ·14 223 | 14580·1 |
| + ·49 | ·41 919 | 7228·01 | + ·64 | ·81 397 | 6372·03 | ·05 963 | 2815·04 | ·00 000 | ·1 | + ·8875 | ·18 230 | 17506·2 |
| + ·50 | ·49 291 | 7476·55 | + ·65 | ·87 141 | 5102·96 | ·09 380 | 4068·60 | ·00 000 | ·3 | + ·8900 | ·22 984 | 20525·2 |
| + ·51 | ·56 763 | 7469·58 | + ·66 | ·91 603 | 3831·81 | ·14 190 | 5588·54 | ·00 001 | 1·0 | + ·8925 | ·28 486 | 23462·2 |
| + ·52 | ·64 142 | 7197·43 | + ·67 | ·94 847 | 2686·01 | ·20 611 | 7267·30 | ·00 003 | 2·7 | + ·8950 | ·34 690 | 26098·2 |
| + ·53 | ·71 099 | 6678·46 | + ·68 | ·97 045 | 1749·29 | ·28 713 | 8908·49 | ·00 007 | 7·6 | + ·8975 | ·41 490 | 28195·7 |
| + ·54 | ·77 431 | 5957·64 | + ·69 | ·98 426 | 1052·89 | ·38 327 | 10245·41 | ·00 020 | 20·4 | + ·9000 | ·48 723 | 29523·1 |
| + ·55 | ·82 968 | 5100·32 | + ·70 | ·99 226 | 582·29 | ·49 007 | 10996·59 | ·00 054 | 52·2 | + ·9025 | ·56 171 | 29891·7 |
| + ·56 | ·87 611 | 4182·53 | + ·71 | ·99 652 | 293·99 | ·60 053 | 10950·67 | ·00 139 | 127·7 | + ·9050 | ·63 580 | 29191·4 |
| + ·57 | ·91 338 | 3278·64 | + ·72 | ·99 857 | 134·54 | ·70 622 | 10051·41 | ·00 340 | 296·5 | + ·9075 | ·70 678 | 27421·2 |
| + ·58 | ·94 194 | 2451·39 | + ·73 | ·99 947 | 55·36 | ·79 917 | 8441·61 | ·00 793 | 650·3 | + ·9100 | ·77 211 | 24702·9 |
| + ·59 | ·96 281 | 1744·08 | + ·74 | ·99 983 | 20·30 | ·87 372 | 6433·48 | ·01 751 | 1336·9 | + ·9125 | ·82 970 | 21271·2 |
| + ·60 | ·97 730 | 1177·75 | + ·75 | ·99 995 | 6·57 | ·92 779 | 4407·96 | ·03 643 | 2557·9 | + ·9150 | ·87 814 | 17445·1 |
| + ·61 | ·98 684 | 752·76 | + ·76 | ·99 999 | 1·85 | ·96 291 | 2686·67 | ·07 115 | 4514·3 | + ·9175 | ·91 688 | 13572·5 |
| + ·62 | ·99 278 | 454·04 | + ·77 | 1·00 000 | ·45 | ·98 312 | 1439·29 | ·12 947 | 7276·6 | + ·9200 | ·94 623 | 9973·7 |
| + ·63 | ·99 626 | 257·60 | + ·78 | 1·00 000 | ·09 | ·99 329 | 668·46 | ·21 856 | 10586·9 | + ·9225 | ·96 718 | 6888·6 |
| + ·64 | ·99 818 | 136·99 | + ·79 | 1·00 000 | ·02 | ·99 772 | 264·93 | ·34 065 | 13717·1 | + ·9250 | ·98 121 | 4448·2 |
| + ·65 | ·99 917 | 68·03 | + ·80 | 1·00 000 | ·00 | ·99 935 | 87·98 | ·48 865 | 15580·0 | + ·9275 | ·98 998 | 2669·3 |
| + ·66 | ·99 965 | 31·41 | + ·81 | 1·00 000 | ·00 | ·99 985 | 23·97 | ·64 475 | 15226·2 | + ·9300 | ·99 505 | 1478·7 |
| + ·67 | ·99 986 | 13·42 | + ·82 | 1·00 000 | ·00 | ·99 997 | 5·22 | ·78 522 | 12526·6 | + ·9325 | ·99 776 | 750·6 |
| + ·68 | ·99 995 | 5·28 | + ·83 | 1·00 000 | ·00 | ·99 999 | ·88 | ·89 067 | 8449·2 | + ·9350 | ·99 907 | 346·2 |
| + ·69 | ·99 998 | 1·90 | + ·84 | 1·00 000 | ·00 | 1·00 000 | ·11 | ·95 483 | 4528·2 | + ·9375 | ·99 965 | 143·7 |
| + ·70 | ·99 999 | ·62 | + ·85 | 1·00 000 | ·00 | 1·00 000 | ·01 | ·98 547 | 1855·5 | + ·9400 | ·99 988 | 53·1 |
| + ·71 | 1·00 000 | ·19 | + ·86 | 1·00 000 | ·00 | 1·00 000 | ·00 | ·99 647 | 554·8 | + ·9425 | ·99 996 | 17·3 |
| + ·72 | 1·00 000 | ·05 | + ·87 | 1·00 000 | ·00 | 1·00 000 | ·00 | ·99 940 | 114·2 | + ·9450 | ·99 999 | 4·9 |
| + ·73 | 1·00 000 | ·01 | + ·88 | 1·00 000 | ·00 | 1·00 000 | ·00 | 1·00 000 | 15·0 | + ·9475 | 1·00 000 | 1·2 |
| + ·74 | 1·00 000 | ·00 | + ·89 | 1·00 000 | ·00 | 1·00 000 | ·00 | 1·00 000 | 1·2 | + ·9500 | 1·00 000 | ·2 |
| + ·75 | 1·00 000 | ·00 | + ·90 | 1·00 000 | ·00 | 1·00 000 | ·00 | 1·00 000 | ·0 | + ·9525 | 1·00 000 | ·0 |
| + ·76 | 1·00 000 | ·00 | + ·91 | 1·00 000 | ·00 | 1·00 000 | ·00 | 1·00 000 | ·0 | + ·9550 | 1·00 000 | ·0 |
| + ·77 | 1·00 000 | ·00 | + ·92 | 1·00 000 | ·00 | 1·00 000 | ·00 | 1·00 000 | ·0 | + ·9575 | 1·00 000 | ·0 |
| + ·78 | 1·00 000 | ·00 | + ·93 | 1·00 000 | ·00 | 1·00 000 | ·00 | 1·00 000 | ·0 | + ·9600 | 1·00 000 | ·0 |
| + ·79 | 1·00 000 | ·00 | + ·94 | 1·00 000 | ·00 | 1·00 000 | ·00 | 1·00 000 | ·0 | + ·9625 | 1·00 000 | ·0 |

$\rho = 0.0 — 0.4$      $n = 400$

| r | ρ = 0·0 Area | Ordinate | ρ = 0·1 Area | Ordinate | ρ = 0·2 Area | Ordinate | ρ = 0·3 Area | Ordinate | ρ = 0·4 Area | Ordinate |
|---|---|---|---|---|---|---|---|---|---|---|
| − ·30 | ·00 000 | ·00 | ·00 000 | ·00 | ·00 000 | ·00 | ·00 000 | ·00 | ·00 000 | ·00 |
| − ·28 | ·00 000 | ·00 | ·00 000 | ·00 | ·00 000 | ·00 | ·00 000 | ·00 | ·00 000 | ·00 |
| − ·26 | ·00 000 | ·00 | ·00 000 | ·00 | ·00 000 | ·00 | ·00 000 | ·00 | ·00 000 | ·00 |
| − ·24 | ·00 000 | ·06 | ·00 000 | ·00 | ·00 000 | ·00 | ·00 000 | ·00 | ·00 000 | ·00 |
| − ·22 | ·00 000 | ·43 | ·00 000 | ·00 | ·00 000 | ·00 | ·00 000 | ·00 | ·00 000 | ·00 |
| − ·20 | ·00 003 | 2·46 | ·00 000 | ·00 | ·00 000 | ·00 | ·00 000 | ·00 | ·00 000 | ·00 |
| − ·18 | ·00 015 | 11·72 | ·00 000 | ·00 | ·00 000 | ·00 | ·00 000 | ·00 | ·00 000 | ·00 |
| − ·16 | ·00 066 | 46·83 | ·00 000 | ·01 | ·00 000 | ·00 | ·00 000 | ·00 | ·00 000 | ·00 |
| − ·14 | ·00 251 | 157·92 | ·00 000 | ·08 | ·00 000 | ·00 | ·00 000 | ·00 | ·00 000 | ·00 |
| − ·12 | ·00 817 | 450·11 | ·00 001 | ·52 | ·00 000 | ·00 | ·00 000 | ·00 | ·00 000 | ·00 |
| − ·10 | ·02 282 | 1087·30 | ·00 003 | 2·78 | ·00 000 | ·00 | ·00 000 | ·00 | ·00 000 | ·00 |
| − ·08 | ·05 507 | 2230·86 | ·00 017 | 12·55 | ·00 000 | ·00 | ·00 000 | ·00 | ·00 000 | ·00 |
| − ·06 | ·11 559 | 3894·53 | ·00 070 | 48·35 | ·00 000 | ·01 | ·00 000 | ·00 | ·00 000 | ·00 |
| − ·04 | ·21 249 | 5792·74 | ·00 259 | 158·93 | ·00 000 | ·07 | ·00 000 | ·00 | ·00 000 | ·00 |
| − ·02 | ·34 503 | 7348·11 | ·00 822 | 446·28 | ·00 000 | ·43 | ·00 000 | ·00 | ·00 000 | ·00 |
| 0·00 | ·50 000 | 7953·88 | ·02 268 | 1071·03 | ·00 003 | 2·31 | ·00 000 | ·00 | ·00 000 | ·00 |
| + ·02 | ·65 497 | 7348·11 | ·05 444 | 2197·24 | ·00 014 | 10·54 | ·00 000 | ·00 | ·00 000 | ·00 |
| + ·04 | ·78 751 | 5792·74 | ·11 418 | 3852·65 | ·00 059 | 41·31 | ·00 000 | ·00 | ·00 000 | ·00 |
| + ·06 | ·88 441 | 3894·53 | ·21 037 | 5770·36 | ·00 222 | 138·97 | ·00 000 | ·04 | ·00 000 | ·00 |
| + ·08 | ·94 493 | 2230·86 | ·34 290 | 7375·53 | ·00 720 | 400·90 | ·00 000 | ·24 | ·00 000 | ·00 |
| + ·10 | ·97 718 | 1087·30 | ·49 897 | 8034·24 | ·02 044 | 990·50 | ·00 002 | 1·37 | ·00 000 | ·00 |
| + ·12 | ·99 183 | 450·11 | ·65 584 | 7445·55 | ·05 030 | 2092·56 | ·00 009 | 6·72 | ·00 000 | ·00 |
| + ·14 | ·99 748 | 157·92 | ·79 011 | 5857·46 | ·10 807 | 3771·75 | ·00 039 | 28·38 | ·00 000 | ·00 |
| + ·16 | ·99 934 | 46·83 | ·88 773 | 3901·54 | ·20 349 | 5785·55 | ·00 156 | 102·92 | ·00 000 | ·01 |
| + ·18 | ·99 985 | 11·72 | ·94 784 | 2193·59 | ·33 770 | 7529·69 | ·00 543 | 319·53 | ·00 000 | ·08 |
| + ·20 | ·99 997 | 2·46 | ·97 914 | 1037·29 | ·49 796 | 8285·39 | ·01 639 | 846·48 | ·00 001 | ·52 |
| + ·22 | 1·00 000 | ·43 | ·99 290 | 410·91 | ·65 987 | 7677·21 | ·04 286 | 1905·87 | ·00 004 | 2·95 |
| + ·24 | 1·00 000 | ·06 | ·99 799 | 135·72 | ·79 765 | 5962·90 | ·09 716 | 3631·31 | ·00 018 | 14·45 |
| + ·26 | 1·00 000 | ·00 | ·99 954 | 37·18 | ·89 589 | 3862·09 | ·19 137 | 5825·37 | ·00 083 | 60·37 |
| + ·28 | 1·00 000 | ·00 | ·99 991 | 8·40 | ·95 425 | 2073·80 | ·32 891 | 7823·66 | ·00 330 | 214·37 |
| + ·30 | 1·00 000 | ·00 | ·99 999 | 1·55 | ·98 302 | 917·19 | ·49 695 | 8740·76 | ·01 121 | 642·95 |
| + ·32 | 1·00 000 | ·00 | 1·00 000 | ·23 | ·99 476 | 331·65 | ·66 770 | 8065·53 | ·03 265 | 1617·70 |
| + ·34 | 1·00 000 | ·00 | 1·00 000 | ·03 | ·99 872 | 97·25 | ·80 964 | 6097·89 | ·08 132 | 3388·74 |
| + ·36 | 1·00 000 | ·00 | 1·00 000 | ·00 | ·99 977 | 22·92 | ·90 919 | 3743·65 | ·17 306 | 5859·49 |
| + ·38 | 1·00 000 | ·00 | 1·00 000 | ·00 | ·99 996 | 4·30 | ·96 379 | 1847·68 | ·31 550 | 8282·92 |
| + ·40 | 1·00 000 | ·00 | 1·00 000 | ·00 | ·99 998 | ·63 | ·98 815 | 724·97 | ·49 592 | 9469·37 |
| + ·42 | 1·00 000 | ·00 | 1·00 000 | ·00 | ·99 999 | ·07 | ·99 691 | 223·30 | ·68 050 | 8649·85 |
| + ·44 | 1·00 000 | ·00 | 1·00 000 | ·00 | 1·00 000 | ·01 | ·99 945 | 53·24 | ·83 117 | 6227·72 |
| + ·46 | 1·00 000 | ·00 | 1·00 000 | ·00 | 1·00 000 | ·00 | ·99 997 | 9·67 | ·92 781 | 3480·35 |
| + ·48 | 1·00 000 | ·00 | 1·00 000 | ·00 | 1·00 000 | ·00 | ·99 998 | 1·32 | ·97 553 | 1483·85 |
| + ·50 | 1·00 000 | ·00 | 1·00 000 | ·00 | 1·00 000 | ·00 | 1·00 000 | ·13 | ·99 341 | 473·26 |
| + ·52 | 1·00 000 | ·00 | 1·00 000 | ·00 | 1·00 000 | ·00 | 1·00 000 | ·01 | ·99 869 | 110·44 |
| + ·54 | 1·00 000 | ·00 | 1·00 000 | ·00 | 1·00 000 | ·00 | 1·00 000 | ·00 | ·99 999 | 18·39 |
| + ·56 | 1·00 000 | ·00 | 1·00 000 | ·00 | 1·00 000 | ·00 | 1·00 000 | ·00 | 1·00 000 | 2·12 |
| + ·58 | 1·00 000 | ·00 | 1·00 000 | ·00 | 1·00 000 | ·00 | 1·00 000 | ·00 | 1·00 000 | ·16 |
| + ·60 | 1·00 000 | ·00 | 1·00 000 | ·00 | 1·00 000 | ·00 | 1·00 000 | ·00 | 1·00 000 | ·08 |
| + ·62 | 1·00 000 | ·00 | 1·00 000 | ·00 | 1·00 000 | ·00 | 1·00 000 | ·00 | 1·00 000 | ·00 |
| + ·64 | 1·00 000 | ·00 | 1·00 000 | ·00 | 1·00 000 | ·00 | 1·00 000 | ·00 | 1·00 000 | ·00 |
| + ·66 | 1·00 000 | ·00 | 1·00 000 | ·00 | 1·00 000 | ·00 | 1·00 000 | ·00 | 1·00 000 | ·00 |
| + ·68 | 1·00 000 | ·00 | 1·00 000 | ·00 | 1·00 000 | ·00 | 1·00 000 | ·00 | 1·00 000 | ·00 |
| + ·70 | 1·00 000 | ·00 | 1·00 000 | ·00 | 1·00 000 | ·00 | 1·00 000 | ·00 | 1·00 000 | ·00 |

n = 400                                                                 ρ = 0.5 — 0.9

| r | ρ=0.5 Area | ρ=0.5 Ordinate | ρ=0.6 Area | ρ=0.6 Ordinate | r | ρ=0.7 Area | ρ=0.7 Ordinate | r | ρ=0.8 Area | ρ=0.8 Ordinate | r | ρ=0.9 Area | ρ=0.9 Ordinate |
|---|---|---|---|---|---|---|---|---|---|---|---|---|---|
| +.26 | .00 000 | .00 | .00 000 | .00 | +.530 | .00 000 | .0 | +.630 | .00 000 | .0 | | | |
| +.27 | .00 000 | .00 | .00 000 | .00 | +.535 | .00 000 | .0 | +.635 | .00 000 | .0 | +.8325 | .00 000 | .0 |
| +.28 | .00 000 | .00 | .00 000 | .00 | +.540 | .00 000 | .0 | +.640 | .00 000 | .0 | +.8350 | .00 000 | .0 |
| +.29 | .00 000 | .00 | .00 000 | .00 | +.545 | .00 000 | .0 | +.645 | .00 000 | .0 | +.8375 | .00 000 | .0 |
| +.30 | .00 000 | .01 | .00 000 | .00 | +.550 | .00 000 | .1 | +.650 | .00 000 | .0 | +.8400 | .00 000 | .1 |
| +.31 | .00 000 | .27 | .00 000 | .00 | +.555 | .00 000 | .1 | +.655 | .00 000 | .0 | +.8425 | .00 000 | .2 |
| +.32 | .00 001 | .73 | .00 000 | .00 | +.560 | .00 000 | .2 | +.660 | .00 000 | .0 | +.8450 | .00 000 | .5 |
| +.33 | .00 002 | 1.86 | .00 000 | .00 | +.565 | .00 000 | .4 | +.665 | .00 000 | .0 | +.8475 | .00 000 | 1.2 |
| +.34 | .00 005 | 4.56 | .00 000 | .00 | +.570 | .00 001 | .8 | +.670 | .00 000 | .0 | +.8500 | .00 001 | 2.6 |
| +.35 | .00 012 | 10.75 | .00 000 | .00 | +.575 | .00 001 | 1.5 | +.675 | .00 000 | .0 | | | |
| | | | | | | | | | | | +.8525 | .00 002 | 5.5 |
| +.36 | .00 029 | 24.26 | .00 000 | .00 | +.580 | .00 002 | 2.8 | +.680 | .00 000 | .0 | +.8550 | .00 004 | 11.8 |
| +.37 | .00 065 | 52.37 | .00 000 | .00 | +.585 | .00 004 | 5.1 | +.685 | .00 000 | .0 | +.8575 | .00 008 | 24.5 |
| +.38 | .00 142 | 108.06 | .00 000 | .00 | +.590 | .00 008 | 9.3 | +.690 | .00 000 | .1 | +.8600 | .00 017 | 49.7 |
| +.39 | .00 298 | 212.77 | .00 000 | .00 | +.595 | .00 014 | 16.5 | +.695 | .00 000 | .2 | +.8625 | .00 035 | 98.7 |
| +.40 | .00 595 | 399.28 | .00 000 | .01 | +.600 | .00 025 | 28.7 | +.700 | .00 000 | .4 | +.8650 | .00 070 | 191.0 |
| +.41 | .01 139 | 713.03 | .00 000 | .02 | +.605 | .00 053 | 49.3 | +.705 | .00 000 | .9 | +.8675 | .00 137 | 359.7 |
| +.42 | .02 082 | 1209.95 | .00 000 | .06 | +.610 | .00 076 | 82.9 | +.710 | .00 001 | 2.1 | +.8700 | .00 261 | 658.1 |
| +.43 | .03 639 | 1947.73 | .00 000 | .20 | +.615 | .00 130 | 136.8 | +.715 | .00 003 | 5.0 | | | |
| | | | | | | | | | | | +.8725 | .00 483 | 1167.3 |
| +.44 | .06 073 | 2969.17 | .00 001 | .60 | +.620 | .00 218 | 221.2 | +.720 | .00 007 | 11.2 | +.8750 | .00 871 | 2004.1 |
| +.45 | .09 674 | 4278.49 | .00 001 | 1.75 | +.625 | .00 388 | 350.1 | +.725 | .00 015 | 24.6 | +.8775 | .01 525 | 3322.8 |
| +.46 | .14 707 | 5815.78 | .00 005 | 4.83 | +.630 | .00 584 | 542.5 | +.730 | .00 034 | 52.2 | +.8800 | .02 588 | 5308.0 |
| +.47 | .21 336 | 7442.27 | .00 013 | 12.69 | +.635 | .00 916 | 821.9 | +.735 | .00 072 | 107.3 | | | |
| | | | | | | | | | | | +.8825 | .04 251 | 8149.4 |
| +.48 | .29 551 | 8945.38 | .00 033 | 31.68 | +.640 | .01 420 | 1216.6 | +.740 | .00 149 | 213.0 | +.8850 | .06 747 | 11993.1 |
| +.49 | .39 102 | 10075.12 | .00 083 | 74.92 | +.645 | .02 168 | 1758.2 | +.745 | .00 300 | 407.8 | +.8875 | .10 334 | 16869.3 |
| +.50 | .49 500 | 10606.00 | .00 199 | 167.53 | +.650 | .03 208 | 2478.7 | +.750 | .00 582 | 751.3 | +.8900 | .16 254 | 22606.9 |
| +.51 | .60 069 | 10406.88 | .00 449 | 353.21 | +.655 | .04 670 | 3404.8 | +.755 | .01 090 | 1388.1 | +.8925 | .21 674 | 28773.3 |
| +.52 | .70 073 | 9490.55 | .00 959 | 700.37 | +.660 | .06 650 | 4553.8 | +.760 | .01 967 | 2418.7 | +.8950 | .29 617 | 34617.7 |
| +.53 | .78 864 | 8018.84 | .01 934 | 1302.37 | +.665 | .09 261 | 5924.1 | +.765 | .03 416 | 3634.6 | +.8975 | .38 897 | 39319.8 |
| +.54 | .86 014 | 6256.50 | .03 684 | 2264.24 | +.670 | .12 606 | 7487.7 | +.770 | .05 698 | 5595.5 | +.9000 | .49 100 | 41873.1 |
| +.55 | .91 377 | 4491.43 | .06 611 | 3668.41 | +.675 | .16 771 | 9185.0 | +.775 | .09 116 | 8178.3 | | | |
| +.56 | .95 074 | 2955.56 | .11 171 | 5518.87 | +.680 | .21 798 | 10922.2 | +.780 | .13 969 | 11311.7 | +.9025 | .59 601 | 41650.7 |
| +.57 | .97 408 | 1775.28 | .17 755 | 7680.63 | +.685 | .27 679 | 12574.9 | +.785 | .20 480 | 14751.3 | +.9050 | .69 672 | 38500.2 |
| +.58 | .98 750 | 969.06 | .26 535 | 9847.52 | +.690 | .34 335 | 13999.7 | +.790 | .28 701 | 18070.7 | +.9075 | .78 645 | 32888.9 |
| +.59 | .99 451 | 478.42 | .37 305 | 11579.98 | +.695 | .41 616 | 15048.6 | +.795 | .38 437 | 20708.0 | +.9100 | .86 000 | 25806.1 |
| +.60 | .99 782 | 212.53 | .49 399 | 12429.40 | +.700 | .49 300 | 15598.3 | +.800 | .49 200 | 22098.5 | | | |
| +.61 | .99 922 | 84.47 | .61 772 | 12113.78 | +.705 | .57 117 | 15565.8 | +.805 | .60 261 | 21854.0 | +.9125 | .91 528 | 18478.4 |
| +.62 | .99 975 | 29.86 | .73 244 | 10659.61 | +.710 | .64 765 | 14931.6 | +.810 | .70 771 | 19922.1 | +.9150 | .95 310 | 11984.9 |
| +.63 | .99 993 | 9.33 | .82 827 | 8416.93 | +.715 | .71 956 | 13743.0 | +.815 | .79 957 | 16644.2 | +.9175 | .97 646 | 6985.0 |
| +.64 | .99 998 | 2.56 | .89 994 | 5923.78 | +.720 | .78 435 | 12116.4 | +.820 | .87 297 | 12664.0 | +.9200 | .98 941 | 3626.0 |
| +.65 | .99 999 | .61 | .94 762 | 3688.70 | +.725 | .84 026 | 10212.4 | +.825 | .92 626 | 8714.7 | | | |
| +.66 | 1.00 000 | .13 | .97 561 | 2016.25 | +.730 | .88 632 | 8212.0 | +.830 | .96 116 | 5382.9 | +.9225 | .99 576 | 1660.1 |
| +.67 | 1.00 000 | .02 | .99 001 | 958.82 | +.735 | .92 251 | 6287.4 | +.835 | .98 162 | 2959.7 | +.9250 | .99 851 | 663.1 |
| +.68 | 1.00 000 | .00 | .99 644 | 392.89 | +.740 | .94 955 | 4572.5 | +.840 | .99 227 | 1435.4 | +.9275 | .99 954 | 228.3 |
| +.69 | 1.00 000 | .00 | .99 891 | 137.25 | +.745 | .96 872 | 3150.6 | +.845 | .99 714 | 607.8 | +.9300 | .99 988 | 66.8 |
| +.70 | 1.00 000 | .00 | .99 971 | 40.04 | +.750 | .98 159 | 2052.1 | +.850 | .99 909 | 222.2 | | | |
| +.71 | 1.00 000 | .00 | .99 992 | 9.89 | +.755 | .98 975 | 1259.5 | +.855 | .99 975 | 69.3 | +.9325 | .99 997 | 16.4 |
| +.72 | 1.00 000 | .00 | .99 997 | 1.98 | +.760 | .99 462 | 726.6 | +.860 | .99 994 | 18.2 | +.9350 | .99 998 | 3.3 |
| +.73 | 1.00 000 | .00 | .99 998 | .32 | +.765 | .99 735 | 392.6 | +.865 | .99 999 | 3.9 | +.9375 | .99 999 | .5 |
| +.74 | 1.00 000 | .00 | 1.00 000 | .04 | +.770 | .99 878 | 198.1 | +.870 | 1.00 000 | .7 | +.9400 | 1.00 000 | .1 |
| +.75 | 1.00 000 | .00 | 1.00 000 | .00 | +.775 | .99 948 | 93.0 | +.875 | 1.00 000 | .1 | | | |
| +.76 | 1.00 000 | .00 | 1.00 000 | .00 | +.780 | .99 980 | 40.5 | +.880 | 1.00 000 | .0 | +.9425 | 1.00 000 | .0 |
| +.77 | 1.00 000 | .00 | 1.00 000 | .00 | +.785 | .99 993 | 16.2 | +.885 | 1.00 000 | .0 | +.9450 | 1.00 000 | .0 |
| +.78 | 1.00 000 | .00 | 1.00 000 | .00 | +.790 | .99 999 | 6.0 | +.890 | 1.00 000 | .0 | +.9475 | 1.00 000 | .0 |
| +.79 | 1.00 000 | .00 | 1.00 000 | .00 | +.795 | 1.00 000 | 2.0 | +.895 | 1.00 000 | .0 | +.9500 | 1.00 000 | .0 |
| +.80 | 1.00 000 | .00 | 1.00 000 | .00 | +.800 | 1.00 000 | .0 | +.900 | 1.00 000 | .0 | +.9525 | 1.00 000 | .0 |
| | | | | | | | | | | | +.9550 | 1.00 000 | .0 |
| | | | | | | | | | | | +.9575 | 1.00 000 | .0 |
| | | | | | | | | | | | +.9600 | 1.00 000 | .0 |

# CONFIDENCE BELTS

## CHART I. CHANCE OF REJECTING THE HYPOTHESIS WHEN TRUE = ·05 + ·05 = ·10

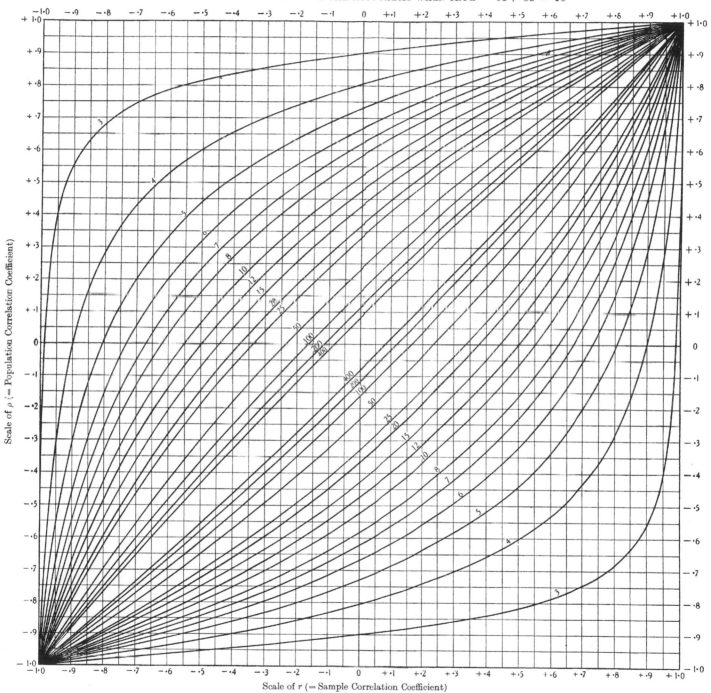

Scale of $r$ (= Sample Correlation Coefficient)

The numbers on the curves indicate sample size

# CONFIDENCE BELTS

## CHART II. CHANCE OF REJECTING THE HYPOTHESIS WHEN TRUE = ·025 + ·025 = ·05

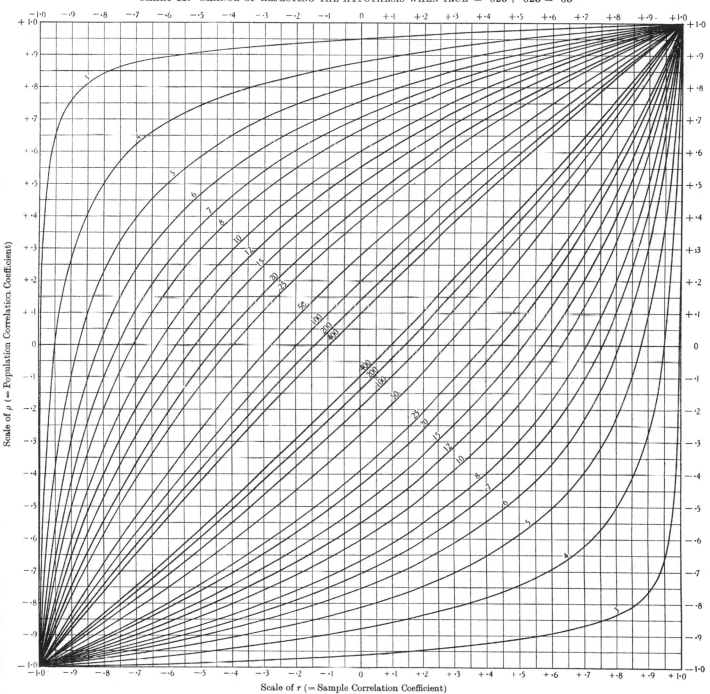

Scale of $r$ (= Sample Correlation Coefficient)

The numbers on the curves indicate sample size

# CONFIDENCE BELTS

CHART III. CHANCE OF REJECTING THE HYPOTHESIS WHEN TRUE $= \cdot01 + \cdot01 = \cdot02$

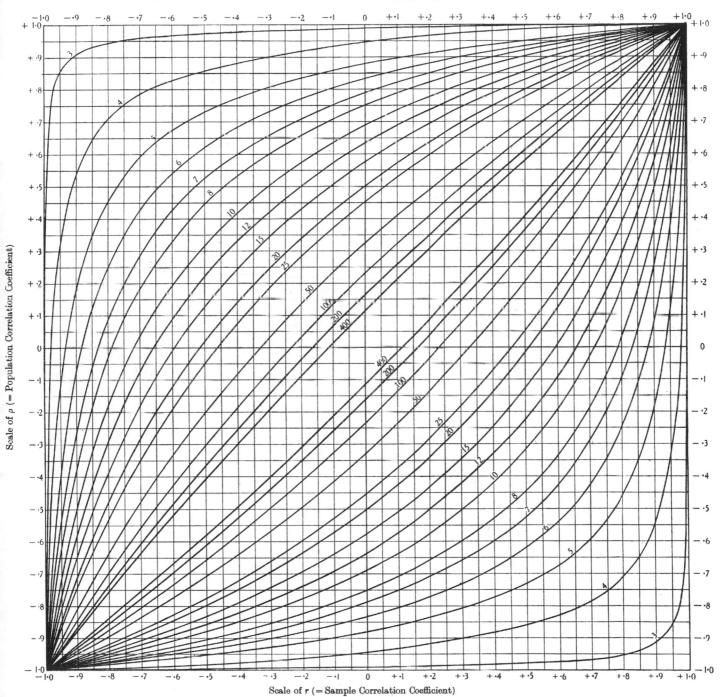

Scale of $r$ (= Sample Correlation Coefficient)

The numbers on the curves indicate sample size

# CONFIDENCE BELTS

## CHART IV. CHANCE OF REJECTING THE HYPOTHESIS WHEN TRUE $= \cdot 005 + \cdot 005 = \cdot 01$

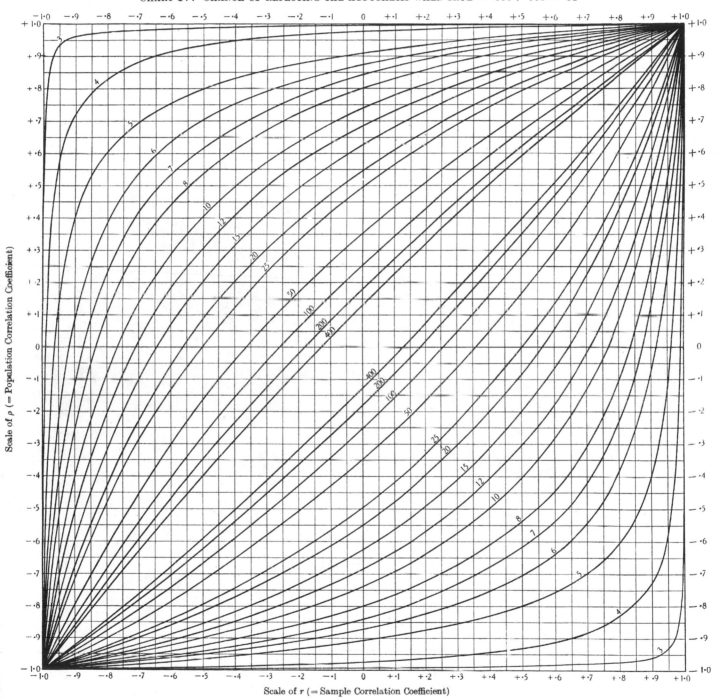

Scale of $r$ (= Sample Correlation Coefficient)

The numbers on the curves indicate sample size